Lecture Notes in Mathematics

Edited by A. Dold, B. Eckmann and F. Takens

Subseries: Fondazione C.I.M.E., Firenze
Adviser: Roberto Conti

1403

B. Simeone (Ed.)

Combinatorial Optimization

Lectures given at the 3rd Session of the Centro Internazionale Matematico Estivo (C.I.M.E.) held at Como, Italy, August 25–September 2, 1986

Springer-Verlag
Berlin Heidelberg New York London Paris Tokyo Hong Kong

Editor

Bruno Simeone
Dipartimento di Statistica,
Probabilità e Statistiche Applicáte
Università di Roma "La Sapienza"
Piazzale Aldo Moro 5, 00185 Roma, Italy

Mathematics Subject Classification (1980): 90C27; 68R99

ISBN 3-540-51797-9 Springer-Verlag Berlin Heidelberg New York
ISBN 0-387-51797-9 Springer-Verlag New York Berlin Heidelberg

Printing and binding: Druckhaus Beltz, Hemsbach/Bergstr.
2146/3140-543210 – Printed on acid-free paper

PREFACE

The present volume contains the proceedings of the CIME International Summer School on "Combinatorial Optimization", which was held at Villa Olmo, Como, Italy, from August 25 to September 2, 1986.

This was the first CIME Summer School specifically devoted to this quickly developing area, although the Varenna School on "Matroid Theory and its Applications" organized by Prof. Barlotti in 1980 did already include some lectures on matroid optimization. As a matter of fact, the idea of the present School came up for the first time there.

Combinatorial Optimization has a peculiar location in the map of Applied Mathematics, being placed in an interzone in the middle of Combinatorics, Computer Science and Operations Research. From a mathematical point of view, it draws on pure combinatorics, including graphs and matroids, on Boolean algebras and switching functions, partially ordered sets, group theory, linear algebra, convex geometry and probability theory, as well as other tools. Over the past years, a substantial amount of research has been devoted to the connections between Combinatorial Optimization and theoretical Computer Science, and in particular to computational complexity and algorithmic issues. Quite often the study of combinatorial optimization problems is motivated by real-life applications, such as scheduling, assignment, location, distribution, routing, districting, design and other Operations Research applications.

Although references to actual applications were frequently given, the emphasis of the School has been on theoretical aspects of Combinatorial Optimization. The four invited Lecturers, Prof. Peter L. Hammer, Rutgers University, USA; Prof. Ellis L. Johnson, IBM Scientific Research Center, Yorktown Heights, USA; Prof. Bernhard Korte, University of Bonn, West Germany; and Prof. Eugene L. Lawler, University of California, Berkeley, USA, have given a broad account of recent results and current trends in the area. Special attention has been devoted to the study of important classes of functions (either real- or binary-valued) defined on the binary n-cube (Prof. Hammer); to polyhedral combinatorics and its connections with combinatorial duality theories and min-max identities (Prof. Johnson); to the deep link between greedy algorithms and finite geometries such as matroids and greedoids (Prof. Korte); to the role of submodularity (a discrete analogue of convexity) and to a general decomposition theory leading to linear-time graph algorithms (Prof. Lawler).

Contributed papers were presented by D. Acketa, C. Arbib, J. Bisschop, S. Dragutin, O. Holland, M. Lucertini, S. Pallottino, G. Pirillo, W. Piotrowski, B. Simeone, and P. Winter, and many of them are collected in this volume. We have also included contributions by M. Conforti and P. Hansen, who had planned to attend the School, but at the last moment were not able to come.

It is a pleasure to acknowledge the financial support of CIME, as well as the valuable assistance provided by Prof. Roberto Conti, Director, and Prof. Antonio Moro, Secretary of CIME. I am grateful to Fondazione "A. Volta" and its Director Prof. Giulio Casati for their kind hospitality: the elegant neo-classic architectures of Villa Olmo and the scenic beauty of Lake Como have created a charming atmosphere for the School; and the local staff, in particular Dr. Chiara De Santis and Mrs. Donatella Marchegiano, has efficiently handled even the tiniest logistic details. I also thank my colleagues Prof. Mario Lucertini and Prof. Stefano Pallottino for their personal help in the organization of the School. Finally, my deepest thanks to the invited Lecturers and to the other participants for their individual contributions to the School.

Bruno Simeone,
University of Rome "La Sapienza"

TABLE OF CONTENTS

Courses

Seminars

QUADRATIC FUNCTIONS OF BINARY VARIABLES

by
Peter L. Hammer
RUTCOR, Rutgers University, New Brunswick, NJ, USA

and

Bruno Simeone
RUTCOR, Rutgers University, New Brunswick, NJ, USA
and
Department of Statistics, University of Rome, Italy

Contents

1 Introduction

The present survey is devoted to quadratic functions of n binary variables. Part I (Sec. 2 to 3) deals with binary-valued functions (*boolean functions*, *truth functions*) and its main theme is the efficient solution of quadratic boolean equations; Part II (Sec. 4 to 17) deals with real-valued functions (*pseudo-boolean functions*, *set-functions*) and focuses on the maximization of such functions over the binary n-cube.

Quadratic functions of binary variables deserve attention for a variety of reasons. They naturally arise in modelling *interactions*. Consider a set of n objects, labelled $1, 2, ..., n$, each of which can be either chosen or not. Assume that for any pair (i, j) of objects a real number a_{ij}, measuring the "interaction" between i and j, is given. Also, assume that the global interaction is the sum of the interactions between all pairs of chosen objects. Let $x_i = 1$ or 0 depending on whether object i is chosen or not. Then the global interaction can be written as a quadratic function $\sum_{i=1}^{n} \sum_{j=1}^{n} a_{ij} x_j x_j$ of the n variables $x_1, ..., x_n$.

For example, inter-city traffic [Rhys (1970)] and kinetic energy in spin-glass models [Kirkpatrick, Gelatt and Vecchi (1983)] can be represented in this way.

Quadratic functions of binary variables also naturally arise in *least-square approximation*. Assume that a weight w_i is assigned to each object $i = 1, ..., n$, and that one wants to choose a subset of objects whose total weight is as close as possible to a "target" weight t. This leads to the minimization of the quadratic function $(w_1 x_1 + \cdots + w_n x_n - t)^2$. One nice application deals with the optimal distribution of cargoes among the trips of a space shuttle in the supply support system of a lunar base [Freeman, Gogerty, Graves, and Brooks (1966)]. As another example, consider the *optimal regression* problem [Beale, Kendall, and Wall (1967)]. An endogenous variable Y is approximated by a linear function $a_1 Z_1 + \cdots + a_n Z_n$ of n exogenous variables $Z_1, ..., Z_n$. We assume that the coefficients a_j have been already estimated from a sample of m observations $(y_i, z_{i1}, ..., z_{in})$ of the variables $Y, Z_1, ..., Z_n$ via standard linear regression techniques. However, for practical reasons one often wants to choose only p variables $(p \ll n)$ among $Z_1, ..., Z_n$ and express Y as a linear combination of the chosen variables with the smallest possible loss of information. Let $x_j = 1$ or 0 depending on whether variable Z_j is chosen or not. Then the problem can be formulated as the minimization of the quadratic function

$$\sum_{i=1}^{m} (y_i - a_1 z_{i1} x_1 - \cdots - a_n z_{in} x_n)^2$$

subject to the cardinality constraint $x_1 + \cdots + x_n = p$.

There is a rich and fruitful interplay between the theory of quadratic functions of binary variables and the theory of graphs. As a matter of fact, with any graph $G = (V, E)$, where $V = \{1, ..., n\}$ is the vertex-set and E is the edge-set, one can naturally associate the quadratic monotone boolean function $f_G(x) = \bigvee_{(i,j) \in E} x_i x_j$, and vice-versa. For this reason, quadratic monotone boolean functions are sometimes called *graphic*. One important direction of research attempts to find connections between combinatorial properties of G and functional properties of f_G. This line of research is typified by a theorem of Chvátal and Hammer (1977) stating that a graphic function is threshold if and only if the associated graph does not contain squares, paths of length 3, or parallel edges.

Graph-theoretic methods are very useful in the solution of quadratic boolean equations. Actually, all the fastest known solution algorithms are graph-theoretic in nature (see Sec. 3). Graphs are also a very useful tool in the maximization of quadratic pseudo-boolean functions. For example, a reduction of this problem to finding a maximum weight stable set in a graph is exhibited in Sec. 14. Conversely, many graph optimization problems can be naturally formulated as quadratic $0-1$ optimization problems (see Sec. 5).

One important reason for studying quadratic boolean equations is that, unlike equations of higher order, they can be solved in polynomial – in fact, linear – time. Furthermore, the most common types of logical relations,

"Either P or Q is false"

"Either P or Q is true"

"P implies Q"

are represented by *quadratic* equations, namely

$pq = 0,$

$\overline{p}\,\overline{q} = 0,$

$p\overline{q} = 0,$

respectively.

The survey is structured as follows. Some fundamental concepts of the theory of boolean functions and pseudo-boolean functions are recalled in Sec. 2 and 4, respectively. Sec. 3, which is based on [Petreschi and Simeone (1985)], describes some fast graph-theoretic algorithms for the solution of quadratic boolean equations.

Sec. 5 describes many combinatorial applications of quadratic $0-1$ maximization.

The remaining sections are devoted to a detailed account of the theory of roof-duality in quadratic $0-1$ maximization. They are mainly based on [Hammer, Hansen and Simeone (1984)]; however, new developments are reported in Sections 11, 12 and 13.

The theory of roof-duality allows one to get – with a modest computational effort – upper bounds of the maximum of a quadratic function f over the binary n-cube B^n. This is achieved by considering a special class of linear overestimators of f – the so called roofs – and maximizing them (instead of f) over B^n. "Best" roofs can be generated in polynomial time via maximal flow techniques (Sec. 15). There is no guarantee that the upper bound obtained by maximizing a best roof does coincide with the quadratic optimum; however, one can check whether this is indeed the case in polynomial time by solving a quadratic boolean equation (Sec. 17).

Sec. 16 is devoted to the intriguing *persistency* phenomenon: consider any maximum point x^* of the quadratic function f and any maximum point $\tilde{x}$ of a best roof g; then for every variable x_i with a non-zero coefficient in g one has $\tilde{x}_i = x_i^*$. This phenomenon, deserves, along with many other topics related to quadratic functions of binary variables, further substantial investigation.

PART I
Quadratic boolean functions and equations

2 Boolean functions and boolean equations

It is well known that the set $B = \{0, 1\}$, endowed with the operations

$$\begin{array}{lll} x \vee y = max\{x, y\} & \text{(union)} & \\ x \wedge y = min\{x, y\} = x \cdot y & \text{(intersection or product)} & (2.1) \\ \overline{x} = 1 - x & \text{(complementation)} & \end{array}$$

is a boolean algebra.

Let $x_1, \ldots, x_n$ be boolean variables, i.e. variables taking values in B. A *literal* is either a variable x_i or its complement $\overline{x}_i$. A *term* is a finite product of distinct literals a *boolean expression* (in *disjunctive normal form*, DNF) a finite union of terms.

A *boolean function* is any mapping $f : B^n \rightarrow B$. Given any boolean expression ϕ, one can associate with each $\alpha \in B^n$ the element $\phi(\alpha) \in B$ obtained by replacing in ϕ each variable x_i by α_i and evaluating the resulting expression according to (2.1). Such mapping $\alpha \mapsto \phi(\alpha)$ is a boolean function, and conversely each boolean function is representable in this way.

A boolean expression is called

- *primitive*, if no two distinct terms involve the same variables, no matter whether complemented or not. For example, the presence of the term xyz forbids the presence of terms $\overline{x}yz$, $\overline{x}\,\overline{y}z$, $\overline{x}\,\overline{y}\,\overline{z}$, $x\,\overline{y}z$ and so on, as well as other occurrences of xyz.
- *normal*, if all terms are different and if no two terms of the form xC, $\overline{x}C$ are present. Example: $xyz \vee \overline{x}\,\overline{y}\,\overline{z} \vee x\,\overline{y} \vee y\,\overline{z}$.
- *pure*, if every term contains at least one uncomplemented variable. Example: $xyz \vee \overline{x}y \vee y\,\overline{z}$.
- *mixed*, if every term contains both complemented and uncomplemented variables. Example: $x\,\overline{y}\,\overline{z} \vee \overline{x}y \vee y\,\overline{z}$.

Let ϕ be a boolean expression. A monomial I (i.e. a finite product of distinct literals but not necessarily a term of ϕ) is an *implicant* of ϕ if $I \leq \phi$; that is, when $I = 1$ then $\phi = 1$.

The implicant I is *prime* if there is no implicant $J \neq I$ such that $I \leq J \leq \phi$. The prime implicants of a boolean expression ϕ can be found by the following *consensus method*, due to Quine (1955): Starting from the list of terms of ϕ, execute as many times as possible the following two operations.

CONSENSUS: If two terms xC and $\overline{x}D$ are present in the current list and no variable is complemented in C and uncomplemented in D or vice versa, then add to the current list the *consensus* CD of the two terms after deleting possible repeated literals.

ABSORPTION: If the current list contains two terms C, D such that each literal in D appears in C, then delete D from the list.

Quine (1955) has proved that when no further consensus or absorption operation is possible, then the final list contains all prime implicants of ϕ, and only them.

The consensus method may well take exponential time. However, when the method is applied to *quadratic* boolean expressions, only quadratic or linear terms can be generated at each step. It follows that the number of terms in the list at each step is $O(n^2)$ and this in turn implies that in the quadratic case the consensus method runs in polynomial time.

Let us now turn our attention to boolean equations. Let ϕ be a boolean expression. The boolean equation $\phi = 0$ is said to be *consistent* if there exist some $\alpha \in B^n$ such that $\phi(\alpha) = 0$. Then α is called a *solution* of the equation.

It is easy to see that a boolean equation $\phi = 0$ is consistent if and only if the constant 1 is not a prime implicant of ϕ. Hence one method for checking the consistency of $\phi = 0$ consists in applying the consensus method to ϕ and to verify whether 1 is a prime implicant or not. Many methods for solving general or specific boolean equations have been proposed in the literature [see e.g. Rudeanu (1974)].

Clearly a pure boolean equation is always consistent, since $\alpha = (0, ..., 0)$ is always a solution. This statement can be somehow reversed, as shown by Prop. (2.1) below.

Given the boolean expression $\phi(x_1, \ldots, x_n)$, the *switch* on the variable x_i is the operation which replaces each occurence of x_i in ϕ by $\overline{x}_i$ and vice versa. Similarly one defines a switch on a set S of variables.

Proposition 2.1 *A boolean equation is consistent if and only if it can be transformed into a pure one by a switch on some set S of variables.*

Proof: Given the boolean equation $\phi(x_1, \ldots, x_n) = 0$, assume that $\alpha = (\alpha_1, \ldots, \alpha_n)$ is a solution. If $S = \{x_i : \alpha_i = 1\}$, the switch on S transforms the equation into a pure one. Conversely, suppose that there is some set S of variables such that the switch on S transforms the equation into a pure one. Then $(0, ..., 0)$ is a solution of the transformed equation; hence, if one defines $\alpha_i = 1$ or 0 according as x_i belongs to S or not, the vector $\alpha = (\alpha_1, \ldots, \alpha_n)$ is a solution of the original equation. □

3 Efficient graph-theoretic algorithms for solving quadratic boolean equations

In the present section, we shall describe three fast algorithms for the solution of a quadratic boolean equation $\phi = 0$:

- The Labelling Algorithm of Even, Itai and Shamir (1976) (this paper contains only an outline of the algorithm; more detailed descriptions can be found in Gavril (1977) and Simeone (1985);
- The Switching Algorithm of Petreschi and Simeone (1980);
- The Strong Components Algorithm of Aspvall, Plass and Tarjan (1979).

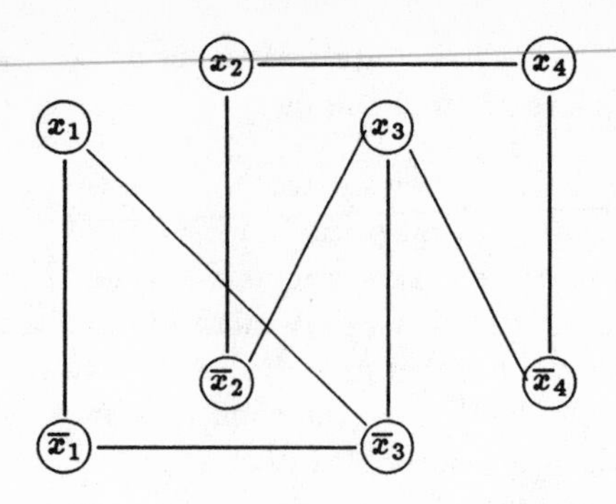

Figure 1: *The graph G associated with $\psi = x_1\overline{x}_3 \vee \overline{x}_1\overline{x}_3 \vee \overline{x}_2 x_3 \vee x_2 x_4 \vee x_3 \overline{x}_4$.*

A common feature of the three above algorithms is their graph-theoretic nature: in the first two algorithms, the quadratic boolean expression ϕ is represented by an undirected graph (the so called "matched graph" or "clause graph"), while the third algorithm exploits a digraph model (the so called "implication graph").

Consider a quadratic boolean equation (in *DNF*) in n variables $x_1, \ldots, x_n$ and m terms

$$\phi \equiv T_1 \vee ... \vee T_n \; = \; 0, \tag{3.1}$$

where, without loss of generality, we may assume that each term is the product of exactly two literals and that the presence of the term $\xi\eta$ forbids the presence of other occurences of $\xi\eta$.

Definition. The *matched-graph* associated with ϕ is the undirected graph $G = (V, E)$, where $V = \{x_1, \ldots, x_n; \overline{x}_1, \ldots, \overline{x}_n\}$ and

$$E = \{\langle \xi, \eta \rangle \text{ for each term } \xi\eta \text{ of } \phi\} \cup \{\langle x_i, \overline{x}_i \rangle : i = 1, ..., n\}$$

The edges of G are classified into *positive*, *negative*, *mixed* or *null* ones according to whether they are of the form $\langle x_i, x_j \rangle$, $\langle \overline{x}_i, \overline{x}_j \rangle$, $\langle x_i, \overline{x}_j \rangle$, $\langle x_i, \overline{x}_i \rangle$, respectively.

Figure 1. shows the matched graph G associated with the following expression

$$\psi = x_1\overline{x}_3 \vee \overline{x}_1\overline{x}_3 \vee \overline{x}_2 x_3 \vee x_2 x_4 \vee x_3 \overline{x}_4 \tag{3.2}$$

As shown by Theorem (3.1) below, the consistency of the quadratic boolean equation $\phi = 0$ has a nice graph-theoretic counterpart in the matched graph G associated with ϕ. We recall that, given an arbitrary graph $G' = (V', E')$, a *matching* M of G' is any set of pairwise non-incident edges; a *transversal* T of G' is any set of vertices such that every edge of G' has at least an endpoint in G'; and that G' is said to have the *König-Egerváry Property* (briefly, the KE Property) if the maximum cardinality of a matching is equal to the minimum cardinality of a transversal.

Theorem 3.1 : *The quadratic boolean equation $\phi = 0$ is consistent iff the matched graph G associated with ϕ has the König-Egerváry Property.*

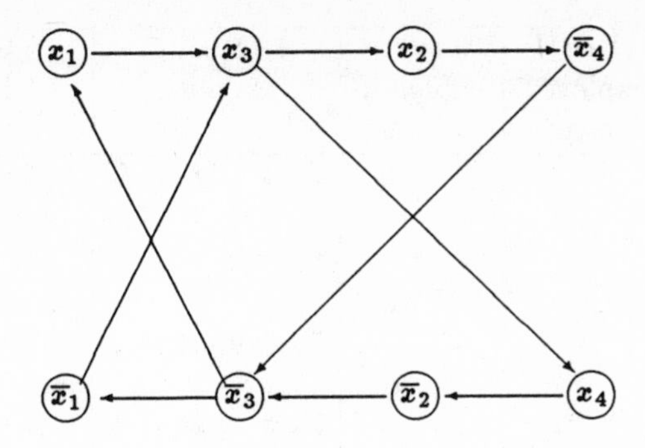

Figure 2: *The implication graph of ψ, as given by (3.2).*

Proof: See Simeone (1985). □

Gavril (1977) has described an $O(m)$ algorithm for recognizing the König-Egerváry Property in a graph with m edges, assuming that a maximum cardinality matching is at hand. One such matching can be found in $O(n^{2.5})$ time, e.g. via the implementation of Edmonds' Blossom Algorithm described in Micali and Vazirani (1980).

In view of Theorem (3.1), one can check the consistency of the equation $\phi = 0$ by executing Gavril's algorithm on the matched graph G associated with ϕ. (Note that the null edges form a maximum cardinality matching of G). The resulting procedure turns out to be essentially Even, Itai and Shamir's method.

Definition. The *implication graph* associated with ϕ is the digraph $D(V, A)$, where V is defined as above and

$$A = \{(\xi, \overline{\eta}), (\overline{\xi}, \eta) \text{ for each term } \xi\eta \text{ in } \phi\}.$$

The digraph D is isomorphic to the digraph $\tilde{D}$ obtained from D by reversing the orientation of every edge and complementing the name of every variable.

The implication graph of ψ (as given by (3.2)) is shown in Figure 2.

We are ready to describe the three above mentioned algorithms.

3.1 Labelling Algorithm

The idea of the algorithm is to *guess* the value of an arbitrary literal ξ in some solution and to *deduce* the possible consequences of this guess on other variables appearing in the expression. Since ξ can take either the value 0 or the value 1, the algorithm analyzes in parallel the consequences of these two alternative guesses on ξ. One keeps track of these consequences by a "red" labelling (corresponding to the guess $\xi = 1$) and by a "green" labelling (corresponding to the guess $\xi = 0$). Initially all terms are declared to be "red-unexplored" and "green- unexplored"; then one selects on arbitrary literal ξ and assigns

Step	Red-explored terms	Green-explored terms	Red labels	Green labels
0	/	/	$x_1 = 1; \overline{x}_1 = 0$ (guess)	$x_1 = 0; \overline{x}_1 = 1$ (guess)
1	$x_1\overline{x}_3$		$\overline{x}_3 = 0; x_3 = 1$	
2		$\overline{x}_1\overline{x}_3$		$\overline{x}_3 = 0; x_3 = 1$
3	$\overline{x}_2 x_3$		$\overline{x}_2 = 0; x_2 = 1$	
4		$x_3\overline{x}_4$		$\overline{x}_4 = 0; x_4 = 1$
5	$x_3\overline{x}_4$		$\overline{x}_4 = 0; x_4 = 1$	
6		$\overline{x}_2 x_3$		$\overline{x}_2 = 0; x_2 = 1$
7	$x_2 x_4$		$x_4 = 0; \overline{x}_4 = 1$ conflict	
8		$x_2 x_4$	conflict	$x_4 = 0; \overline{x}_4 = 1$

Table 1: *Steps for the Labelling algorithm for ψ as given by (3.2)*

to it the red label 1 and the green label 0. Then $\overline{\xi}$ must receive the red label 0 and the green label 1.

Then the two labellings are extended to as many literals as possible by alternately performing for the red labelling and for the green one the following *STEP*.

STEP: Take an arbitrary unscanned term $\eta\zeta$ such that η has the label 1, and assign to ζ and to $\overline{\zeta}$ the labels 0 and 1, respectively, making sure that ζ did not previously get the label 1. Declare the term $\eta\zeta$ scanned. (Of course, terms like "label", "unscanned", "scanned" involved in *STEP* are relative to the color currently under consideration).

If a conflict arises, say, for the red labelling (i.e. a literal which was peviously red-labelled 1 is forced to get the red label 0 or vice versa), the red labelling stops and the red labels are erased. If, at a later stage, a conflict occurs also for the green labelling, the algorithm stops and the equation has no solution. It may happen that a labelling, say the red one, "gets stuck": no conflict has occurred, but there are still literals having no red label. This is possible only when, for each red-unexplored term, the literals appearing in that term are either red-unlabelled or have red label 1. If this situation occurs then red labels are taken for granted and both the red- and the green-labellings are restarted on the reduced expression involving only the red-unlabelled literals.

The algorithm can be shown to run in $O(m)$ time (see Gavril (1977)). As an example, the following Table 1 summarizes the algoritm steps when the input expression is ψ as given by (3.2).

The expression ψ is not satisfiable because both labellings end up in a conflict.

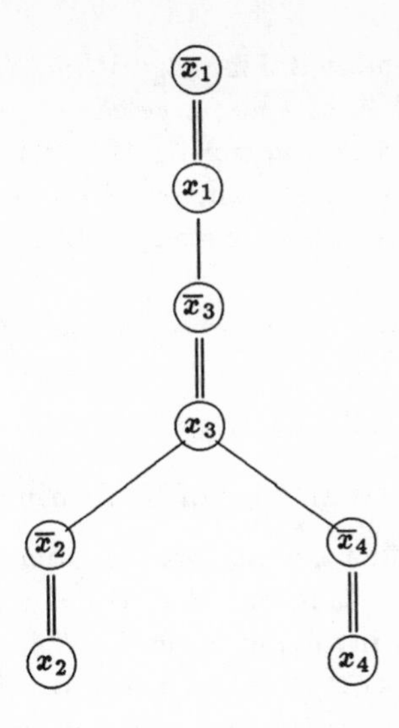

Figure 3: *The alternating tree rooted at* $\overline{x}_1$ *for the graph of Figure 1.*

3.2 Switching algorithm

On the basis of Prop. (2.1), the algoritm tries to transform the given expression ϕ into a pure one, if possible, through a sequence of switches. Before describing the algorithm, it is convenient to introduce some preliminary definitions. (we use the tree-terminology of Aho, Hopcroft and Ullman (1974)).

Definition: A variable x in ϕ is said to be *forced* to the value α ($\alpha = 0$ or 1) if either the equation $\phi = 0$ is inconsistent or x has the value α in all solutions.

Definition: An *alternating tree rooted at* $\overline{x}_i$ is any subgraph of G (the matched graph of ϕ) which is a tree $T(\overline{x}_i)$ rooted at $\overline{x}_i$ and has the following properties:

1. $\overline{x}_i$ is the root;
2. if $\overline{x}_j$ is a vertex of $T(\overline{x}_i)$, then its father in $T(\overline{x}_i)$ is $\overline{x}_j$.
3. if $\overline{x}_j$ is a vertex of $T(\overline{x}_i)$, then its father is a vertex x_r of $T(\overline{x}_i)$, such that $\langle x_r, \overline{x}_j\rangle$ is a mixed edge of G;
4. if x_j is a vertex of $T(\overline{x}_i)$ and $\langle x_j, \overline{x}_r\rangle$ is a mixed edge of G, then $\overline{x}_r$ is a vertex of $T(\overline{x}_i)$.

For the matched graph of Figure 1., an alternating tree rooted at $\overline{x}_1$ is shown in Figure 3.

Definition: The *join* of two vertices ξ and η of $T(\overline{x}_i)$ is their common ancestor which is farthest from the root $\overline{x}_i$.

Let us now briefly describe the switching algorithm. Again the algorithm works on the matched graph G. An endpoint $\overline{x}_i$ of a negative edge $\overline{x}_i\overline{x}_r$ is selected and the alternating tree $T(\overline{x}_i)$ is grown. As soon as a new vertex x_h of $T(\overline{x}_i)$ is generated, one checks whether there is in G any positive edge $x_h x_k$ linking x_h to a previously generated vertex x_k of $T(\overline{x}_i)$. If this is the case, the variable x_j corresponding to the join of x_h and x_k must be forced to 0.

As a consequence, other variables are forced in cascade according to the following rules:

(i) if ξ is forced to 0, then $\overline{\xi}$ is forced to 1;

(ii) if ξ is forced to 1 and $\langle \xi, \eta \rangle$ is an edge of G, then η is forced to 0.

If during this process a conflict occurs (that is, some variableis forced to both 0 and 1), then the algorithm stops pointing out that the equation is inconsistent. Otherwise one obtains a reduced equation involving fewer variables and a new cycle begins. If the construction of $T(\overline{x}_i)$ has been completed and no positive edge between two vertices of $T(\overline{x}_i)$ has been detected, then a switch is performed on all the variables corresponding to vertices of $T(\overline{x}_i)$. In this way one obtains an equivalent expression and a new cycle begins. The procedure is iterated until either a pure equation is obtained or all variables are forced. In both cases a solution of the original equation $\phi = 0$ can be found by inspecting the list of the forced variables and the list of the switched ones.

As an example, given ψ as in (3.2), the alternating tree $T(\overline{x}_1)$ of Figure 3. is grown. Since x_2 and x_4 are linked by a positive edge, their join x_3 is forced to 0 and hence $\overline{x}_3$ is forced to 1. In turn, $\overline{x}_3 = 1$ forces $x_1 = 0$ and $\overline{x}_1 = 1$. On the other hand, because of the edge $\langle \overline{x}_1, \overline{x}_3 \rangle$, $\overline{x}_1$ must also be forced to 0 and thus a conflict arises. Hence the equation $\psi = 0$ is inconsistent.

For a formal description of the algorithm and for a correctness proof, the reader is referred to Petreschi and Simeone (1980), where a worst-case $O(mn)$ bound on the time-complexity of the algorithm is also given.

3.3 Strong components algorithm

This algorithm is based on the following key result.

Theorem 3.2 (Aspvall, Plass and Tarjan (1979)) . *The equation $\phi = 0$ is consistent iff in the implication graph D no vertex x_i is in the same strong component as its complement $\overline{x}_i$ (i.e. no circuit of D contains both x_i and $\overline{x}_i$).* □

The algorithm works on D and preliminarily finds the strong components of D in reverse topological order (see Tarjan (1972)). The isomorphism between D and $\tilde{D}$ implies that for every strong component SC of D there exist a "mirror" component $\tilde{SC}$, the *complement* of SC, induced by the complements of the vertices in SC. Hence Theorem (3.2) can be restated as follows: "ϕ is satisfiable iff in D no strong component coincides with its complement".

The general step of the algorithm consists in processing the strong components of D in the following way.

For each strong component SC, one of the following cases must occur:

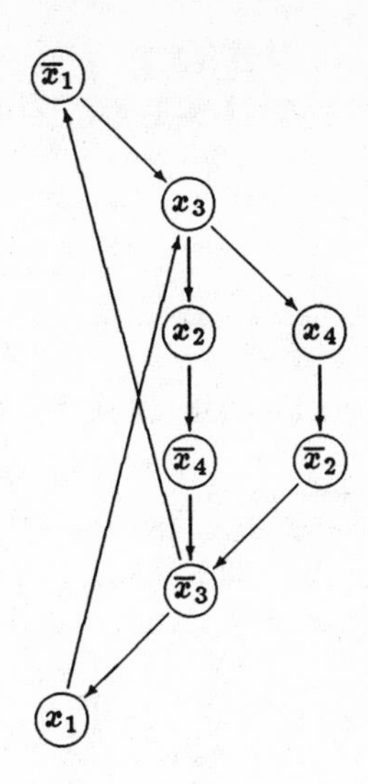

Figure 4: *The spanning arborescence generated by the algorithm of Tarjan (1972).*

(a) SC is already labelled. The algoritm processes another strong component.

(b) $SC = \tilde{SC}$. The algorithm stops. In view of Theorem (3.2) the equation $\psi = 0$ is not consistent.

(c) SC is unlabelled. The algorithm assigns the label 1 to SC and the label 0 to $\tilde{SC}$.

SC_1, is said to be a *predecessor* of SC_2 (and SC_2 a successor of SC_1,) if there exists an edge from some vertex of SC_1, to some vertex of SC_2. It is easy to show that every component labelled 1 has only components with label 1 as successors and every component labelled 0 has only components with label 0 as predecessors. Thus, if we assign to each vertex ξ the label of the component containing ξ_1 we get a solution to $\phi = 0$.

As an example, we consider again ψ and the associated implication graph D of Figure 2.

The algorithm of Tarjan (1972) for finding the strong components produces the arborescence of Figure 4: the digraph D turns out to be strongly connected. For every i, the vertices x_i and $\overline{x}_i$ belong to the same strong component and thus, by Theorem (3.2), the equation $\psi = 0$ is inconsistent.

The complexity of the Strong Components Algorithm is $O(m)$. For further details, see Aspvall,Plass and Tarjan (1979).

A randomized algorithm with expected $O(n)$ time has been described by Hansen, Jaumard and Minoux (1984).

PART II
Quadratic pseudo-boolean functions

4 Generalities on pseudo-boolean functions

4.1 Pseudo-boolean functions and posiforms

A *pseudo-boolean function* is any mapping $f : B^n \to R$ where B^n, as usual, denotes the n-dimensional binary cube $\{0,1\}^n$ and R is the set of reals.

It has been known for a long time [cfr. Hammer (1968)] that any pseudo-boolean function f can be uniquely represented as a multilinear (or affine) function of n binary variables $x_1, \ldots, x_n$:

$$f(x) = \sum_{T \in \mathcal{T}} a_T \prod_{j \in T} x_j, \tag{4.1}$$

where $\mathcal{T}$ is a collection of subsets of $\{1, ..., n\}$ and $a_T \neq 0$ for all $T \in \mathcal{T}$; we make the convention that $\prod_{j \in \emptyset} x_j = 1$: this allows for the presence of a constant term in the r.h.s. of (4.1).

A *monomial* is a product aT, where T is a *term*, i.e. a finite product of distinct literals (cfr. Section 1) and a is a real number, the *coefficient* of the monomial.

A *positive pseudo-boolean form* (briefly, a *posiform*) is any sum of monomials with positive coefficients:

$$\phi = a_1 T_1 + \cdots + a_m T_m, \quad (a_1, \ldots, a_m > 0). \tag{4.2}$$

By definition, a posiform is allowed to include a constant term: then the posiform is said to be *inhomogeneous*. Otherwise the posiform is called *homogeneous*. The notions of primitive, normal, pure and mixed boolean expressions (see Section 1) can be extended to posiforms without any difficulty.

The *boolean frame* $\hat{\phi}$ of a posiform ϕ is the boolean expression (in DNF) obtained from ϕ by disregarding the coefficients a_i and replacing the ordinary sums of (4.2) by (boolean) unions:

$$\hat{\phi} = T_1 \vee \cdots \vee T_m. \tag{4.3}$$

We shall exploit later on the property that $\min_{x \in B^n} \phi(x, \overline{x}) = 0$ if and only if the boolean equation $\hat{\phi} = 0$ is consistent.

Hammer and Rosenberg (1974) have observed that, for any given pseudo-boolean function f, there exist a constant c and a posiform ϕ such that

$$f(x) = c + \phi(x, \overline{x}), \quad \forall x \in B^n. \tag{4.4}$$

This is an easy consequence of the identity

$$-x_1 x_2 \cdots x_p = \overline{x}_1 x_2 \cdots x_p + \overline{x}_2 x_3 \cdots x_p + \cdots + \overline{x}_{p-1} x_p + \overline{x}_p - 1.$$

The representation (4.4) is by no means unique. For example, if

$$f(x) = -5x_1x_2x_3 + 6x_1x_3 - 4x_2x_3 + 5x_1$$

one has identically

$$\begin{aligned} f(x) &= -9 + 5\bar{x}_1 x_2 x_3 + 6x_1 x_3 + 9\bar{x}_2 x_3 + 5x_1 + 9\bar{x}_3 \\ &= -4 + 5x_1 x_2 \bar{x}_3 + 5x_1 \bar{x}_2 + 6x_1 x_3 + 4x_2 \bar{x}_3 + 4\bar{x}_2 \\ &= -6 + 5x_1 \bar{x}_2 x_3 + \bar{x}_1 \bar{x}_3 + 4\bar{x}_2 x_3 + 6x_1 + 5\bar{x}_2 + 2x_3 \\ &= \cdots . \end{aligned}$$

A pair (c, ϕ) such that (4.4) holds will be called here a *positive representation* of f (in earlier papers, the term "pseudo-disjunctive normal form" was used).

The concept of positive representation turns out to be very useful in the maximization of pseudo-boolean functions. Let $(-c, \phi)$ be a positive representation of $-f$. Then

$$f + \phi = c.$$

Since $\phi \geq 0$ the constant c is an upper bound of $z = \max_{x \in B^n} f(x)$, and actually coincides with z iff the boolean equation $\hat{\phi} = 0$ is consistent.

Simeone (1979) has shown that there is always a posiform ϕ^* such that $f + \phi^* = z$, and has described a "squeezing" algorithm for constructing such ϕ^*, and determining z. Unlike most algorithms for the maximization of pseudo-boolean functions, the squeezing algorithm is not enumerative. However, its computational complexity is still exponential in the worst case.

4.2 Conflict graphs

Again, let f be an arbitrary pseudo-boolean function and (c, ϕ) a positive representation of f. Maximizing f over B^n is equivalent to

$$\max_{x \in B^n} \phi(x, \bar{x}). \tag{4.5}$$

In its turn, (4.5) can be formulated as a maximum weighted stable set problem in a graph (recall that a stable set of the graph G is a set of pairwise non-adjacent vertices of G). This reduction relies on the concept of conflict graph. Let α be any boolean expression (in DNF). The *conflict graph* G_α of α is the graph whose vertices are the terms of α and where two vertices T and T' are adjacent if and only if,as terms, they have at least one conflict variable x, i.e. a variable which is complemented in T and uncomplemented in T', or vice versa.

Conversely, given a graph G, a (conflict) *code* of G is any boolean expression a having G as its conflict graph. Every graph has always a conflict code, and generally many of them. For example, K_4, the complete graph on 4 vertices has, among others, the codes shown below.

Let us consider now a posiform $\phi = a_1 T_1 + \cdots + a_m T_m$. We can associate with ϕ a weighted graph (a, G), where G is the conflict graph of $\hat{\phi}$, the boolean frame of ϕ, and where each vertex T_i has the weight a_i.

The basic observation is that problem (4.5) is actually the problem of determining a stable set of G having maximum weight. Indeed,if we evaluate ϕ at an arbitrary point $x \in B^n$, some terms T_i will have the value 0 and some others the value 1. By the definition of a conflict graph, the latter terms form a stable set of G, whose weight is precisely $\phi(x, \bar{x})$.

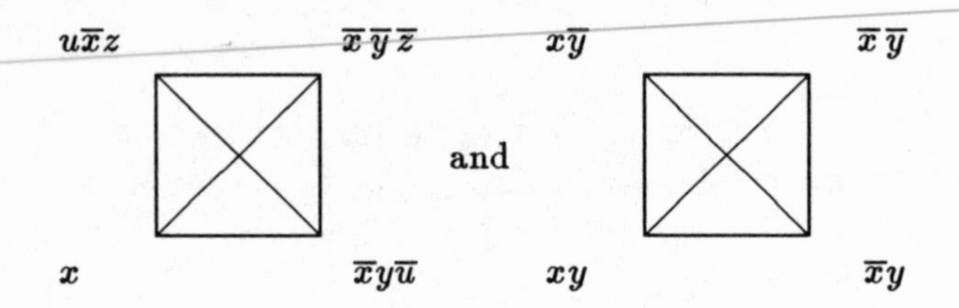

Figure 5:

It is not true that, conversely, every stable set of G arises in this way; however, it is not hard to see that every *maximal* stable set (and in particular every maximum weight stable set) does. For example, the problem of maximizing

$$\begin{aligned} \phi &= 2\overline{x}_1x_2 + 5\overline{x}_2x_3 + \overline{x}_2x_4 + 5\overline{x}_5x_6 + 9\overline{x}_7x_8 + \\ &\quad +2\overline{x}_3\overline{x}_4x_5 + 7\overline{x}_3x_5x_8 + 15\overline{x}_3x_4\overline{x}_6x_7 \end{aligned} \tag{4.6}$$

is equivalent to finding a maximum weighted stable set in the graph of Figure 6.

Now, let α be a different code of G and let ψ be the posiform whose boolean frame is α and whose coefficients are the same as those of ϕ. The weighted graph associated with ϕ and ψ is the same; hence the two problems $\max_{x\in B^n}\phi(x,\overline{x})$ and $\max_{y\in B^q}\psi(y,\overline{y})$ are equivalent.

It is easy to obtain the optimal solutions of the first problem from those of the second one.

The interest of this transformation lies in the fact that one might be able to replace ϕ by another ψ which is easier to maximize. For example, the graph in Figure 6 has also the code as in Figure 7.

Hence maximizing (4.6) is equivalent to the problem of maximizing

$$\psi(x,\overline{x}) = 5x_1 + 5x_1\overline{x}_4 + 2\overline{x}_1\overline{x}_2 + 15\overline{x}_1\overline{x}_3 + x_1x_5 + 7x_2x_4 + 9\overline{x}_2x_3 + 2x_4\overline{x}_5.$$

5 Maximization of quadratic pseudo-boolean functions

5.1 Combinatorial applications of quadratic $0-1$ optimization

From now on, we shall deal with the *quadratic* $0-1$ *maximization* problem

$$z = \max_{x\in B^n} f(x) \tag{5.1}$$

where $f(x) = x^TQx = \sum_{i=1}^n\sum_{j=1}^n q_{ij}x_ix_j$ is a quadratic form. Without loss of generality, we may assume that the $n\times n$ matrix Q is upper triangular, i.e. that $q_{ij}=0$ whenever $i>j$. Note that, since $x_i^2=x_i$ when x_i is a binary variable, linear terms $q_{ii}x_i$ may appear in f.

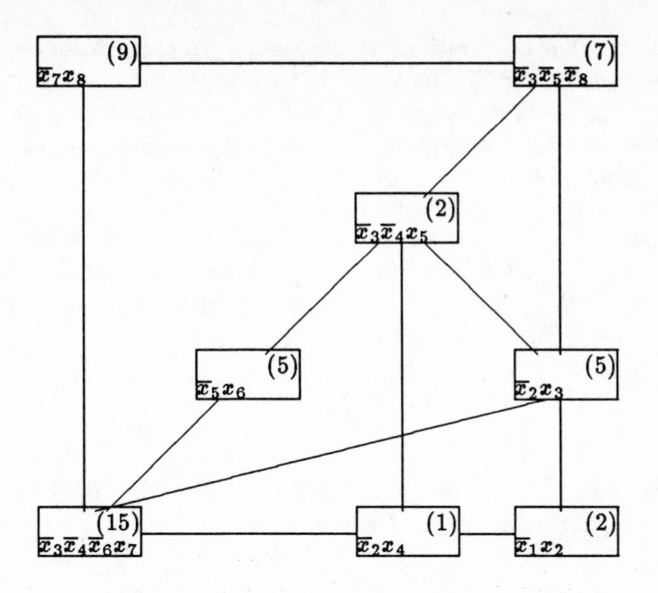

Figure 6:

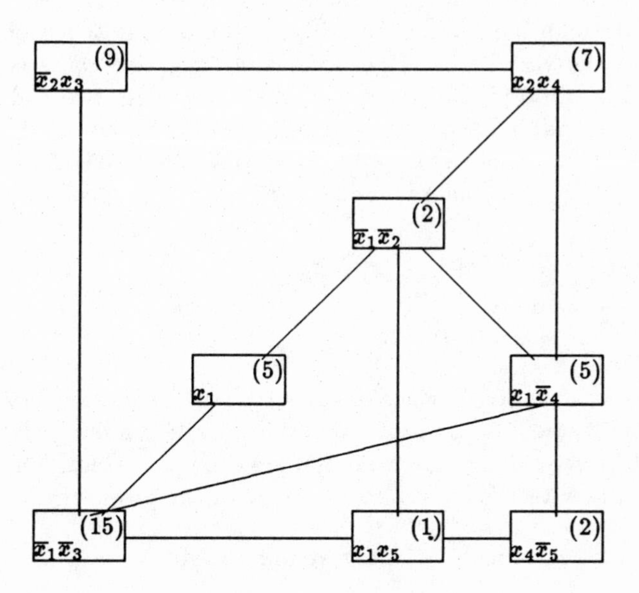

Figure 7:

The quadratic $0-1$ maximization problem is known to be NP-complete.

As a matter of fact, most of the problems in the "Gotha" of combinatorial optimization can be "naturally"[1] formulated as quadratic $0-1$ optimization problems.

In certain cases the formulation as a quadratic $0-1$ optimization problem is immediate. Here are some examples.

1. *The minimum cut problem.* Given a digraph $D = (V, A)$ with a unique source s and a unique sink t, let a capacity c_{ij} be assigned to every edge (i, j). A *cut* is any bipartition $\tau = \{S, T\}$ of the vertex set V such that $s \in S$ and $t \in T$. The *capacity* of the cut $\tau = \{S, T\}$ is defined by

$$c(\tau) = \sum_{i \in S} \sum_{j \in T} c_{ij}.$$

Defining $x_j = 0$ or 1 depending on whether $j \in S$ or $j \in T$, the problem of finding a minimum capacity cut can be formulated as

$$\min_{x \in B^n} \sum_{i=1}^{n} \sum_{j=1}^{n} c_{ij}(1 - x_i)x_j.$$

2. *Balancing a signed graph.* A signed graph is a graph G together with a bipartition of its edges into "positive" and "negative" ones. The signed graph G is balanced if none of its cycles contains an odd number of negative edges; a well-known necessary and sufficient condition for this to happen [Harary (1953)] is that there exists a bipartition of the vertices into "positive" and "negative" ones, such that the sign of each edge is equal to the product of the signs of its two endpoints. Given an arbitrary signed graph G, consider the problem of determining the smallest number of edges $\beta(G)$ to be removed in order to make G balanced. As shown by Hammer (1972), $\beta(G)$ is equal to the minimum of the quadratic pseudo-boolean function

$$f(x) = \sum_{\substack{\text{all positive} \\ \text{edges } (i,j)}} (x_i \overline{x}_j + \overline{x}_i x_j) + \sum_{\substack{\text{all negative} \\ \text{edges } (i,j)}} (x_i x_j + \overline{x}_i \overline{x}_j).$$

3. *Maximum 2-satisfiability.* Given a quadratic boolean equation $T_1 \vee T_2 \vee \cdots \vee T_m = 0$, the maximum 2-satisfiability problem consists in determining the smallest number of terms to be removed, so as to make the equation consistent. This number is easily seen to coincide with the minimum of the quadratic posiform $T_1 + T_2 + \cdots + T_m$.

Broad classes of constrained $0-1$ programming problems, e.g.

(a) maximization of a linear function subject to a quadratic boolean equation,

[1] Of course every problem in the class NP is reducible in polynomial time to a quadratic $0-1$ maximization problem, since the latter one is NP- complete. By a "natural" reduction we mean one that is 1) concise and 2) easy to figure out. For example, we are not aware of any "natural" reduction of the set covering problem to quadratic $0-1$ maximization.

(b) maximization of a linear function subject to a system of linear equations,

(c) maximization of a quadratic function subject to a system of linear equations,

can be easily reformulated as (unconstrained) quadratic $0-1$ maximization problems.

(a) Consider the problem

$$\begin{aligned} max \quad & c_1x_1 + \cdots + c_nx_n \\ s.t. \quad & \phi(x) = 0, \\ & x \in B^n, \end{aligned} \tag{5.2}$$

where $\phi = T_1 \vee \cdots \vee T_m$ is a quadratic boolean expression. Note that the equation $\phi = 0$ is equivalent to a system of "degree-two" inequalities (cfr. Johnson and Padberg (1980)), i.e. inequalities of the form

$$x_i + x_j \leq 1, \quad x_i \leq x_j \quad or \quad x_i + x_j \geq 1,$$

corresponding to $x_ix_j = 0$, $x_i\overline{x}_j = 0$ and $\overline{x}_i\overline{x}_j = 0$, respectively. This class contains the maximum weighted stable set problem on a graph, which can be formulated as

$$\begin{aligned} max \quad & c_1x_1 + \cdots + c_nx_n \\ s.t. \quad & x_i + x_j \leq 1, \quad \text{for all edges } \langle i,j \rangle, \\ & x \in B^n. \end{aligned}$$

If M is a large constant, problem (5.2) can be formulated as

$$\max_{x \in B^n} c_1x_1 + \cdots + c_nx_n - M(T_1 + \cdots + T_m).$$

(b) The problem

$$\begin{aligned} max \quad & c_1x_1 + \cdots + c_nx_n \\ s.t. \quad & Ax = b, \\ & x \in B^n, \end{aligned} \tag{5.3}$$

can be re-formulated as the quadratic $0-1$ maximization problem

$$\max c_1x_1 + \cdots + c_nx_n - M(Ax-b)^T(Ax-b),$$

where M is a large constant. The *set partitioning* problem is a special case of (5.3), with $b = e = (1, ..., 1)$ and A being any $0-1$ matrix. By further specializing A, b and c, one obtains, in particular, the classical assignment problem.

(c) A similar technique leads to the re-formulation of

$$\begin{aligned} max \quad & x^TQx \\ s.t. \quad & Ax = b, \\ & x \in B^n, \end{aligned} \tag{5.4}$$

as

$$\max_{x \in B^n} x^TQx - M(Ax-b)^T(Ax-b)$$

with the usual meaning of M. Important special cases of (5.4) are

c.1. the quadratic assignment problem,

c.2. the quadratic semiassignment problem,

c.3. the cardinality constrained quadratic $0-1$ maximization problem.

c.1. The quadratic assignment problem is

$$\begin{aligned} min \quad & \sum_{h=1}^{p}\sum_{k=1}^{p}\sum_{i=1}^{n}\sum_{j=1}^{n} d_{ijhk}x_{ih}x_{jk} \\ s.t. \quad & \sum_{j=1}^{n} x_{jk} = 1, \quad k = 1, ..., p, \\ & \sum_{k=1}^{p} x_{jk} = 1, \quad j = 1, ..., n, \\ & x_{jk} \in \{0,1\} \quad \text{for all } j, k. \end{aligned}$$

Layout problems and travelling salesman problems can be formulated in this way [see Burkard and Rendl (1985)].

c.2. The quadratic semiassignment problem is

$$\begin{aligned} min \quad & \sum_{k=1}^{p}\sum_{i=1}^{n}\sum_{j=1}^{n} c_{ij}x_{ik}x_{jk} \\ s.t. \quad & \sum_{k=1}^{p} x_{jk} = 1, \quad j = 1, ..., n \\ & x_{jk} \in \{0,1\}, \quad \text{for all } j, k. \end{aligned} \tag{5.5}$$

Special cases of (5.5) are

- the *clustering* problem: given n objects and an $n \times n$ dissimilarity matrix $[c_{ij}]$, find a partition of the objects into p classes or *clusters* which minimizes the sum of the dissimilarities between objects belonging to a same cluster.
- the *equipartition* problem: given n objects with weights w_i, $i = 1, ..., n$, find a partition of the objects into p classes so as to minimize the variance of the class weights. As noticed by Simeone (1986), this problem can be formulated as (5.5) with $c_{ij} = w_i w_j$ for all i, j.
- *chromatic number*: a graph $G = (V, E)$ admits a coloration of its vertices with p colors (adjacent vertices carrying different colors) iff the optimal value of

$$\begin{aligned} min \quad & \sum_{k=1}^{p}\sum_{(i,j)\in E} x_{ik}x_{jk} \\ s.t. \quad & \sum_{k=1}^{p} x_{jk} = 1, \quad j = 1, ..., n, \\ & x_{jk} \in \{0,1\} \quad \text{for all } j, k \end{aligned}$$

  is zero.

c.3. The cardinality constrained quadratic $0-1$ maximization problem is

$$\begin{aligned} & max \quad x^T Q x \\ & s.t. \quad x_1 + \cdots + x_n = p, \\ & \qquad\;\; x \in B^n, \end{aligned} \tag{5.6}$$

where p is an integer between 0 and n (see Witzgall(1975)). In particular, when Q is the adjacency matrix of a graph G, Gallo, Hammer and Simeone (1980) have observed that G has a clique of order p (i.e. a set of p pairwise adjacent vertices) if and only if the optimal value of (5.6) is $p(p-1)$.

5.2 Reducibility of non-linear $0-1$ optimization to quadratic $0-1$ optimization

Rosenberg (1975) has shown that the problem of maximizing an arbitrary pseudo-boolean function f is reducible to the maximization of a *quadratic* pseudo-boolean function. Basically, his idea consists in replacing a product $x_i x_j$ (appearing in at least one term of f) by a new variable y_{ij}. A quadratic penalty term $-M\,(x_i x_j + (3 - 2x_i - 2x_j) y_{ij})$ is then added to the objective function in order to force the relation $y_{ij} = x_i x_j$ to hold.

5.3 "Virtually quadratic" $0-1$ optimization problems and quadratic graphs

In Sect. 4.2 we have seen that the problem of maximizing a pseudo- boolean function f over B^n can be reduced to that of finding a maximum weight stable set in a weighted graph, which is naturally associated with a given positive representation (c, ϕ) of f.

If the graph happens to have a quadratic code then, using the procedure described in Sect. 4.2, one can find a quadratic posiform ψ *with the same number of terms* as ϕ, in such a way that the optimal solutions to

$$\max_{x \in B^n} \phi(x, \overline{x}) \tag{5.7}$$

can be easily determined, once the optimal solutions to

$$\max_{y \in B^q} \psi(y, \overline{y})$$

are known. In this sense, one might say that the maximization problem (5.7) is "virtually quadratic". Such procedure may be contrasted with Rosenberg's technique (see Sec. 5.2), which is applicable to arbitrary pseudo-boolean functions but usually leads to a considerable increase in the number of terms.

A graph G is said to be *quadratic* if it admits a quadratic conflict code α. *Dummies*, i.e. variables which appear only complemented or only uncomplemented in α, are allowed. If x is any variable appearing in α, we call "color x" the set of all edges $\langle T, T' \rangle$ such that x appears uncomplemented in T and complemented in T', or vice versa. Clearly, a color spans a complete bipartite subgraph of G (if x is a dummy, the color is empty).

Thus, a quadratic graph has the property that its edge set can be covered by complete bipartite graphs (the colors) in such a way that at most two different colors meet at each vertex. Moreover, it is readily seen that this property *characterizes* quadratic graphs. If one further requires that the colors be stars, then the graph is called *bistellar*. A quadratic graph is called *primitive*, *pure*, *mixed* when it admits a primitive, pure or mixed quadratic code, respectively. Note that a quadratic graph is primitive if and only if the colors form a partition of the edge-set. The question of characterizing quadratic graphs appears to be still open. Recently Crama and Hammer (1985) have proved the following negative result.

Theorem 5.1 *Recognizing quadratic primitive graphs is an NP-complete problem.* □

Benzaken, Hammer and Simeone (1980) remarked that the quadratic mixed graphs are precisely the adjoints of directed graphs (recall that the adjoint of the digraph D is the undirected graph whose vertices are the edges of D and where two vertices u and v are adjacent if and only if the head of v coincides with the tail of u or vice versa).

Benzaken, Boyd, Hammer and Simeone (1983) proved that a graph is quadratic primitive pure iff it has an edge-orientation in which ten special forbidden configurations $H_1, ..., H_{10}$ are absent.

Finally, Hammer and Simeone (1980) characterized the bistellar graphs as those for which the subgraph H induced by all vertices of degree ≥ 3 is injective. (i.e. every connected component of H has at most one cycle).

6 Upper planes

Consider the "primal" problem

$$z = \max_{x \in B^n} f(x) \tag{6.1}$$

where f is an arbitrary pseudo-boolean function.

An *upper plane* (or *linear overestimator*) of f is any linear function $p(x) = v_0 + v_1 x_1 + \cdots + v_n x_n$ such that $p(x) \geq f(x)$ for all $x \in B^n$. Replacing f by p results in a *linear relaxation* of (6.1):

$$\max_{x \in B^n} p(x) \tag{6.2}$$

Clearly the optimal value of (6.2) is an upper bound of z Let $\mathcal{U}$ be a set of upper planes of f. We shall make the assumption – usually satisfied in most cases of interest – that the set $\mathcal{U}$, regarded as a subset of R^{n+1}, is convex and closed. We are interested in those upper planes p in $\mathcal{U}$ for which the optimal value of (6.2) is as close as possible to z: this leads to the problem

$$W = \min_{p \in \mathcal{U}} \max_{x \in B^n} p(x) \tag{6.3}$$

the *plane-dual* of (6.1) (with respect to $\mathcal{U}$). An upper plane p^* such that $\max_{x\in B^n} p^*(x) = w$ is called a *best* upper plane (in $\mathcal{U}$).

One has always $w \geq z$, and the difference $w - z$ is called the *plane-duality gap* (with respect to $\mathcal{U}$).

The set of upper planes $\mathcal{U}$ is said to be *complete* if

$$f(x) = \min_{p\in\mathcal{U}} p(x), \quad \text{for all } x \in B^n.$$

In this case then one has

$$w = \min_{p\in\mathcal{U}} \max_{x\in B^n} p(x) \geq \max_{x\in B^n} \min_{p\in\mathcal{U}} p(x) = z. \tag{6.4}$$

7 Roofs

Perhaps the simplest way to obtain upper planes for a quadratic pseudo-boolean function

$$f(x) = x^T Q x = \sum_{i=1}^{n}\sum_{j=1}^{n} q_{ij}x_i x_j \tag{7.1}$$

(where we assume *w.l.o.g.* that $q_{ij} = 0$ whenever $i > j$) consists in generating, for each term $q_{ij}x_i x_j$, a *local* upper plane

$$p(x_i, x_j) = a_{ij}x_i + b_{ij}x_j + c_{ij} \tag{7.2}$$

and then adding up all these local upper planes.

Note that $a_{ij}x_i + b_{ij}x_j + c_{ij}$ is an upper plane of $q_{ij}x_i x_j$ if and only if

$$\begin{array}{c} c_{ij} \geq 0, \;\; a_{ij} + c_{ij} \geq 0, \;\; b_{ij} + c_{ij} \geq 0, \\ a_{ij} + b_{ij} + c_{ij} \geq q_{ij} \end{array} \tag{7.3}$$

Any upper plane of the form

$$\sum_{i=1}^{n}\sum_{j=1}^{n}(a_{ij}x_i + b_{ij}x_j + c_{ij}),$$

where the a_{ij}, b_{ij}, c_{ij} satisfy (7.3), will be called a *paved* upper plane.

Let $P = \{(i,j) \,:\, 1 \leq i < j \leq n,\; q_{ij} > 0\}$ and $N = \{(i,j) \,:\, 1 \leq i < j \leq n,\; q_{ij} < 0\}$.

The above inequalities (7.3) are satisfied if, in particular, one chooses

$$p_{ij}(x_i, x_j) = \begin{cases} \lambda_{ij}x_i + (q_{ij} - \lambda_{ij})x_j, & (i,j) \in P \\ \lambda_{ij}(1 - x_i - x_j), & (i,j) \in N \end{cases} \tag{7.4}$$

where

$$0 \leq \lambda_{ij} \leq |q_{ij}|, \quad \text{for all } (i,j) \in P \cup N. \tag{7.5}$$

Any upper plane of the form

$$p(x, \lambda) = \sum_{(i,j)\in P} \lambda_{ij}x_i + (q_{ij} - \lambda_{ij})x_j + \sum_{(i,j)\in N} \lambda_{ij}(1 - x_i - x_j) \tag{7.6}$$

where the λ_{ij}'s satisfy (7.5), is called a *roof* of f. Grouping terms, one can write

$$p(x,\lambda) = v_0(\lambda) + v_1(\lambda)x_1 + \cdots + v_n(\lambda)x_n, \tag{7.7}$$

where

$$v_0(\lambda) = \sum_{(i,j)\in N} \lambda_{ij}$$

and

$$v_i(\lambda) = q_{ii} + \sum_{j|(j,i)\in P} q_{ji} + \sum_{j|(i,j)\in P} \lambda_{ij} - \sum_{j|(i,j)\in P} \lambda_{ji} - \tag{7.8}$$
$$- \sum_{j|(i,j)\in N} \lambda_{ij} - \sum_{j|(j,i)\in N} \lambda_{ji}.$$

The family of roofs properly includes several known classes of upper planes, such as the *selections* [Hammer and Rosenberg (1974), Hammer,Peled and Sorensen (1977)] and the *penalty - relaxation* upper planes [Hansen (1975), Hammer and Hansen (1981)].

One nice property of roofs is their completeness.

Theorem 7.1 *The set of roofs is complete.*

Proof: It is enough to proof that, for all $x \in B^n$ and for each term $q_{ij}x_ix_j$ of $f(x)$, there is a λ^*_{ij}, $0 \le \lambda^*_{ij} \le |q_{ij}|$, such that

$$q_{ij}x_ix_j = \begin{cases} \lambda^*_{ij}x_i + (q_{ij} - \lambda^*_{ij})x_j, & (i,j) \in P \\ \lambda^*_{ij}(1 - x_i - x_j), & (i,j) \in N. \end{cases}$$

But it is easy to exhibit one such λ^*_{ij}: just take $\lambda^*_{ij} = |q_{ij}|x_i$. □

When f is quadratic and $\mathcal{U}$ is the set of roofs of f, the plane- dual (6.3) will be called the *roof-dual.* As noticed above, the optimal value of the roof-dual is an upper bound of z, the optimal value of the primal.

8 Complementation and the height

We shall introduce now another upper bound of a quadratic pseudo- boolean function f - the "height" of f. As we have seen in Sect. 4.1, there are always a constant c_0 and a quadratic posiform ϕ such that

$$f(x) + \phi(x,\overline{x}) = c_0. \tag{8.1}$$

Obviously, c_0 is an upper bound of $z = \max_{x\in B^n} f(x)$ and is actually equal to z if and only if the quadratic boolean equation

$$\hat{\phi}(x,\overline{x}) = 0$$

is consistent (as in Section 4.1, $\hat{\phi}$ is the boolean frame of ϕ). A word of caution: in general it is not possible to find a *quadratic* posiform ϕ such that $f(x)+\phi(x,\overline{x}) = z$ for all $x \in B^n$.

For example, if $f(x) = -x_1 + x_1x_2 + x_1x_3 - x_2x_3$, then one has $\max_{x\in B^n} f(x) = 0$, but there is no quadratic posiform ϕ such that $f(x) + \phi(x, \overline{x}) = 0$ for all $x \in B^3$.

The above considerations lead to the following question: what is the smallest constant c_0 for which (8.1) holds for some quadratic posiform ϕ? Such smallest constant will be called the *height* of f and will be denoted by $H(f)$. Clearly, one has $H(f) \geq z$. A pseudo-boolean function g will be called a *posifunction* if there is a posiform ϕ such that $g(x) = \phi(x, \overline{x})\ \forall\ x \in B^n$. If in addition ϕ is quadratic then g is called a quadratic posifunction.

For a given quadratic pseudo-boolean function f, there is a unique quadratic posifunction f^* such that

$$f(x) + f^*(x) = H(f). \tag{8.2}$$

Such function f^* will be called the (quadratic) *complement* of f. Some elementary properties of complementation follow.

A quadratic posifunction g is said to be *flat* if there are no quadratic posifunction g' and no constant $k > 0$ such that $g = k + g'$.

Lemma 8.1 : *The complement f^* of any quadratic pseudo-boolean function f is flat.*

Proof: If $f^* = f' + k$ for some quadratic posifunction f' and some $k > 0$ then $f + f' = H(f) - k$, contradicting the definition of height. □

Theorem 8.2 *For any quadratic posifunction g, one has $g^{**} \leq g$, equality holding if and only if g is flat.*

Proof: One has

$$g + g^* = H(g), g^* + g^{**} = H(g^*). \tag{8.3}$$

Notice that $H(g) = g + g^* \geq H(g^*)$ by definition of $H(g^*)$.

From (8.3) one gets $g = g^{**} + k$, where $k = H(g) - H(g^*)$. If g is flat then $k = 0$ and $g^{**} = g^*$. Conversely if $g^{**} = g$ then g is the complement of g^* and hence is flat by Lemma 8.1. □

Corollary 8.3 *For every quadratic pseudo-boolean function f, one has $f^{***} = f^*$.* □

9 Linearization

In order to establish a third upper bound on the optimal value of the primal, we shall re-formulate the problem using on idea originally proposed by Rhys (1970).

Thus, let us replace in $f(x)$ each quadratic term $q_{ij}x_ix_j$ with negative coeffient by $q_{ij}x_j - q_{ij}\overline{x}_ix_j$. Then the objective function takes the form

$$f(x) = \sum_{(i,j)\in P} q_{ij}x_ix_j - \sum_{(i,j)\in N} q_{ij}\overline{x}_ix_j + \sum_{i=1}^{n}(q_{ii} + \sum_{j|(j,i)\in N} q_{ji})x_i. \tag{9.1}$$

Introducing new $0-1$ variables associated with the quadratic terms $x_i x_j$ and constraining them to take the value of the products of the two literals in these terms, we arrive at the following linear $0-1$ program

$$\max \sum_{(i,j)\in P} q_{ij}y_{ij} - \sum_{(i,j)\in N} q_{ij}y_{ij} + \sum_{i=1}^{n}(q_{ii} + \sum_{j|(j,i)\in N} q_{ji})x_i \tag{9.2}$$

s.t.

$$\begin{aligned} y_{ij} &\le x_i, & (i,j) &\in P & (9.3)\\ y_{ij} &\le x_j, & (i,j) &\in P & (9.4)\\ y_{ij} &\le 1 - x_i, & (i,j) &\in N & (9.5)\\ y_{ij} &\le x_j, & (i,j) &\in N & (9.6)\\ x_i &\in \{0,1\}, & i &= 1,...,n & (9.7)\\ y_{ij} &\in \{0,1\}, & (i,j) &\in P \cup N. & (9.8) \end{aligned}$$

This is the *discrete Rhys form* (drf) of the primal and it can be easily seen to be equivalent to the primal. Rhys (1970) has noticed that when $N = \emptyset$, the constraints (9.7) and (9.8) can be replaced by

$$x_i \le 1 \tag{9.9}$$

$$x_i \ge 0, y_{ij} \ge 0 \tag{9.10}$$

because of the total unimodularity of the constraint matrix (clearly, the constraints $y_{ij} \ge 0$ are inessential).

For $N \ne \emptyset$, the linear program (9.2)-(9.6), (9.9),(9.10) is a relaxation of the primal and will be called the *continuous Rhys form* (crf) of the primal. Its optimal value z_{crf} is obviously an upper bound of the primal optimum z.

Example. The quadratic function

$$\begin{aligned} f(x) = \;& 6x_3 + 10x_4 + 6x_1x_2 - 2x_1x_3 - \\ & -10x_1x_4 + 2x_2x_3 - 8x_2x_4 \end{aligned} \tag{9.11}$$

can also be written as

$$\begin{aligned} f(x) = \;& -9 + 4x_3 + 8\overline{x}_4 + 6x_1x_2 + 2\overline{x}_1x_3 + \\ & +10\overline{x}_1x_4 + 2x_2x_3 + 8\overline{x}_2x_4 \end{aligned} \tag{9.12}$$

The crf of f is

$$\begin{array}{lll} max & 4x_3 - 8x_4 + 6y_{12} + 2y_{13} + 10y_{14} + 2y_{23} + 8y_{24} & \\ s.t. & y_{12} \le x_1, & y_{12} \le x_2 \\ & y_{13} \le 1 - x_1, & y_{13} \le x_3 \\ & y_{14} \le 1 - x_1, & y_{14} \le x_4 \\ & y_{23} \le x_2, & y_{23} \le x_3 \\ & y_{24} \le 1 - x_2, & y_{24} \le x_4 \\ & 0 \le x_i \le 1, & i = 1, ..., 4 \\ & y_{ij} \ge 0 & \text{for all } (i,j) \end{array} \tag{9.13}$$

The expression (9.1) of $f(x)$ involves a somewhat arbitrary choice of the variables to be complemented (e.g

$$-x_i x_j = -x_i + \overline{x}_i x_j = -x_j + x_i \overline{x}_j = \frac{1}{2}(-x_i - x_j + \overline{x}_i x_j + x_i \overline{x}_j)$$

and so on).

Obviously similar linear programs would be obtained if one would start by any other expression of f in which the quadratic terms have positive coefficients.

Theorem 9.2 below states that the optima of all such linear programs actually coincide. A key tool for proving this result is the representation theorem 9.1, which provides an explicit description of all expressions of f with positive quadratic terms.

Theorem 9.1 *Let $f(x)$ be given by (7.1) and $\Phi(x, \overline{x})$ be a homogeneous quadratic normal posiform and l a linear function. Then one has*

$$f(x) = \Phi(x, \overline{x}) + l(x) \quad \text{for all } x \in B^n \tag{9.14}$$

if and only if there exist λ_{ij}, w_{ij} $((i,j) \in P \cup N)$, such that

$$\Phi(x, \overline{x}) = \sum_{(i,j)\in P} (\lambda_{ij} x_i x_j + w_{ij} \overline{x}_i \overline{x}_j) + \sum_{(i,j)\in N} (\lambda_{ij} \overline{x}_i x_j + w_{ij} x_i \overline{x}_j). \tag{9.15}$$

and

$$l(x) = l_0 + \sum_{i=1}^{n} l_i x_i \tag{9.16}$$

where

$$\lambda_{ij} \ge 0, \; w_{ij} \ge 0, \quad (i,j) \in P \cup N, \tag{9.17}$$

$$\lambda_{ij} + w_{ij} = |q_{ij}|, \quad (i,j) \in P \cup N, \tag{9.18}$$

and

$$l_i = q_{ii} + \sum_{j|(j,i)\in P} w_{ji} + \sum_{j|(j,i)\in P} w_{ij} - \sum_{j|(j,i)\in N} \lambda_{ji} - \sum_{j|(i,j)\in N} w_{ij}, \; i = 1, 2, ..., n, \tag{9.19}$$

$$l_0 = - \sum_{(i,j)\in P} w_{ij}. \tag{9.20}$$

Proof: If (9.15)-(9.20) hold, then, by replacing $\overline{x}_i$ by $1 - x_i$ everywhere in the r.h.s. of (9.14) and simplifying, we actually get $f(x)$. Conversely, assume that (9.14) holds and that Φ is given by

$$\Phi(x, \overline{x}) = \sum(\alpha_{ij}x_ix_j + \beta_{ij}\overline{x}_i\overline{x}_j + \gamma_{ij}\overline{x}_ix_j + \delta_{ij}x_i\overline{x}_j) \tag{9.21}$$

where the $\alpha_{ij}, \beta_{ij}, \gamma_{ij}$ and δ_{ij} are nonnegative. By the normality assumption, we see that the set of index pairs (i, j) with $i < j$ is partitioned into three sets

$$S_1 \equiv \{(i,j) |\ \alpha_{ij} + \beta_{ij} > 0, \gamma_{ij} = \delta_{ij} = 0\},$$

$$S_2 \equiv \{(i,j) |\ \gamma_{ij} + \delta_{ij} > 0, \alpha_{ij} = \beta_{ij} = 0\},$$

$$S_3 \equiv \{(i,j) |\ \alpha_{ij} = \beta_{ij} = \gamma_{ij} = \delta_{ij} = 0\}.$$

Therefore, after replacement of $\overline{x}_i$ by $1 - x_i$ $(i = 1, ..., n)$, the sum of the quadratic terms of $f(x) - \Phi(x, \overline{x})$ becomes

$$\sum_{(i,j)\in S_1} (q_{ij} - \alpha_{ij} - \beta_{ij})x_{ij} + \sum_{(i,j)\in S_2} (q_{ij} + \gamma_{ij} + \delta_{ij})x_{ij}. \tag{9.22}$$

From the uniqueness of the polynomial representation (without complemented variables) of a pseudo-boolean function, it follows that $q_{ij} - \alpha_{ij} - \beta_{ij} = 0$ for all $(i,j) \in S_1$ and $q_{ij} + \gamma_{ij} + \delta_{ij} = 0$ for all $(i,j) \in S_2$. From the nonnegativity of $\alpha, \beta, \gamma, \delta$ it follows now that $S_1 = P$, $S_2 = N$ and thus (9.15), (9.17), (9.18) hold. Moreover,

$$\begin{aligned} l(x) &= l_0 + \sum_{i=1}^n l_ix_i = \\ &= f(x) - \Phi(x, \overline{x}) = \\ &= \sum_{i=1}^n q_{ii}x_i + \sum_{(i,j)\in P} q_{ij}x_ix_j - \sum_{(i,j)\in P} \lambda_{ij}x_ix_j - \\ &- \sum_{(i,j)\in P} w_{ij}(x_ix_j - x_i - x_j + 1) + \sum_{(i,j)\in N} q_{ij}x_ix_j - \\ &- \sum_{(i,j)\in N} \lambda_{ij}(x_j - x_ix_j) - \sum_{(i,j)\in N} w_{ij}(x_i - x_ix_j). \end{aligned}$$

Because of (9.18), all quadratic terms cancel out, and (9.19), (9.20) then follow from the uniqueness of the polynomial representation of $l(x)$. □

Let us consider the general expression of f given by (9.14)-(9.20) and associate with it the general form of the *crf* (the *gcrf*).

$$\max \sum_{(i,j)\in P\cup N} (\lambda_{ij}y_{ij} + w_{ij}t_{ij}) + \sum_{i=1}^n l_ix_i + c \tag{9.23}$$

s.t.

$$\left.\begin{aligned} y_{ij} &\le x_i, \\ y_{ij} &\le x_j, \\ t_{ij} &\le 1 - x_i, \\ t_{ij} &\le 1 - x_j, \end{aligned}\right\} \quad (i,j) \in P, \tag{9.24}$$

$$\left.\begin{array}{l} y_{ij} \leq 1 - x_i, \\ y_{ij} \leq x_j, \\ t_{ij} \leq x_i, \\ t_{ij} \leq 1 - x_j, \end{array}\right\} \quad (i,j) \in N, \tag{9.25}$$

$$0 \leq x_i \leq 1, \quad i = 1, \ldots, n, \tag{9.26}$$

$$y_{ij} \geq 0, \;\; t_{ij} \geq 0, \quad (i,j) \in P \cup N. \tag{9.27}$$

where λ, W and l satisfy (9.17)-(9.20).

Clearly, the Rhys linear program of f is obtained from the general Rhys linear program by setting all $w_{ij} = 0$.

Theorem 9.2 *The optimum value of z_{crf} of the crf of f is equal to the optimum value $\hat{z}$ of the $gcrf$ of f.*

Proof: The linear program (9.23)-(9.27) always has an optimal solution (x^*, y^*, z^*) such that

$$\left.\begin{array}{l} y^*_{ij} = \min\{x^*_i, x^*_j\}, \\ t^*_{ij} = \min\{1 - x^*_i, 1 - x^*_j\}, \end{array}\right\} \quad (i,j) \in P,$$

$$\left.\begin{array}{l} y^*_{ij} = \min\{1 - x^*_i, x^*_j\}, \\ t^*_{ij} = \min\{x^*_i, 1 - x^*_j\}, \end{array}\right\} \quad (i,j) \in N.$$

Hence, if $U^n = [0,1]^n$ is the full unit n-cube, we have

$$\begin{aligned} \hat{z} \;&= \max_{x \in U^n} \left\{ \sum_{(i,j)\in P} \lambda_{ij} \min\{x_i, x_j\} + \right. \\ &+ \sum_{(i,j)\in P} w_{ij} \min\{1 - x_i, 1 - x_j\} + \\ &+ \sum_{(i,j)\in N} \lambda_{ij} \min\{1 - x_i, x_j\} + \\ &+ \sum_{(i,j)\in N} w_{ij} \min\{x_i, 1 - x_j\} + \\ &+ \sum_{i=1}^{n} \left(q_{ii} + \sum_{j|(i,j)\in P} w_{ij} + \sum_{j|(j,i)\in P} w_{ji} - \right. \\ &\left.\left. - \sum_{j|(j,i)\in N} \lambda_{ji} - \sum_{j|(i,j)\in N} w_{ij} \right) x_i - \sum_{(i,j)\in P} w_{ij} \right\} \end{aligned} \tag{9.28}$$

and in particular

$$
\begin{aligned}
z_{crf} &= \max_{x \in U^n} \Bigg\{ \sum_{(i,j)\in P} q_{ij} \min\{x_i, x_j\} - \\
&\quad - \sum_{(i,j)\in N} q_{ij} \min\{1 - x_i, x_j\} + \\
&\quad + \sum_{i=1}^{n} \left(q_{ii} + \textstyle\sum_{j|(i,j)\in N} q_{ij} \right) x_i \Bigg\}.
\end{aligned} \tag{9.29}
$$

For all $x \in U^n$ and for all $(i, j) \in P$, one has

$$
\begin{aligned}
&\lambda_{ij} \min\{x_i, x_j\} + w_{ij} \min\{1 - x_i, 1 - x_j\} + w_{ij}x_j + w_{ij}x_i - w_{ij} \\
&\quad = \lambda_{ij} \min\{x_i, x_j\} + w_{ij}(1 - \max\{x_i, x_j\}) + w_{ij}x_j + w_{ij}x_i - w_{ij} \\
&\quad = \lambda_{ij} \min\{x_i, x_j\} + w_{ij}(x_i + x_j - \max\{x_i, x_j\}) \\
&\quad = \lambda_{ij} \min\{x_i, x_j\} + w_{ij} \min\{x_i, x_j\}
\end{aligned}
$$

and thus

$$
\begin{aligned}
&\lambda_{ij} \min\{x_i, x_j\} + w_{ij} \min\{1 - x_i, 1 - x_j\} + \\
&\quad + w_{ij}x_j + w_{ij}x_i - w_{ij} \\
&\quad = q_{ij} \min\{x_i, x_j\}.
\end{aligned} \tag{9.30}
$$

Similarly, for all $x \in U^n$ and $(i, j) \in N$, one has

$$
\begin{aligned}
&\lambda_{ij} \min\{1 - x_i, x_j\} + w_{ij} \min\{x_i, 1 - x_j\} - v_{ij}x_ix_j - w_{ij}x_i \\
&\quad = \lambda_{ij} \min\{1 - x_i, x_j\} + w_{ij}(1 - \max\{1 - x_i, x_j\}) + \\
&\qquad + q_{ij}x_j + w_{ij}(1 - x_i) - w_{ij} \\
&\quad = \lambda_{ij} \min\{1 - x_i, x_j\} + w_{ij}(1 - x_i + x_j - \max\{1 - x_i, x_j\}) + q_{ij}x_j \\
&\quad = \lambda_{ij} \min\{1 - x_i, x_j\} + w_{ij} \min\{1 - x_i, x_j\} + q_{ij}x_j
\end{aligned}
$$

and thus

$$
\begin{aligned}
&v_{ij} \min\{1 - x_i, x_j\} + w_{ij} \min\{x_i, 1 - x_j\} - \\
&\quad - v_{ij}x_j - w_{ij}x_i \\
&\quad = -q_{ij} \min\{1 - x_i, x_j\} + q_{ij}x_j.
\end{aligned} \tag{9.31}
$$

Adding up all identities (9.30) for $(i, j) \in P$ and all identities (9.31) for $(i, j) \in N$ and taking into account (9.28) and (9.29) we finally get $\hat{z} = z_{crf}$. □

10 Equivalence between roof duality, complementation and linearization

In the previous section it was shown that each of there different approaches under consideration, namely roof duality, complementation and linearization, leads to an upper bound on the primal optimum $z = \max_{x \in B^n} f(x)$. Actually, the three upper bounds are

(i) w_R, the optimum value of the roof dual (Section 6)

(ii) $H(f)$, the height of f (Section 7)

(iii) z_{crf}, the optimum value of the continuous Rhys form of f (Section 8).

The main result of the present section is that $w_R = H(f) = z_{crf}$.

10.1 Upper planes with quadratic residual

In order to prove the identity $w_R = H(f)$, we shall first prove the fact that the roofs are precisely those upper planes which have a quadratic residual.

Let ϕ be a quadratic posiform such that $f + \phi = c_0$ for some constant c_0. We can write

$$\phi(x,\bar{x}) = \phi_0 + \phi_l(x,\bar{x}) + \phi_q(x,\bar{x}), \tag{10.1}$$

where ϕ_0 is a constant, ϕ_l is the sum of the linear terms of ϕ and ϕ_q is the sum of the quadratic terms. Then

$$f(x) + \phi_q(x,\bar{x}) = c_0 - \phi_0 - \phi_l(x,\bar{x}) \tag{10.2}$$

Clearly, the r.h.s. of (10.2) is an upper plane of f. An upper plane $p(x)$ of $f(x)$ such that there exists a "purely" quadratic normal posiform ϕ_q satisfying

$$f(x) + \phi_q(x,\bar{x}) = p(x), \quad x \in B^n \tag{10.3}$$

is said to have a *quadratic residual.* For example, if $f = x_1, x_2$ and $p(x) = \frac{1}{2}x_1 + \frac{1}{2}x_2$, then $\phi_q(x,\bar{x}) = \frac{1}{2}\bar{x}_1 x_2 + x_1\bar{x}_2$.

Theorem 10.2 below characterizes the roofs as those upper planes having a quadratic residual.

Lemma 10.1 *Let $\lambda = [\lambda_{ij}]$ be such that $0 \le \lambda_{ij} \le |q_{ij}|$ for all $(i,j) \in P \cup N$, and let*

$$\begin{aligned} Q(x,\lambda) &= \sum_{(i,j)\in P} [\lambda_{ij} x_i \bar{x}_j + (q_{ij} - \lambda_{ij})\bar{x}_i x_j] \\ &\quad + \sum_{(i,j)\in N} [\lambda_{ij}\bar{x}_i\bar{x}_j + (q_{ij} + \lambda_{ij}) x_i x_j]. \end{aligned} \tag{10.4}$$

Then $Q(x,\lambda)$ is a quadratic residual of f and all quadratic residuals of f are of this form.

Proof: The quadratic normal posiform Q is a quadratic residual of f if and only if there is a linear function p such that $-f = Q - p$. The statement then immediately follows by applying Theorem 9.1 to $-f$. We now prove the announced result. □

Theorem 10.2 *An upper plane of the quadratic function f is a roof if and only if it has a quadratic residual. Moreover, if a roof $p(x,\lambda)$ is given by (7.6), then its own quadratic residual $Q(x,\lambda)$ is given by (10.4) with the same values of the parameters λ_{ij}, and vice versa.*

Proof: Let $p(x, \lambda)$ be a roof given by (7.6). Writing

$$\lambda_{ij}x_i + (q_{ij} - \lambda_{ij})x_j - q_{ij}x_ix_j = \lambda_{ij}x_i\overline{x}_j + (q_{ij} - \lambda_{ij})\overline{x}_ix_j$$

for each $(i, j) \in P$ and

$$\lambda_{ij}(1 - x_i - x_j) - q_{ij}x_ix_j = \lambda_{ij}\overline{x}_i\overline{x}_j - (q_{ij} - \lambda_{ij})x_ix_j$$

for each $(i, j) \in N$; and adding up all such identities we get

$$p(x, \lambda) - f(x) = Q(x, \lambda)$$

and hence p has a quadratic residual.

Conversely, let $p(x)$ be an arbitrary upper plane with a quadratic residual $Q(x, \lambda)$. Substituting everywhere $1 - x_i$ for $\overline{x}_i$ $(i = 1, ..., n)$ in the l.h.s. of the relation $f(x) + Q(x, \lambda) = p(x)$, we find that $p(x)$ is precisely the roof $p(x)$ given by (7.6). □

10.2 Equivalence between roof duality and complementation

Theorem 10.3 *The optimal value w_R of the roof-dual is equal to the height $H(f)$ of f.*

Proof: Let $p(x) = v_0 + v_1x_1 + \cdots + v_nx_n$ be any roof. By Theorem 10.2, p has a quadratic residual Q and

$$f + Q = p \tag{10.5}$$

On the other hand, let

$$\begin{aligned} L(x, \overline{x}) &= \textstyle\sum_{j=1}^n v_j^+ x_j - \sum_{j=1}^n v_j^- x_j \\ &\text{and} \\ c &= v_0 + \textstyle\sum_{j=1}^n v_j^+ \end{aligned} \tag{10.6}$$

where, as customary, $z^+ = \max\{0, z\}$ and $z^- = \min\{0, z\}$. L is a linear posiform and

$$p + L = c \tag{10.7}$$

Adding up (10.5) and (10.7) yields

$$f + h = c, \quad \text{where} \quad h = Q + L.$$

Since h is a quadratic posifunction, we have $c \geq H(f)$ by the definition of height. On the other hand, by (10.6), $\max_{x \in B^n} p(x) = c$. Hence $w_R \geq H(f)$. Let us prove the reverse inequality. One has

$$f + f^* = H(f), \tag{10.8}$$

f^* being the complement of f. Since f^* is a quadratic posifunction, there is a quadratic normal posiform ϕ such that $\phi = f^*$. Since f^* is flat by Lemma 8.1, ϕ must be homogeneous. Let us denote by Q and by L the sum of the quadratic terms and of the linear terms of ϕ, respectively. Then

$$f^* = Q + L. \tag{10.9}$$

Define the linear function p by

$$p(x) = H(f) - L(x, \overline{x}). \tag{10.10}$$

Then from (10.8), (10.9) and (10.10) one obtains

$$f + Q = p. \tag{10.11}$$

The last identity shows that p has a quadratic residual. But then, by Theorem 10.2, p is a roof. Moreover, in view of (10.10), $\max_{x \in B^n} p(x) = H(f)$. Hence $w_R \leq H(f)$, and the proof is completed. □

10.3 Equivalence between roof-duality and linearization

In this subsection we prove that the optimum w_R of the roof-dual coincides with the optimum z_{crf} of the continuous Rhys form of f. We preliminarly show that the roof-dual

$$\min_{p \in \mathcal{R}} \max_{x \in B^n} p(x) \tag{10.12}$$

where $\mathcal{R}$ is the set of roofs, can be formulated as a linear program with special structure.

Theorem 10.4 *The roof dual (10.12) can be expressed as a (specially structured) linear program.*

Proof: Denoting by Λ the set

$$\{\lambda = [\lambda_{ij}] : 0 \leq \lambda_{ij} \leq |q_{ij}|, (i,j) \in P \cup N\}$$

we have

$$\begin{aligned} w_R &= \min\nolimits_{p \in \mathcal{R}} \max\nolimits_{x \in B^n} p(x, \lambda) \\ &= \min\nolimits_{\lambda \in \Lambda} \max\nolimits_{x \in B^n} \{v_0(\lambda) + v_1(\lambda)x_1 + \cdots + v_n(\lambda)x_n\} \\ &= \min\nolimits_{\lambda \in \Lambda} \{v_0(\lambda) + v_1^+(\lambda) + \cdots + v_n^+(\lambda)\}. \end{aligned} \tag{10.13}$$

where $a^+ = \max\{0, a\}$ Hence

$$\begin{aligned} w_R = \quad \min \quad & v_0(\lambda) + \textstyle\sum_{i=1}^n u_i \\ \text{s.t.} \quad & \left.\begin{aligned} u_i &\geq v_i(\lambda) \\ u_i &\geq 0 \end{aligned}\right\} \quad i = 1, ..., n \\ & 0 \leq \lambda_{ij} \leq |q_{ij}|, \quad (i,j) \in P \cup N. \end{aligned} \tag{10.14}$$

Since the $v_i(\lambda)$, $i = 0, 1, ..., n$, are linear functions of the λ_{ij}'s (see (7.8)), it follows that (10.14) is a linear program in the variables u_i and λ_{ij}. □

Notice that all coefficients in the constraints of (10.14), are $0, 1$, or -1. Disregarding the upper bounds $\lambda_{ij} \leq |q_{ij}|$ there are at most two non-zero elements per column, both being equal to 1, if the column corresponds to a variable λ_{ij}, $(i,j) \in N$; and being equal to 1 and -1, respectively, if the column corresponds to some λ_{ij}, with $(i,j) \in P$; finally, columns corresponding to variables u_i are unit columns. In fact, (10.14) is a bidirected flow problem (cfr Lawler (1976)).

Example. Consider again the quadratic pseudo-boolean function

$$f(x) = 6x_1x_2 - 2x_1x_3 - 10x_1x_4 + 2x_2x_3 - 8x_2x_4 + 6x_3 + 10x_4,$$

to be maximized. Then the roof-dual can be written as

$$\begin{array}{ll} \min & \lambda_{13} + \lambda_{14} + \lambda_{24} + u_1 + u_2 + u_3 + u_4 \\ \text{s.t} & u_1 - \lambda_{12} + \lambda_{13} + \lambda_{14} \geq 0, \\ & u_2 + \lambda_{12} - \lambda_{23} + \lambda_{24} \geq 6, \\ & u_3 + \lambda_{13} + \lambda_{23} \geq 8, \\ & u_4 + \lambda_{14} + \lambda_{24} \geq 10, \\ & u_1, u_2, u_3, u_4 \geq 0, \end{array} \tag{10.15}$$

$0 \leq \lambda_{12} \leq 6$, $0 \leq \lambda_{13} \leq 2$, $0 \leq \lambda_{14} \leq 10$, $0 \leq \lambda_{23} \leq 2$, $0 \leq \lambda_{24} \leq 8$.

The second main result of this section is the following.

Theorem 10.5 *The roof-dual is the linear programming dual of the continuous Rhys form* crf *of f. Hence $w_R = z_{crf}$.*

Proof: After the introduction of slack variables $\mu_{ij} = |q_{ij}| - \lambda_{ij}$, the roof-dual takes the following *symmetric form*:

$$\begin{array}{lll} \min & \sum_{i=1}^{n} u_i + \sum_{(i,j)\in N} \lambda_{ij} & \\ \text{s.t.} & u_i - \sum_{j|(i,j)\in P} \lambda_{ij} - \sum_{j|(j,i)\in P} \mu_{ji} & \\ & + \sum_{j|(i,j)\in N} \lambda_{ij} - \sum_{j|(j,i)\in N} \mu_{ij} & \\ & \geq q_{ii} + \sum_{j|(j,i)\in N} q_{ji}, & i = 1, ..., n \\ & \left.\begin{array}{l} \lambda_{ij} + \mu_{ij} = |q_{ij}| \\ \lambda_{ij}, \mu_{ij} \geq 0 \end{array}\right\}, & (i,j) \in P \cup N. \end{array} \tag{10.16}$$

Let us now consider the crf (9.2) - (9.6), (9.9), (9.10), and let us associate dual variables λ_{ij} with the constraints (9.3) and (9.5), μ_i with the constraints (9.4) and (9.6), and u_i with the constraints (9.9). It is easily checked that the dual of the crf is precisely the roof-dual in its symmetric form 10.16.

Remark 10.6 *The above theorem implies that the roof duality gap is the same as the integrality gap between the continuous and the discrete Rhys form.*

11 Elementary boolean operations

Following Bourjolly, Hammer and Simeone (1983), we shall outline in this section a fourth interpretation of the roof-duality bound.

Let us start from the simple observation that, if ξ and η are literals, then the following identities hold:

$$\xi + \overline{\xi} = 1 \tag{11.1}$$

$$\xi\eta + \overline{\xi}\eta = \eta \tag{11.2}$$

$$\xi + \overline{\xi}\eta = \eta + \xi\overline{\eta} \tag{11.3}$$

Accordingly, we can define three *elementary boolean operations* mapping quadratic posiforms into quadratic posiforms:

$\mathcal{O}_1$ (FUSION): If both linear terms $c\xi$ and $c\overline{\xi}$ are present, then replace them by the constant $c > 0$.

$\mathcal{O}_2$ (CONDENSATION): If both terms $c\xi\eta$ and $c\overline{\xi}\eta$ are present, then replace them by the linear term $c\eta$.

$\mathcal{O}_3$ (EXCHANGE): If the terms $c\xi$ and $c\overline{\xi}\eta$ are present, then replace them by the pair of terms $c\eta$ and $c\xi\overline{\eta}$.

In addition to the three above boolean operations, we shall make use of the following ordinary algebra operation:

$\mathcal{O}_4$ (SPLIT/MERGE): Replace a term cM, where c is a positive constant and M is a monomial, by the sum $c_1M + c_2M$, where $c_1 + c_2 = c$ and $c_1, c_2 > 0$; and vice versa replace $c_1M + c_2M$ by cM.

Starting from a homogeneous quadratic posiform ϕ_0 and executing any of the above four operations – and hence also any finite sequence of them – we obtain another homogeneous quadratic posiform ϕ and a constant $k \geq 0$ such that $\phi_0 = \phi + k$. We shall then say that the constant k *has been squeezed out of* ϕ_0.

Now, given an arbitrary quadratic pseudo-boolean function f, we know that we can always write $f + \phi_0 = c_0$, where c_0 is a constant and ϕ_0 is a homogeneous quadratic posiform. What is the largest constant k^* that can be squeezed out of ϕ_0? Certainly, k^* can never exceed $c_0 - H(f)$, where $H(f)$ is the height of f, because from the relation $\phi_0 = \phi + k^*$ one gets $f + \phi = c_0 - k^* \geq H(f)$ by the definition of height. Actually, the following striking result holds.

Theorem 11.1 [Bourjolly,Hammer,Simeone (1983)] *Let f be a quadratic pseudo-boolean function, ϕ_0 homogeneous quadratic posiform and c_0 a constant such that*

$$f + \phi_0 = c_0.$$

Then the largest constant k^* which can be squeezed out from ϕ_0 by executing a finite sequence of the elementary operations $\mathcal{O}_1$, $\mathcal{O}_2$, $\mathcal{O}_3$, $\mathcal{O}_4$ is precisely equal to $c_0 - H(f)$.

The proof of the above result is quite lengthy and hence will be omitted here.

Bourjolly,Hammer and Simeone (1983) describe a "squeezing" algorithm, based on the above ideas, for getting an upper bound on the maximum of a quadratic pseudo-boolean function f.

12 Equivalence between roof-duality and paved duality

We have already mentioned (Sec. 6) that roofs are a special case of paved upper planes. Let us denote by R the set of all roofs and by Π the set of all paved upper planes. In analogy with the roof-dual

$$w(R) = \min_{p \in R} \max_{x \in B^n} p(x) \tag{12.1}$$

one can introduce the *paved-dual*

$$w(\Pi) = \min_{p \in \Pi} \max_{x \in B^n} p(x). \tag{12.2}$$

Clearly $z \leq w(\Pi) \leq w(R)$, where z, as usual, is the maximum of f over B^n. One might think that, in general, the upper bound $w(\Pi)$ is sharper than $w(R)$. Surprisingly, as recently shown by Lu and Simeone (1987), the two bounds turn out to be always equal.

Theorem 12.1 *One has* $w(\Pi) = w(R)$.

Proof: See Lu and Simeone (1987). □

The above theorem implies that roof-duality and paved-duality are in fact equivalent – in the sense that they yield the same upper bound of the quadratic optimum z – and thus provides a further interpretation of roof-duality. Essentially, the theorem says that by considering arbitrary upper planes obtained by "termwise bounding" one cannot hope to do any better than by using only roofs.

13 "Local" vs "Global" concave envelopes

Still another perspective on roof duality is provided by the work of Hammer and Kalantari (1986). Their viewpoint is related to – but conceptually different from – the linearization approach described in Section 9. Given the quadratic pseudo-boolean function $f(x) = x^T Q x$ (where $q_{ij} = 0$ whenever $i > j$), they introduce the function

$$\begin{aligned} r(x) = & \sum_{(i,j) \in P} q_{ij} \min\{x_i, x_j\} - \sum_{(i,j) \in N} q_{ij} \min\{1 - x_i, x_j\} \\ & + \sum_{i=1}^{n} \left(q_{ii} + \sum_{j|(j,i) \in N} q_{ji} \right) x_i \end{aligned} \tag{13.1}$$

defined for all x in the full unit n-cube U^n. We shall call r the *tent* of f.

The function $r(x)$ is

(i) piecewise-linear,

(ii) concave,

(iii) an upper function (or overestimator) of f in U^n; that is,

$$r(x) \geq f(x), \quad \text{for all } x \in U^n,$$

(iv) coincides with $f(x)$ when $x \in B^n$.

Properties (i) and (ii) are easy to check. Properties (iii), (iv) follow from the expression (9.1) of f (which holds for all $x \in U^n$) and from the fact that $0 \leq \xi, \eta \leq 1$ implies $\xi\eta \leq \min\{\xi, \eta\}$, with equality when $\xi, \eta \in \{0, 1\}$. Moreover, the following result holds.

Theorem 13.1 [Hammer,Hansen and Simeone (1984), formula (1.54); Hammer and Kalantari (1986), Thm 2.1] *The maximum of $r(x)$ over U^n is equal to the optimum z_{crf} of the continuous Rhys form.*

We have already proved this theorem: cfr (9.29). □

After numbering the extreme points of U^n (i.e. the points of B^n) as x^1, x^2, ..., x^q, where $q = 2^n$, let us define, for all $x \in U^n$,

$$\begin{aligned} e(x) = \quad \max \quad & \sum_{i=1}^{q} \alpha_i f(x^i) \\ s.t. \quad & \sum_{i=1}^{q} \alpha_i x^i = x \\ & \sum_{i=1}^{q} \alpha_i = 1 \\ & \alpha_i \geq 0, \quad i = 1, ..., q. \end{aligned} \tag{13.2}$$

The function $e(x)$ is the concave envelope of f over U^n [Falk and Hoffman (1976)]. It can be shown that $e(x)$ is a concave upper function of f in U^n: actually $e(x)$ is a tightest concave upper function of f, in the sense that $e(x) \leq g(x)$ for all concave upper functions g of f. Furthermore, $e(x) = f(x)$ for all $x \in B^n$ and

$$\max_{x \in U^n} e(x) = \max_{x \in B^n} f(x).$$

While one has $e(x) \leq r(x)$ for all $x \in U^n$, equality does not have to hold in general. However, as shown by Hammer and Kalantari (1986), $r(x)$ does coincide with $e(x)$ in U^n in the special case when $q_{ij} \geq 0$ for all i, j. On the other hand, it is worth pointing out the following property.

Proposition 13.2 *The concave envelope of $q_{ij}x_ix_j$ in U^2 is given by*

$$f_{ij}(x_i, x_j) = \begin{cases} q_{ij} \min\{x_i, x_j\}, & (i,j) \in P \\ q_{ij}x_j - q_{ij} \min\{1 - x_i, x_j\}, & (i,j) \in N \end{cases}$$

The proof is left to the reader. □

If we re-write $r(x)$ as

$$\begin{aligned} r(x) &= \sum_{(i,j)\in P} q_{ij}\min\{x_i, x_j\} + \sum_{i=1}^{n} q_{ii}x_i + \\ &\quad + \sum_{(i,j)\in N} (q_{ij}x_j - q_{ij}\min\{1 - x_i, x_j\}) \\ &= \sum_{(i,j)\in P\cup N} f_{ij}(x_i, x_j) + \sum_{i=1}^{n} q_{ii}x_i, \end{aligned}$$

we see that $r(x)$ is precisely the sum of the concave envelopes of the individual terms in f.

Thus we might say that $r(x)$ is a "termwise" or "locally" tightest concave overestimator of f. The fact that in general $r(x)$ does not coincide with the "globally" tightest concave overestimator $e(x)$ explains the presence of a gap

$$\left(\max_{x\in U^n} r(x) - \max_{x\in U^n} e(x)\right),$$

which turns out to be exactly the roof-duality gap, as shown above.

14 Weighted stability in graphs and the König-Egerváry property

In this section, we recall some graph-theoretic results which will be exploited in the subsequent sections. Let $G = (V, E)$ an undirected graph with $|V| = n$ vertices and $|E| = m$ edges. Recall that a stable set of G is a set S of pairwise nonadjacent vertices.

The *stability number* $\alpha(G)$ of G is the maximum cardinality of a stable set of G. The binary points of the polytope

$$\mathcal{P} = \{x : x \in R^n,\ x \geq 0,\ x_i + x_j \leq 1 \text{ for all } \langle i,j\rangle \in E\}$$

are the characteristic vectors of the stable sets of G. By analogy, the extreme points of $\mathcal{P}$ are called *fractional stable sets* of G.

Theorem 14.1 [Balinski (1968)], *Each extreme point of* $\mathcal{P}$ *has components* $0, 1$ *or* $\frac{1}{2}$.

□

Assume the vertices of G to have weights w_i, $i = 1, ..., n$. We shall make the assumption that the weights w_i are integers. This assumption is not too restrictive: in practice, one can always take the w_i to be rational numbers, and multiplying them by a suitable integer one gets integral weights. Consider the "maximum weighted stable set problem" on G

$$(WS)\qquad \begin{array}{ll} \max & \sum_{i=1}^{n} w_i x_i \\ \text{s.t.} & x_i + x_j \leq 1, \quad \langle i,j\rangle \in E \\ & x \in B^n \end{array} \tag{14.1}$$

and its continuous relaxation,

$$(CWS)\qquad \begin{array}{ll} \max & \sum_{i=1}^{n} w_i x_i \\ \text{s.t.} & x_i + x_j \leq 1, \quad \langle i,j\rangle \in E \\ & x \geq 0 \end{array} \tag{14.2}$$

By Theorem 14.1, all optimal basic solutions to CWS have components $0, 1$ or $\frac{1}{2}$.

Theorem 14.2 [Nemhauser,Trotter (1975)] *If $x_i = \alpha$ ($\alpha = 0$ or 1) in* some *optimal solution to CWS, then $x_i = \alpha$ in* some *optimal solution to WS.* □

Theorem 14.3 [Hammer,Hansen,Simeone (1982)] *If $x_i = \alpha$ ($\alpha = 0$ or 1) in* all *optimal solutions to CWS, then $x_i = \alpha$ in* all *optimal solutions to WS.* □

A *transversal* of G is any set T of vertices such that every edge of G has at least one endpoint in T. The complement of transversal is a stable set, and vice versa. Finding a "maximum weighted transversal" of G can be formulated as a $0-1$ linear program,

$$(WT)\qquad \begin{array}{ll} \min & \sum_{i=1}^{n} w_i x_i \\ \text{s.t.} & x_i + x_j \geq 1 \\ & x \in B^n \end{array} \tag{14.3}$$

Replacing the constraints $x \in B^n$ by the non-negativity constraints $x \geq 0$, one obtains the continuous relaxation CWT of WT. The basic feasible solutions to CWT are called the *fractional transversals.* Again they can be shown to have components $0, 1$ or $\frac{1}{2}$. The linear programming dual of CWT is

$$(CWM)\qquad \begin{array}{lll} \max & \sum_{\langle i,j\rangle \in E} \lambda_{ij} & \\ \text{s.t} & \sum_{j \in N(i)} \lambda_{ij} \leq w_i, & i = 1, \ldots, n \\ & \lambda_{ij} \geq 0, & \langle i,j\rangle \in E \end{array} \tag{14.4}$$

where $N(i) \equiv \{j \in V : \langle i,j\rangle \in E\}$ is the *neighborhood* of vertex i.

The feasible solutions to (CWM) are called *fractional w-matchings.* If, in addition, they have integral components, they are simply called w-matchings. We shall denote by WM the problem obtained by adding to CWM the constraints:

$$\lambda_{ij} \text{ integer}, \quad \langle i,j\rangle \in E.$$

The *value* of a (fractional or integral) w-matching λ is $\sum_{\langle i,j\rangle \in E} \lambda_{ij}$.

An edge $\langle i,j\rangle$ is said to be *active* (in λ) if $\lambda_{ij} > 0$; otherwise $\langle i,j\rangle$ is said to be *passive.* Finally, a vertex i is *insaturated* (in λ) if $\sum_{j\in N(i)} \lambda_{ij} < w_i$.

As pointed out by Fulkerson,Hoffman and Mc Andrews (1965), the maximum fractional w-matching problem CWM given by (14.4) can be formulated as a maximal flow problem in a suitable network. To see this, let us introduce the network $\mathcal{N}$ whose nodes are

$$s(\text{source}),\quad t(\text{sink}),\quad 1,2,\ldots n; 1',2',\ldots,n'$$

and whose arcs are

$$\begin{array}{lll} (s,1), & \cdots, & (s,n) \\ (1',t), & \cdots, & (n',t) \\ (i,j') & \text{and} & (i',j) \quad \text{for all } \langle i,j\rangle \in E \end{array}$$

Furthermore, capacities w_i are assigned to the arcs (s,i) and (i',t) and infinite capacities are assigned to all remaining arcs.

Then it can be seen that, if Φ is a maximal flow in $\mathcal{N}$, the vector λ defined by $\lambda_{ij} = \frac{1}{2}(\Phi_{ij'} + \Phi_{i'j})$, $\langle i,j\rangle \in E$ is an optimal solution to (CWM).

Let us denote by $v(P)$ the optimal value of an optimization problem P. The following relations hold

$$v(WS) + v(WT) = \quad v(CWS) + v(CWT) \quad = \sum_{i=1}^{n} w_i \tag{14.5}$$

$$v(WM) \leq \quad v(CWM) = v(CWT) \quad \leq v(WT) \tag{14.6}$$

The weighted graph (G,w) is said to have the *w–König-Egerváry property* (briefly, G is a $w-KEG$) if the minimum weight of a transversal is equal to the maximum value of a w-matching; that is, if all inequalities in (14.6) are satisfied as equalities.

When all weights are 1, one obtains the usual König-Egerváry property of graphs.

Theorem 14.4 [Bourjolly,Hammer,Simeone (1984)] *For a weighted graph (G,w), the following statements are equivalent:*

(i) *G is a $w-KEG$*

(ii) $v(WS) = v(CWS)$

(iii) $v(WT) = v(CWT)$

(iv) *Given an arbitrary maximum fractional w-matching λ^*, there exists $S \subseteq V$ such that*

> **(a)** *every insaturated node belongs to S*
>
> **(b)** *every active edge has an endpoint in S*
>
> **(c)** *every edge has an endpoint in $T = V - S$.* □

An interesting consequence of Thm 14.4 (iv) is that the w-König-Egerváry property is equivalent to the consistency of an appropriate quadratic boolean equation. In order to *generate* such equation, one needs to solve a max-flow problem. Here are the details. Let λ be a maximum fractional w-matching of G. Let I be the set of those vertices which are insaturated in λ^*. Let A be the set of those edges which are active in λ^*. Define the following quadratic boolean expression

$$\Psi = \left(\bigvee_{i\in I} \overline{x}_i\right) \vee \left(\bigvee_{\langle i,j\rangle \in A} \overline{x}_i \overline{x}_j\right) \vee \left(\bigvee_{\langle i,j\rangle \in E} x_i x_j\right) \tag{14.7}$$

Corollary 14.5 *G is a $w-KEG$ if and only if the quadratic boolean equation $\Psi = 0$ is consistent.*

Proof: Let S be any subset of V and let x be the characteristic vector of S, i.e. $x_j = 1$ for all $j \in S$ and $x_j = 0$ for all $j \in T = V - S$. Then in Thm 14.4 (iv)

- condition (a) is equivalent to $\overline{x}_i = 0$ for all $i \in I$
- condition (b) is equivalent to $\overline{x}_i\overline{x}_j = 0$ for all $\langle i, j\rangle \in A$
- condition (c) is equivalent to $x_i x_j = 0$ for all $\langle i, j\rangle \in E$

The thesis easily follows. □

Remark 14.6 *In view of Corollary 14.5, one can recognize the w-König-Egerváry property by first solving CWM and then by checking the consistency of the quadratic boolean equation $\psi = 0$. Since both steps can be executed in time polynomial in n and m, it follows that w-KEGs can be recognized in time polynomial in n and m.*

Further structural characterizations and recognition algorithms for w-KEGs are given in Bourjolly,Hammer and Simeone (1984).

15 Weighted stability in graphs and efficient computation of best roofs

We have mentioned in Section 5.1 that the problem of finding maximum weighted stable sets of graphs can be reduced to the maximization of a quadratic pseudo-boolean function. In this section we show that the converse reduction, for a quadratic pseudo-boolean function in n variables and with m terms, is possible in a "succinct" way, namely building a graph with $2n + m$ vertices and $n + 2m$ edges. Such reduction will lead us to

1. a further interpretation of the roof duality gap as the integrality gap in the stable set problem,
2. a computationally efficient procedure for finding best roofs, and
3. the persistency results of the next section.

15.1 SAM graphs

We now introduce the basic concept for proving the announced equivalence between quadratic maximization and stability. We shall call G a *SAM* (Stable and Matching) graph if the vertex set of G can be partitioned into two (possibly empty) subsets V_1 and V_2 such that

(i) V_1 is stable,

(ii) the set of edges with both endpoints in V_2 is a matching.

Given the quadratic function $f(x) = \sum_{i=1}^{n} \sum_{j=1}^{n} q_{ij} x_i x_j$, let P and N be the sets defined by (2.15) and let M be a large constant: it is enough to take $M > 2(U_f - L_f)$ where

$$L_f = \sum_{i=1}^{n} q_{ii}^{-} + \sum_{(i,j)\in N} q_{ij} \tag{15.1}$$

$$U_f = \sum_{i=1}^{n} q_{ii}^{+} + \sum_{(i,j)\in P} q_{ij} - \sum_{(i,j)\in N} q_{ij} \tag{15.2}$$

and where $q^+ = \max(q,0)$ and $q^- = \min(q,0)$. Let us associate with f a weighted *SAM* graph $S_f = (V, E)$ as follows:

$$V \equiv \{(i,j) : q_{ij} \neq 0\} \cup \{1, 2, ..., n\} \cup \{\bar{1}, \bar{2}, ..., \bar{n}\}$$

$$E = \{\langle (i,j), \bar{i}\rangle : (i,j) \in P\} \cup \{\langle (i,j), \bar{j}\rangle : (i,j) \in P\} \cup \{\langle (i,j), i\rangle : (i,j) \in N\}$$
$$\cup \{\langle (i,j), \bar{j}\rangle : (i,j) \in N\} \cup \{\langle i, \bar{i}\rangle : i = 1, ..., n\}.$$

For all $(i,j) \in P$, vertex (i,j) has the weight q_{ij}. For all $(i,j) \in N$, vertex (i,j) has the weight $-q_{ij}$. For all $i = 1, ..., n$, vertex i has the weight

$$q_{ii} + \sum_{j|(j,i)\in N} q_{ji} + M$$

and vertex $\bar{i}$ has the weight M.

The bipartition $\{V_1, V_2\}$ of V required by the definition of a *SAM* graph is as follows:

$$V_1 = \{(i,j) : q_{ij} \neq 0\} \tag{15.3}$$

$$V_2 = \{1, ..., n\} \cup \{\bar{1}, ..., \bar{n}\} \tag{15.4}$$

Depending on the choice of the posiform, different *SAM* graphs can be associated with the same function.

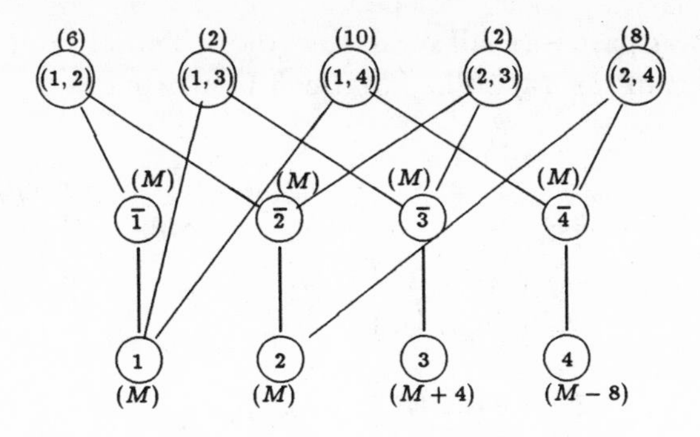

Figure 8:

Example (continuation). The weighted SAM graph S_f associated with the function f given by (9.11), is shown in Figure 8.

Let us associate binary variables with the vertices of S_f as follows.

$$i \to x_i, \quad \bar{i} \to x_i', \quad (i,j) \to y_{ij}.$$

Then the maximum weighted stable set problem (WS) corresponding to S_f (with the given weights) is

$$\max \sum_{i=1}^{n} \left(q_{ii} + \sum_{j|(j,i)\in N} q_{ji} \right) x_i + \sum_{(i,j)\in P} q_{ij} y_{ij} - \sum_{(i,j)\in N} q_{ij} y_{ij} + M \sum_{i=1}^{n} (x_i + x_i' - 1)$$

$$\begin{array}{lll} \text{s.t.} & \left. \begin{array}{l} y_{ij} + x_i' \leq 1, \\ y_{ij} + x_j' \leq 1, \end{array} \right\} & (i,j) \in P, \\ & \left. \begin{array}{l} y_{ij} + x_i \leq 1, \\ y_{ij} + x_j' \leq 1, \end{array} \right\} & (i,j) \in N, \\ & x_i + x_i' \leq 1, & i = 1, ..., n, \\ & x_i, x_i', y_{ij} \in \{0,1\}. & \end{array} \tag{15.5}$$

We denote by CWS the continuous relaxation of WS, obtained by replacing the constrains $x_i, x_i', y_{ij} \in \{0,1\}$ by $x_i, x_i', y_{ij} \geq 0$ (the constraints $x_i, x_i', y_{ij} \leq 1$ are redundant).

Theorem 15.1 *The optimum value of the discrete weighted stability problem (15.5) is equal to the optimum value of the primal problem while the optimum value of the corresponding continuous weighted stability problem is equal to the optimum value of the roof dual.*

Proof: To prove the first part we shall recall that the optimum value of the primal is equal to the optimum value of the drf. The latter can be written as

$$\max \sum_{i=1}^{n} \left(q_{ii} + \sum_{j|(j,i)\in N} q_{ji} \right) x_i + \sum_{(i,j)\in P} q_{ij} y_{ij} + \sum_{(i,j)\in N} (-q_{ij}) y_{ij}$$

$$\text{s.t.} \quad \left. \begin{array}{l} y_{ij} + \overline{x}_i \le 1, \\ y_{ij} + \overline{x}_j \le 1, \end{array} \right\} \quad (i,j) \in P,$$

$$\left. \begin{array}{l} y_{ij} + x_i \le 1, \\ y_{ij} + \overline{x}_j \le 1, \end{array} \right\} \quad (i,j) \in N, \tag{15.6}$$

$$x_i + \overline{x}_i = 1,$$

$$x_i, \overline{x}_i, y_{ij} \in \{0,1\}.$$

Then the drf is similar to the WS problem (15.5) except for the penalty term $M\sum_{i=1}^{n}(x_i + x_i' - 1)$ appearing in the objective function and the inequality constraint $x_i + x_i' \le 1$ compared to the equality constraint $x_i + \overline{x}_i = 1$. If the values given by the $\overline{x}_i$ and x_i' are the same, any optimal solution to (15.5) corresponds to a feasible solution in (15.6) with the same value, and such value belongs to $[L_f, U_f]$. Now any feasible solution of (15.5) which is not feasible for (15.6) must have $x_i + x_i' < 1$ for at least one i and a value $\le U_f - M < U_f - 2(U_f - L_f) \le U_f - (U_f - L_f) = L_f$, and cannot be optimal.

The proof of the second part is similar: we relax the constraints $x_i, \overline{x}_i, x_i', y_{ij} \in \{0,1\}$ to $x_i, \overline{x}_i, x_i', y_{ij} \ge 0$ for $i,j = 1,2,...,n$. Again with any optimum solution of (15.6) we can associate a feasible solution of (15.5) with the same value belonging to $[L_f, U_f]$. Once more any basic feasible solution of (15.5) which is not feasible for (15.6) must have $x_i + x_i' < 1$, *i.e.* $x_i + x_i' \le \frac{1}{2}$ in view of Theorem 14.1 and a value $\le U_f - \frac{1}{2}M < U_f - (U_f - L_f) = L_f$ and hence cannot be optimal. □

Remark 15.2 *From the above proof it follows that, if x^* is a maximizing point f in B^n and one defines $e = (1, ..., 1)$*

$$\begin{aligned} \overline{x}^* &= e - x^*, \\ y_{ij}^* &= min\{x_i^*, x_j^*\}, \quad (i,j) \in P, \\ &= min\{\overline{x}_i^*, x_j^*\}, \quad (i,j) \in N, \end{aligned}$$

then $(x^*, \overline{x}^*, y^*)$ is an optimal solution to WS, and conversely if $(x^*, \overline{x}^*, y^*)$ is an optimal solution to WS, then $f(x^*) = \max_{x\in B^n} f(x)$. Similarly, if (x_0, y_0) is an optimal solution to crf and $\overline{x}_0 = e - x_0$ then $(x_0, \overline{x}_0, y_0)$ is an optimal solution to CWS. Conversely, if $(x_0, \overline{x}_0, y_0)$ is an optimal solution to CWS then (x_0, y_0) is an optimal solution to the crf.

It is worth summarizing in a table the relationships between the different optimization problems examined in Sections 6 to 13, as well as the different interpretations of the roof-duality gap.

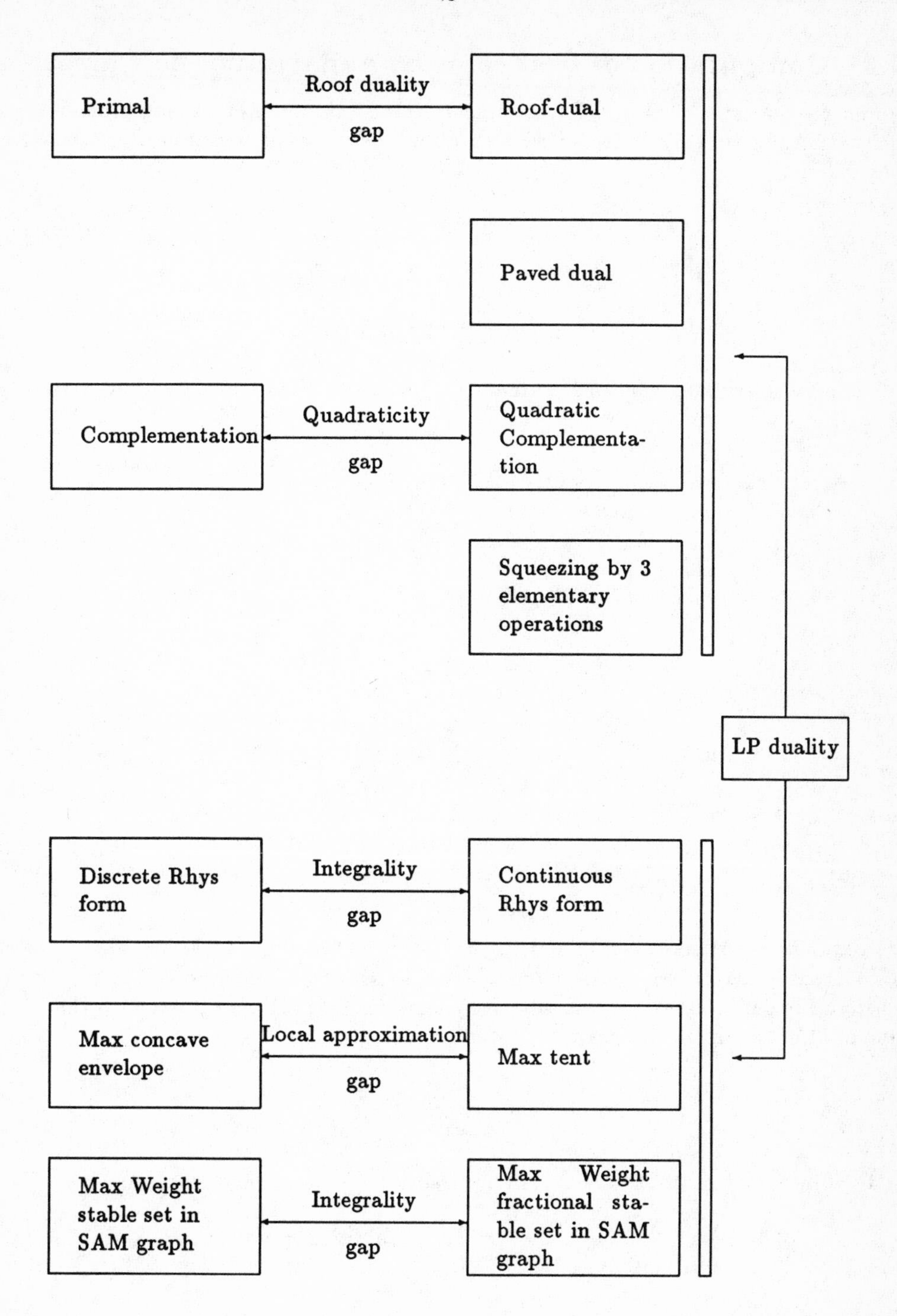

Primal
Roof duality
gap
Roof-dual
Paved dual
Complementation
Quadraticity
gap
Quadratic Complementation
Squeezing by 3 elementary operations
LP duality
Discrete Rhys form
Integrality
gap
Continuous Rhys form
Max concave envelope
Local approximation
gap
Max tent
Max Weight stable set in SAM graph
Integrality
gap
Max Weight fractional stable set in SAM graph

Table 2:

15.2 Computation of best roofs by a maximum flow algorithm

In order to compute best roofs, let us consider the maximum w-matching problem on the weighted SAM graph S_f of the previous section. Let us assign variables to the edges of S_f as follows:

$$(i,\bar{i}) \rightarrow u_i;$$

$$\text{if } (i,j) \in P: \ \langle (i,j),\bar{i}\rangle \rightarrow \lambda_{ij}, \ \langle (i,j),\bar{j}\rangle \rightarrow \lambda_{ji} \ ;$$

$$\text{if } (i,j) \in N: \ \langle (i,j),i\rangle \rightarrow \lambda_{ij}, \ \langle (i,j),\bar{j}\rangle \rightarrow \lambda_{ji} \ .$$

Then the maximum fractional w-matching problem (WM) on S_f can be written as

$$z_M = \max \quad \sum_{(i,j)\in P\cup N} (\lambda_{ij} + \lambda_{ji}) + \sum_{i=l}^{n} \tilde{u}_i - nM \tag{15.7}$$

$$\text{subject to} \quad \lambda_{ij} + \lambda_{ji} \le |q_{ij}| \qquad \forall (i,j) \in P \cup N, \tag{15.8}$$

$$\tilde{u}_i + \sum_{j|(i,j)\in N} \lambda_{ij} \le M + q_{ii} + \sum_{j|(j,i)\in N} q_{ij}, \qquad (i = 1,\ldots,n), \tag{15.9}$$

$$\tilde{u}_i + \sum_{j|(j,i)\in P} \lambda_{ij} + \sum_{j|(i,j)\in P} \lambda_{ij} + \sum_{j|(j.i)\in N} \lambda_{ij} \le M, \ (i = 1,\ldots,n), \tag{15.10}$$

$$\lambda_{ji} \ge 0, \ \lambda_{ij} \ge 0,; \ \tilde{u}_i \ge 0, \qquad \begin{array}{l}(i,j) \in P \cup N, \\ i = 1,\ldots,n.\end{array} \tag{15.11}$$

(the constant $-nM$ has been added to the objective function (15.7) in order to balance the constant $-nM$ in the objective function of (15.5). From Theorem 15.1, identity (14.5) and from the definition of the weights on S_f, and noticing that z_M is equal to the optimal value of (15.5) by linear programming duality, one gets

$$w_R + z_M = \sum_{i=1}^{n} q_{ii} + \sum_{(i,j)\in P} q_{ij}.$$

For the function f of (9.11) the maximum fractional w-matching problem is

$$\begin{aligned} \max \ & \lambda_{12} + \lambda_{21} + \lambda_{13} + \lambda_{31} + \lambda_{14} + \lambda_{41} + \lambda_{23} + \lambda_{32} + \lambda_{24} + \lambda_{42} + \\ & + u_1 + u_2 + u_3 + u_4 - 4M \end{aligned}$$

$$\begin{array}{lll}
\text{s.t.} & \lambda_{ij} \geq 0 \quad \forall (i,j), & \tilde{u}_i \geq 0 \quad \forall i, \\
& \lambda_{12} + \lambda_{21} \leq 6, \quad \lambda_{13} + \lambda_{31} \leq 2, & \lambda_{14} + \lambda_{41} \leq 10, \\
& \lambda_{23} + \lambda_{32} \leq 2, \quad \lambda_{24} + \lambda_{42} \leq 8, & \\
& \tilde{u}_1 + \lambda_{13} + \lambda_{14} \leq M, & \tilde{u}_1 + \lambda_{12} \leq M, \\
& \tilde{u}_2 + \lambda_{24} \leq M, & \tilde{u}_2 + \lambda_{21} + \lambda_{23} \leq M, \\
& \tilde{u}_3 \leq M + 4, & \tilde{u}_3 + \lambda_{31} + \lambda_{32} \leq M, \\
& \tilde{u}_4 \leq M - 8, & \tilde{u}_4 + \lambda_{41} + \lambda_{42} \leq M.
\end{array}$$

The next theorem establishes a simple and useful relation between the optimal solutions of the roof dual (10.14) and those of the maximum fractional w-matching problem (CWM) (15.7) - (15.11)

Theorem 15.3 *(a) There is an optimal solution of (CWM) satisfying (15.8) with equality; (b) If* $[\tilde{u}_i, \lambda_{ij}, \lambda_{ji}]$ *is an optimal solution of (CWM) satisfying (15.8) with equality, and if*

$$\begin{aligned} u_i = \; & \max\{0, \; q_{ii} + \sum_{j|(j,i)\in P} q_{ji} + \sum_{j|(i,j)\in P} \lambda_{ij} - \\ & - \sum_{j|(i,j)\in N} \lambda_{ij} - \sum_{j|(j,i)\in P} \lambda_{ji} - \sum_{j|(j,i)\in N} \lambda_{ji}\}, \end{aligned}$$

$(i = 1, ..., n)$, then $[u_i, \lambda_{ij}]$ is an optimal solution of the roof-dual (10.14).

Proof: (a) Let $(\tilde{u}, \lambda)$ be an optimal solution of (CWM); then equality must hold either in (15.9) or in (15.10). In any case one has $u_i \geq M - (U_f - L_f) > U_f - L_f$. Let $i < j$, be a pair of indices such that

$$\lambda_{ij} + \lambda_{ji} < |q_{ij}|.$$

Let $\epsilon = |q_{ij}| - \lambda_{ij} - \lambda_{ji}$. We observe that the variable λ_{ij} appears only in one of the constraints (15.9),(15.10). Hence if we define a new vector $(\tilde{u}^*, \lambda^*)$ by

$$\begin{array}{ll} \tilde{u}_i^* = \tilde{u}_i - \epsilon, & \lambda_{ij}^* = \lambda_{ij} + \epsilon, \\ \tilde{u}_h^* = \tilde{u}_h \text{ for } h \neq i, & \lambda_{hk}^* = \lambda_{hk} \text{ for } (h,k) \neq (i,j), \end{array}$$

we obtain a new optimal solution to (CWM) such that

$$\lambda_{ij}^* + \lambda_{ji}^* = |q_{ij}|.$$

Iterating this procedure, we obtain an optimal solution to (CWM) satisfying (15.8) with equality for all $i < j$ and nonnegative values for the $\tilde{u}_i$.

(b) Let us assume that (15.8) holds with equality. Then, eliminating all λ_{ji}, $i < j$, we obtain

$$z_M = \max \sum_{(i,j)\in P\cup N} |q_{ij}| + \sum_{i=1}^{n} \tilde{u}_i - nM$$

subject to

$$0 \leq \lambda_{ij} \leq |q_{ij}|; \quad \forall i, j,$$

$$\tilde{u}_i \leq M + q_{ii} + \sum_{j|(i,j)\in N} q_{ji} - \sum_{j|(i,j)\in N} \lambda_{ij},$$

$$\tilde{u}_i \leq M - \sum_{j|(j,i)\in P} q_{ji} + \sum_{j|(j,i)\in N} q_{ji} - \sum_{j|(i,j)\in P} \lambda_{ij} + \sum_{j|(j,i)\in P} \lambda_{ji} + \sum_{j|(j,i)\in N} \lambda_{ji},$$

$$\tilde{u}_i \geq 0,; \quad i = 1, ..., n.$$

Then, denoting, as before, $\Lambda = \{\lambda_{ij} | 0 \leq \lambda_{ij} \leq |q_{ij}|\}$, we have

$$\begin{aligned}
z_M &= \max_{\lambda\in\Lambda} \sum_{(i,j)\in P\cup N} |q_{ij}| - nM + \sum_{i=1}^{n} \min \Bigg\{ M + q_{ii} + \sum_{j|(j,i)\in N} q_{ij} - \sum_{j|(i,j)\in N} \lambda_{ij}, \\
&\quad M - \sum_{j|(j,i)\in P} q_{ij} + \sum_{j|(j,i)\in N} q_{ji} - \sum_{j|(i,j)\in P} \lambda_{ij} + \sum_{j|(j,i)\in P} \lambda_{ji} + \sum_{j|(j,i)\in N} \lambda_{ji} \Bigg\} = \\
&= \sum_{(i,j)\in P\cup N} |q_{ij}| + \max_{\lambda\in\Lambda} \sum_{i=1}^{n} \min \Bigg\{ q_{ii} + \sum_{j|(j,i)\in N} q_{ji} - \sum_{j|(i,j)\in N} \lambda_{ij}, \\
&\quad - \sum_{j|(j,i)\in P} q_{ji} + \sum_{j|(j,i)\in N} q_{ji} - \sum_{j|(i,j)\in P} \lambda_{ij} + \sum_{j|(j,i)\in P} \lambda_{ji} + \sum_{j|(j,i)\in N} \lambda_{ji} \Bigg\} = \\
&= \sum_{(i,j)\in P\cup N} |q_{ij}| - \min_{\lambda\in\Lambda} \sum_{i=1}^{n} \max \Bigg\{ -q_{ii} - \sum_{j|(j,i)\in N} q_{ji} + \sum_{j|(i,j)\in N} \lambda_{ij}, \\
&\quad \sum_{j|(j,i)\in P} q_{ji} + \sum_{j|(j,i)\in N} q_{ji} + \sum_{j|(i,j)\in P} \lambda_{ij} - \sum_{j|(j,i)\in P} \lambda_{ji} - \sum_{j|(j,i)\in N} \lambda_{ji} \Bigg\} = \\
&= \sum_{(i,j)\in P\cup N} |q_{ij}| + \sum_{i=1}^{n} q_{ii} + \sum_{(i,j)\in N} q_{ij} - \min_{\lambda\in\Lambda} \Bigg(\sum_{(i,j)\in N} \lambda_{ij} + \\
&\quad \sum_{i=1}^{n} \max \Bigg\{ 0, q_{ii} + \sum_{j|(j,i)\in N} q_{ji} - \sum_{j|(i,j)\in N} \lambda_{ij} + \sum_{j|(j,i)\in P} q_{ji} - \sum_{j|(j,i)\in N} q_{ji} + \\
&\quad + \sum_{j|(i,j)\in P} \lambda_{ij} - \sum_{j|(j,i)\in P} \lambda_{ji} - \sum_{j|(j,i)\in N} \lambda_{ji} \Bigg\} \Bigg) \\
&= \sum_{i=1}^{n} q_{ii} + \sum_{(i,j)\in P} q_{ij} - \min_{\lambda\in\Lambda} \Bigg(\sum_{(i,j)\in N} \lambda_{ij} + \\
&\quad + \sum_{i=1}^{n} \max \Bigg\{ 0, \quad q_{ii} + \sum_{j|(j,i)\in P} q_{ji} + \sum_{j|(i,j)\in P} \lambda_{ij} - \\
&\quad - \sum_{j|(i,j)\in N} \lambda_{ij} - \sum_{j|(j,i)\in P} \lambda_{ji} - \sum_{j|(j,i)\in N} \lambda_{ji} \Bigg\} \Bigg)
\end{aligned}$$

$$= \sum_{i=1}^{n} q_{ii} + \sum_{(i,j)\in P} q_{ij} - \min\left(\sum_{(i,j)\in N} \lambda_{ij} + \sum_{i=1}^{n} u_i \right)$$

s.t.

$$\begin{aligned} 0 \leq\ & \lambda_{ij} \leq |q_{ij}|, \quad (i,j) \in P \cup N \\ u_i \geq\ & q_{ii} + \sum_{j|(j,i)\in P} q_{ji} + \sum_{j|(i,j)\in P} \lambda_{ij} - \sum_{j|(j,i)\in P} \lambda_{ji} - \\ & - \sum_{j|(i,j)\in N} \lambda_{ij} - \sum_{j|(j,i)\in N} \lambda_{ji}, \\ u_i \geq\ & 0 \end{aligned}$$

Comparing the last expression of z_M with (10.14) and (7.8) the statement follows. □

A max-flow problem on a network with N nodes can be solved in $O(N^3)$ time, e.g. by Karzanov's algorithm. Since the SAM graph has $2n + m$ nodes ($m = |P| + |N|$) and the associated network has $N = 4n + 2m + 2$ nodes, it follows that a best roof can be computed in $O(\max\{m^3, n^3\})$ time.

16 Persistency

In this section we exploit the equivalence of the primal to a stability problem of a SAM graph, and of the roof dual to the continuous relaxation of the stability problem, to derive 'persistency' theorems which allow us to determine the value of some variables in one or in all optimal solutions of the primal.

16.1 Strong and weak persistency theorems

Let us consider again the primal quadratic optimization problem

$$\max_{x\in B^n} f(x) = \max_{x\in B^n} \sum_{i=1}^{n}\sum_{j=1}^{n} q_{ij}x_i x_j \tag{16.1}$$

and a linear relaxation of it,

$$\max_{x\in B^n} p(x). \tag{16.2}$$

We shall say that the upper plane $p(x)$ has the *persistency property* if for every j the fact that $x_j = \alpha$ for all optimal solutions x of (5.2) implies that $x_j = \alpha$ for all optimal solutions x of (5.1) ($\alpha = 0$ or 1)

The main result of this section states that all best roofs have the persistency property.

If $p(x) = v_0 + v_1x_1 + ... + v_nx_n$, is a best roof of f, all optimal solutions $\tilde{x}$ of the problem (16.2) are given by

$$\begin{aligned}\tilde{x}_i &= 0, && \text{if } v_i < 0,\\ &= 1, && \text{if } v_i > 0,\\ &= 0 \text{ or } 1, && \text{if } v_i = 0,\end{aligned} \tag{16.3}$$

so that the simple inspection of the sign of the coefficient $v_i \neq 0$ makes it possible to predict, as a consequence of persistency, the value of the variable x_i in all optimal solutions of the quadratic problem (16.1).

The practical significance of persistency as a *tool for reducing the problem size* in enhanced by the fact (mentioned at the end of the previous section) that best roofs can be computed in polynomial time.

As before, let us denote by R the set of all roofs and by $\text{Best}(R)$ the set of best roofs, i.e.

$$\text{Best}(R) \equiv \left\{p^*(R) : \max_{x \in B^n} p^*(x) = \min_{p \in R} \max_{x \in B^n} p(x)\right\}$$

Theorem 16.1 *For all* $p(x) = v_0 + v_1 x_1 + ... + v_n x_n \in \text{Best}(R)$ *and all optimal solutions* $(\tilde{x}, \tilde{y})$ *to the continuous Rhys linear program crf,*

$$v_i > 0 \Rightarrow \tilde{x}_i = 1, \quad v_i < 0 \Rightarrow \tilde{x}_i = 0. \tag{16.4}$$

Proof: Consider the linear programming formulation (10.14) of the roof dual:

$$\begin{aligned} \min \quad & \sum_{i=1}^{n} u_i + v_0(\lambda) \\ \text{s.t.} \quad & u_i \geq v_i(\lambda), \quad u_i \geq 0, \quad i = 1, ..., n. \\ & 0 \leq \lambda_{ij} \leq |q_{ij}|, \qquad\quad i \in P \cup N. \end{aligned} \tag{16.5}$$

where $v_i(\lambda), i = 0, 1..., n$, are defined by (7.8). Let $p(x) = v_0 + v_1 x_1 + ... + v_n x_n \in \text{Best}(R)$, and let $\tilde{x}$ be an arbitrary optimal solution to crf. Then there is certainly an optimal solution (u^*, λ^*) to (16.5) such that $v_i = v_i(\lambda^*)$.

Since (u^*, λ^*) is an optimal solution to (16.5), $v_i(\lambda^*) > 0 \Rightarrow u_i^* - v_i(\lambda^*) = 0$. By (weak) complementary slackness applied to the primal-dual pair crf (as given by (9.2)-(9.6), (9.9),(9.10)) and (16.5), $u_i^* > 0 \Rightarrow \tilde{x}_i = 1$. On the other hand, if $v_i(\lambda^*) < 0$ then $u_i^* = 0 > v_i(\lambda^*)$. Again, by complementary slackness, one must then have $\tilde{x}_i = 0$. □

An immediate consequence of the above result is the following *sign invariance property* for best roofs.

Corollary 16.2 *Each coefficient* v_i *has constant sign in all* $p \in \text{Best}(R)$ *(i.e. it is either always nonnegative or always nonpositive).* □

The announced persistency result follows.

Theorem 16.3 *(Strong Persistency Theorem). If there is some* $p(x) = v_0 + v_1x_1 + ... + v_nx_n \in \text{Best}(R)$ *such that* $v_i > 0[v_i < 0]$, *one must have* $x_i = 1[x_i = 0]$ *in all maximizing points* x *of* f.

Proof: Let $v_i > 0$ (the proof for $v_i < 0$ is similar). By Theorem 16.1 one must have $x_i = 1$ for all optimal solutions (x, y) of crf. In view of Remark (15.2), one must have $x_i = 1$ in all optimal solutions $(x, \overline{x}, y)$ to CWS. By Theorem 14.3, one must also have $x_i = 1$ in all optimal solutions $(x, \overline{x}, y)$ to WS. Hence,again by Remark 15.2, one has $x_i = 1$ for all maximum points x of f. □

Even when $v_i = 0$, it may happen that there is some optimal solution (x, y) to the crf, such that x_i is equal to 0 or to 1.

Reasoning as in the proof of Theorem 16.3, but invoking Theorem 14.2 rather than Theorem 14.3, one obtains the following result.

Theorem 16.4 *(Weak persistency Theorem). If there is some optimal solution* (x, y) *to the crf such that* $x_i = 1[x_i = 0]$, *then there is a maximum point* x^* *of* f *such that* $x_i^* = 1[x_i^* = 0]$.

When implementing Theorem 16.3 one does not have to compute all best roofs. In fact, the next result shows that one is enough.

Theorem 16.5 *There exist a best roof* $\mu(x) = \mu_0 + \mu_1x_1 + ... + \mu_nx_n$ *such that one has* $x_i = 1[x_i = 0]$ *for all optimal solutions* (x, y) *to crf if and only if* $\mu_i > 0[\mu_i < 0]$.

Proof: By strong complementary slackness applied to the primal-dual pair crf and (16.5) there exist an optimal solution $(\tilde{x}, \tilde{y})$ to crf and an optimal solution $(\tilde{u}, \tilde{\lambda})$ to (16.5) such that

$$\tilde{u}_i > 0 \quad \Leftrightarrow \quad \tilde{x}_i = 1, \tag{16.6}$$

$$\tilde{u}_i > v_i(\tilde{\lambda}) \quad \Leftrightarrow \quad \tilde{x}_i = 0 \tag{16.7}$$

Let $\mu(x) = \mu_0 + \mu_1x_1 + ... + \mu_nx_n$ where $\mu_i = v_i(\tilde{\lambda})$, $i = 0, 1, ..., n$. If $x_i = 1$ for all optimal solutions to crf, then in particular $\tilde{x}_i = 1$ and thus $\tilde{\mu}_i > 0$ by (16.6). Since $(\tilde{u}, \tilde{\lambda})$ is an optimal solution to (16.5) one must have $\tilde{u}_i = v_i(\tilde{\lambda})$ and thus $\mu_i > 0$. On the other hand, if $x_i = 0$ for all optimal solutions to crf then by (16.7) $\tilde{u}_i > v_i(\lambda)$, and therefore $v_i(\tilde{\lambda}) < 0$, since $(\tilde{u}, \tilde{\lambda})$ is optimal. It follows that $\mu_i < 0$. The reverse implications, $\mu_i > 0 \Rightarrow x_i = 1$ and $\mu_i < 0 \Rightarrow x_i = 0$ for all optimal solutions to the crf, follow from the fact that $\mu(x)$ is a best roof and from Theorem 16.3. □

A best roof $\mu(x)$ satisfying the conditions of the above Theorem 16.4 will be called a *master roof*.

Example (continued). For the function f given by (9.11), some best roofs are

$$8 + 6x_3 + 2x_4, \qquad 6 + 8x_3,$$
$$10 - 2x_2 + 4x_3 + 2x_4, \quad 12 - 2x_1 - 2x_2 + 4x_3.$$

Averaging the two last ones one obtains the master roof

$$11 - x_1 - 2x_2 + 4x_3 + x_4.$$

By persistency, one must have $x_1 = x_2 = 0$ and $x_3 = x_4 = 1$ in all maximizing points of f. The optimal value of the roof dual is 16. Since 16 is also the maximum of f in B^4, f is gap-free.

17 Extreme cases

In the present section we give several characterizations of those 'bad' functions for which our methods (strong persistency) allow to force no variables (although weak persistency might still apply).

At the opposite end of the scale, there are the 'gap-free' functions f, i.e. those functions whose maximum in B^n is equal to the optimum value w_R of the roof dual. The maximum of such functions can be computed in polynomial time by using a maximum flow algorithm, as shown in Section 15.2. In this section, we show that there is a polynomial-time recognition algorithm for gap-free functions and exhibit an interesting class of gap-free functions–the unimodular functions– which properly includes all supermodular ones.

17.1 A lower bound on the optimum of the roof dual

We shall first establish a lower bound on the optimum value of the roof dual. This result will be exploited in Theorem 17.1, which provides several characterizations of those functions for which no variable can be fixed by strong persistency.

Let

$$k^* = \frac{1}{2}\left\{\sum_{i=1}^{n} q_{ii} + \sum_{(i,j)\in P} q_{ij}\right\}. \tag{17.1}$$

From now on a quadratic function $f(x)$ will be called *irreducible* (for reasons which will be apparent below) if and only if the linear system

$$\begin{array}{ll} v_i(\lambda) = 0, & i = 1, 2, ..., n, \\ 0 \le \lambda_{ij} \le |q_{ij}|, & i, j = 1, 2, ..., n \end{array} \tag{17.2}$$

is consistent (the linear functions $v_i(\lambda)$ being defined by (7.8)).

Theorem 17.1 *A lower bound of the optimal value w_R of the roof-dual (16.5) is given by k^*; this lower bound is actually equal to w_R if and only if f is irreducible.*

Proof. Let (u^*, λ^*) be an optimal solution to (16.5) and let $I = \{i : v_i(\lambda^*) < 0\}$. (note that I may be empty, in which case all summations over I will be equal to 0). Then

$$\begin{aligned} w_r &= v_0(\lambda^*) + \sum_{i\notin I} v_i(\lambda^*) \\ &\ge v_0(\lambda^*) + \sum_{i\notin I} v_i(\lambda^*) + \tfrac{1}{2}\sum_{i\in I} v_i(\lambda^*) \end{aligned}$$

(with equality holding only if $I = \emptyset$)

$$\begin{aligned}
&= \sum_{(i,j)\in N} \lambda^*_{ij} + \tfrac{1}{2}\sum_{i=1}^{n} v_i(\lambda^*) + \tfrac{1}{2}\sum_{i\notin 1} v_i(\lambda^*) \\
&= \sum_{(i,j)\in N} \lambda^*_{ij} + \tfrac{1}{2}\sum_{i=1}^{n} \left(q_{ii} + \sum_{j|(j,i)\in P} q_{ji} + \sum_{j|(i,j)\in P} \lambda^*_{ij} \right. \\
&\quad \left. - \sum_{j|(j,i)\in P} \lambda^* - \sum_{j|(i,j)\in N} \lambda^*_{ij} - \sum_{j|(j,i)\in N} \lambda^*_{ji} \right) + \tfrac{1}{2}\sum_{i\notin I} v_i(\lambda^*) \\
&= +\tfrac{1}{2}\sum_{i=1}^{n} \left(q_{ii} + \sum_{j|(i,j)\in P} q_{ij} \right) + \tfrac{1}{2}\sum_{i\notin I} v_i(\lambda^*) \\
&= k^* + \tfrac{1}{2}\sum_{i\notin I} v_i(\lambda^*)
\end{aligned}$$

which is greater than or equal to k^*, the equality holding if and only if $I = 0$ and $v_i(\lambda^*) = 0$ for all $i \notin I$. □

17.2 Irreducible functions

Let us now turn our attention to problems for which strong persistency allows us to force no variables.

Theorem 17.2 *The following statements are equivalent:*

(1) *No variable has a constant binary value in all optimal solutions of crf.*

(2) *The only best roof of f is the constant upper plane*

$$p(x) = k^* = \frac{1}{2}\left(\sum_{i=1}^{n} q_{ii} + \sum_{(i,j)\in P} q_{ij}\right).$$

(3) *f has a constant best roof.*

(4) *f is irreducible.*

(5) *$w_R = k^*$.*

(6) *CWS has an optimal solution $(x_0, \overline{x}_0, y_0)$ such that $x_0 = (\frac{1}{2}, \frac{1}{2}, ..., \frac{1}{2})$*

Proof.

(1) ⇒ (2) Let $p(x) = v_0 + v_1x_1 + ... + v_nx_n \in \text{Best}(R)$. If for some index i one had $v_i \neq 0$, then by Theorem 16.1 the variable x_i would take a constant binary value in all optimal solutions of crf. Hence $v_i = 0$ for all j and thus $p(x) = v_0$. Since $v_0 = \max_{x\in B^n} p(x) = w_R$, $p(x)$ is the unique element of Best(R). Moreover, there is an optimal solution $(\tilde{u}, \tilde{\lambda})$ of (16.5) such that $\tilde{u}_i = v_i(\tilde{\lambda}) = 0$. Hence, by Theorem 17.1 $w_R = k^*$ and thus $v_0 = k^*$.

$(2) \Rightarrow (3)$. Obvious.

$(3) \Rightarrow (4)$. Obvious.

$(4) \Rightarrow (5)$. Already proved (see Theorem 17.1)

$(5) \Rightarrow (6)$. Note that $x_i = x'_i = y_{ij} = \frac{1}{2}$ is a feasible solution of CWS giving to the objective function the value k^*. By (5) and Theorem 15.1 this solution is optimal.

□

17.3 Gap-free functions

A quadratic pseudo-boolean function f will be called *gap-free* if $\max_{x \in B^n} f(x) = w_R$, i.e. if there is no roof-duality gap.

Proposition 17.3 *f is a gap-free if and only if the SAM graph S_f has the König-Egerváry property relative to the weights defined in Section 15.1.*

Proof. Follows immediately from Theorems 15.1 and 14.4 □

Corollary 17.4 *Gap-free functions f can be recognized in time polynomial in n, the number of variables, and m, the number of terms in f.*

Proof. Follows from Corollary 14.5 and the subsequent Remark 14.6. □

Remark 17.5 Technically speaking, the definition of W-KEG requires that the weights w_i be integers. Under the (not too restrictive) hypothesis that the q_{ij} are integers, the weights w_i associated with the nodes of the SAM graph are indeed integers and the proof of Prop. 17.3. carries through. Furthermore, even when the w_i are arbitrary real numbers conditions (*ii*) and (*iv*) in Theorem 14.4 can be shown to be equivalent. If follows that one can still determine whether $v(WS) = v(CWS)$ in polynomial time by solving the quadratic boolean equation $\psi = 0$, with ψ given by (14.3). Hence the conclusion of Prop. 17.3 remains true even when the q_{ij} are arbitrary real numbers.

We exhibit now a combinatorially interesting class of functions for which there is no roof-duality gap.

Let us recall that a pseudo-boolean function f is *supermodular* if, for all $x, y \in B^n$

$$f(x \vee y) + f(x \wedge y) \geq f(x) + f(y).$$

As remarked in [Fisher,Nemhauser and Wolsey (1978)] a quadratic pseudo-boolean function $f(x) = x^T Q x$ is supermodular iff $q_{ij} \geq 0$ for all $i \neq j$. In Section 8 we have seen that these functions are gap-free. A more general class of gap-free functions, the *unimodular functions*, has been introduced by Hansen and Simeone (1986). A function f is called unimodular if the matrix of its continuous Rhys form is totally unimodular.

Given $f(x) = x^T Q x$, one can naturally associate with f a signed graph G_f as follows:

- the vertex set of G_f is $\{1, ..., n\}$;
- there is a positive edge between vertices i and j whenever $q_{ij} > 0$;
- there is a negative edge between vertices i and j whenever $q_{ij} < 0$;

We recall that a signed graph is said to be *balanced* if no cycle has an odd number of negative edges.

Theorem 17.6 [Hansen and Simeone (1986)] *For a quadratic function f, the following conditions are equivalent:*

(i) *f is unimodular;*

(ii) *the signed graph G_f is balanced;*

(iii) *there exists a subset S of $\{1, ..., n\}$ such that the switch on S*

$$y_i = \begin{cases} \overline{x}_i, & i \in S \\ x_i, & i \notin S \end{cases}$$

transforms f into a supermodular function of the variables $y_1, ..., y_n$. □

In view of the above theorem, unimodular functions can be recognized by testing G_f for balance. This can be done in $O(m)$ time (see Hansen (1978)).

Lu and Williams (1986) have shown that many concepts and results of roof-duality can be naturally extended to the general case in which one deals with arbitrary pseudo-boolean functions.

References

A. V. Aho, J. E. Hopcroft, J. D. Ullman: *The design and analysis of computer algorithms* (Addison-Wesley, Reading, Mass.,1974).

B. Aspvall, M. F. Plass, R. E. Tarjan: "A linear-time algorithm for testing the truth of certain quantified Boolean formulas", *Inform. Process. Letters* **8** (1979) 121-123.

M. L. Balinski: "Notes on a constructive approach to linear programming", in: G. B. Dantzig and A. F. Veinott,Jr. eds., *Mathematics of decision sciences*, Part I. (Amer. Math. Soc., Providence, 1968) 179-256.

E. M. L. Beale, M. G. Kendall, D. W. Wann: "The discarding of variables in multivariate analysis", *Biometrika* **54** (1967) 357-366.

C. Benzaken, S.C. Boyd, P. L. Hammer, B. Simeone: "Adjoints of bidirected graphs", *Congressus numerantium* **39** (1983) 123-144.

C. Benzaken, P. L. Hammer, B. Simeone: "Some remarks on conflict graphs of quadratic pseudo-boolean functions", in: L. Collatz, G. Meinardus, W. Wetterling eds., *Konstruktive Methoden der finiten nichtlinearen Optimierung* (Birkhäuser, Basel, 1980) 9-30.

C. Berge: *Grahps and hypergraphs* (North-Holland, Amsterdam, 1973).

J. M. Bourjolly, P. L. Hammer, B. Simeone: "A boolean simplex method for computing lower bounds of quadratic pseudo-boolean functions", Research Report CORR 83-29, Univ. of Waterloo (1983).

J. M. Bourjolly, P. L. Hammer, B. Simeone: "Node-weighted graphs having the König-Egerváry property", *Math. Programming Studies* **22**(1984) 44-63.

R. Burkard, F. Rendl: "Quadratic assignment problems", Technical Report, Technische Universität Graz (1985).

V. Chvátal, P. L. Hammer: "Aggregation of inequalities in integer programming", *Ann. Discr. Math.* **1** (1977) 145-162.

Y. Crama, P. L. Hammer: "The complexity of recognizing partition-quadratic graphs", RRR 3-85, Rutgers University (1985).

S. Even, A. Itai, A. Shamir: "On the complexity of time-table and multi-commodity flow problems", *SIAM J. on Computing* **5**(1976) 691-703.

J. E. Falk, K. L. Hoffman: "A successive underestimation method for concave minimization problems", *Math. Oper. Res.* **1**(1976), 251-259.

M. L. Fisher, G. L. Nemhauser, L. A. Wolsey: "An analysis of approximations for maximizing submodular set functions - I", *Math. Programming* **14** (1978) 265-294.

R. J. Freeman, D. C. Gogerty, G. W. Graves, R. B. S. Brooks: "A mathematical model of supply support for space operations", *Oper. Research* **14** (1966) 1-15.

D. R. Fulkerson, A. J. Hoffman, M. H. McAndrew: "Some properties of graphs with multiple edges", *Canad. J. of Math.* **17**(1965) 166-177.

G. Gallo, P. L. Hammer, B. Simeone: "Quadratic knapsack problems", *Math. Programming Studies* **12**(1980) 132-149.

M. R. Garey, D. S. Johnson: *Computers and intractability: a guided tour to the theory of NP-completeness.* (Freeman, S. Francisco, 1979).

F. Gavril: "Testing for equality between maximum matching and minimum node covering", *Inform. Process. Letters* **6**(1977) 199-202.

P. L. Hammer: "Pseudo-boolean remarks on balancing signed graphs", *Graph Theory and Integer Programming* International Series of Numerical Mathematics, **36** (1977) 69-78, Birkhäuser Verlag, Basel, Switzerland.

P. L. Hammer, P. Hansen: "Logical relations in quadratic 0-1 programming", *Revue Roumaine de Math. pures et appliqueés* **26**(1981) 421-429.

P. L. Hammer, S. Rudeanu: *"Boolean methods in operations research and related areas"* (Springer, Berlin, 1968).

P. L. Hammer, P. Hansen, B. Simeone: "Upper planes of quadratic 0-1 functions and stability in graphs", in: O. L. Mangasarian, R. R. Meyer and S. M. Robinson eds., *Nonlinear Programming* 4. (Academic Press, New York, 1981) 395-414.

P. L. Hammer, P. Hansen, B. Simeone: "Vertices belonging to all or to no maximum stable sets of a graph", *SIAM J. Algebr. Discr. Methods* **3**(1982) 511-522.

P. L. Hammer, P. Hansen, B. Simeone: "Roof-duality, complementation and persistency in quadratic 0-1 optimization", *Math. Programming* **28** (1984) 121-155.

P. L. Hammer, B. Kalantari: "Worst-case analysis of the roof-dual gap in quadratic zero-one optimization", Working paper, Rutgers University (1986).

P. L. Hammer, U. N. Peled, S. Sorensen: "Pseudo-boolean functions and game theory, I: Core elements and Shapley value", *Cahiers du Centre d'Etudes de Rech. Operationnelle* **19**(1977) 159-176.

P. L. Hammer, I. G. Rosenberg: "Linear decomposition of a positive group-boolean function", in: L. Collatz, W. Wetterling, eds., *Numerische Methoden bei Optimierung*, vol. II. (Birkhäuser, Basel, 1974) 51-62.

P. L. Hammer, B. Simeone: "Quasimonotone boolean functions and bistellar graphs", *Ann. Discr. Math.* **9**(1980) 107-119.

P. Hansen: "Fonctions d'evaluation et pénalités pour les programmes quadratiques en variables 0-1", in: B. Roy, ed., *Combinatorial programming, methods and applications* (Reidel, Dordrecht, 1975) 361-370.

P. Hansen: "Labelling algorithms for balance in signed graphs", in: J. - C. Bermond et al. eds., Problemes combinatoires et theorie des graphes (Editions CNRS, Paris, 1978) 215-217.

P. Hansen, B. Jaumard, M. Minoux: "A linear expected-time algorithm for deriving all logical conclusions implied by a set of boolean inequalities", *Math. Programming* **34** (1986) 223-231.

P. Hansen, B. Simeone: "Unimodular functions", *Discr. Appl. Math.*, **14** (1986) 269-281.

F. Harary: "On the notion of balance of a signed graph", *Michigan Math. J.* **2**(1953) 143-146.

E. L. Johnson, M. W. Padberg: "Degree-two inequalities, clique facets and biperfect graphs", *Ann. Discr. Math.* **16** (1982) 169-187.

S. Kirkpatrick, C. D. Gelatt, M. P. Vecchi: "Optimization by simulated annealing", *Science* **220** (1983) 671-680.

E. L. Lawler: *Combinatorial optimization: networks and matroids* (Holt, Rinehart and Winston, New York, 1976).

S. H. Lu, B. Simeone: "On the equivalence between roof-duality and paved-duality in quadratic $0-1$ optimization", Technical Report, RRR# 22-87, Rutgers University, 1987.

S. H. Lu, A. C. Williams: "Roof-duality for non-linear 0-1 programming", RRR 2-85, Rutgers University (1985).

S. Micali, Vazirani: "An $O(V^{\frac{1}{2}}E)$ algorithm for finding maximum matchings in general graphs", Proc. 21st Annual Symposium on Foundations of Computer Science, Syracuse 1980 (IEEE Computer Society Press, New York, 1980) 17-27.

G. L. Nemhauser, L. E. Trotter: "Vertex packings: structural properties and algorithms", *Math. Program.* **8**(1975) 235-248.

R. Petreschi, B. Simeone: "A switching algorithm for the solution of quadratic boolean equations", *Inform. Process. Letters.* **11**(1980) 193-198.

R. Petreschi, B. Simeone: "Numerical comparison of 2-satisfiability algorithms", Working paper, Math. Department, University of Rome (1985).

W. V. Quine: "A way of simplifying truth functions", *Amer. Math Monthly* **52**(1952) 627-631.

J. Rhys: "A selection problem of shared fixed costs and networks", *Manag. Science* **17**(1970) 200-207.

I. G. Rosenberg: "Reduction of unconstrained nonlinear 0-1 programming to the quadratic case", *Cahiers du Centre d'Etudes de Rech. Operationnelle* **17**(1975) 71-74.

S. Rudeanu: *Boolean functions and equations* (North-Holland, Amsterdam, 1974).

B. Simeone: "Quadratic 0-1 programming, boolean functions and graphs", Doctoral dissertation, Univ. of Waterloo (1979).

B. Simeone: "Consistency of quadratic boolean equations and the König- Egerváry property for graphs", *Ann. Discr. Math.* **25**(1985) 281-290.

B. Simeone: "An asymptotically exact polynomial algorithm for equipartition problems", *Discr. Appl. Math.*, **14** (1986) 283-293.

R. E. Tarjan: "Depth-first search and linear graph algorithms", *SIAM J. Computing* **1**(1972) 146-160.

C. Witzgall: "Mathematical methods of site selection for Electronic Message Systems (EMS)", NBS Internal Report (1975).

On Binary Group Problems Having the Fulkerson Property

Ellis L. Johnson
IBM Thomas J. Watson Research Center
Yorktown Heights, NY 10598
and
Department of Applied Mathematics and Statistics
State University of New York at Stony Brook
Stony Brook, NY 11794

1. Introduction

Although Gomory's work on the group problem [9] was contemporary with the work of Fulkerson on blocking polyhedra [7], no connection was made until much later. We show that corresponding to blocking pairs of binary clutters there is a pair of binary groups problems that we call a blocking pair of binary group problems. The focus of this paper is to determine when Gomory's corner polyhedra for a blocking pair of binary group problems are a blocking pair of polyhedra in Fulkerson's sense.

Lehman's work on the Shannon switching game [15] established a matroid framework for a class of problems. This important work on the length-width inequality [16] led to Fulkerson's results on polyhedra consequences [7]. Seymour characterized the binary matroids having the "max-flow min-cut" property [18], the stronger form of the condition, and gave some results on the weaker, but apparently more difficult to characterize, condition under study here.

A result of Gomory [9] giving a subadditive characterization of facets is generalized and related to Seymour's characterization of matroids having the "sums of circuits" property [19]. It is shown that the cone Seymour studies includes among its extreme rays the facets of an associated binary group polyhedra. In this way, a result is derived on binary group problems having the Fulkerson property. A stronger result is obtained using a technique of Barahona and Groetschel [2]. This latter theorem requires results on lifting of facets, and we develop a symmetric dual framework for lifting facets and vertices of blocking pairs of binary group problems. This framework has been extended to arbitrary group problems [4].

Two early examples were known of when the Fulkerson property holds: problems over regular matroids [6] and the Chinese postman problem [5]. The former has been extended to arbitrary group problems [3]. The latter fits well into the framework developed here and is a prototype [8] of the blocking pairs of binary group problems having the Fulkerson property. A related problem is the co-postman problem [14] that gives a pair of blocking optimization problems one of which is NP-complete and the other polynomially solvable [10].

The polytope characterizations of Barahona and Groetschel [2] are related to this blocking framework. Our framework has the advantage of having pairs of problems in some sense dual to each other. The vertices and facets of such a pair of problems are more or less interchangeable depending on whether the Fulkerson property holds or not, and it holds for one problem if and only if it holds for the other.

2. Binary Clutters and Matroids

Let $J=\{1,\ldots,n\}$ represent any finite, non-empty set. A clutter is a non-empty family of subsets of J whose members are non-nested. That is, if $\mathcal{Q}$ is a clutter and $A\in\mathcal{Q}, B\in\mathcal{Q}$, then $A\not\subseteq B$. We exclude as a clutter the family whose only member is the empty set and, thereby, exclude the empty set from being a member of a clutter.

Given a clutter $\mathcal{Q}=\{S_1,\ldots,S_m\}$ of subsets of $J=\{1,\ldots,n\}$, the matrix

$$Q = (q_{ij}), \text{ where } q_{ij} = \begin{cases} 1, & \text{if } j\in S_i \\ 0, & \text{otherwise} \end{cases}$$

will be used to represent the clutter, and we refer to "the clutter Q" when actually we mean "the clutter whose subsets have incidence vectors that are the rows of Q."

For binary vectors r and p, define the operations:

$s = r \vee p$ (union) by $s_j = \begin{cases} 0 & \text{if } r_j = 0 \text{ and } p_j = 0 \\ 1 & \text{otherwise} \end{cases}$

$s = r \wedge p$ (intersection) by $s_j = \begin{cases} 1 & \text{if } r_j = 1 \text{ and } p_j = 1 \\ 0 & \text{otherwise} \end{cases}$

$s = r \hat{+} p$ (sum or symmetric difference) by $s_j = \begin{cases} 1 & \text{if } r_j + p_j = 1 \\ 0 & \text{if } r_j + p_j \text{ is 0 or 2} \end{cases}$

$s = r \backslash p$ (set difference) by $s_j = \begin{cases} 1 & \text{if } r_j = 1 \text{ and } p_j = 0 \\ 0 & \text{otherwise} . \end{cases}$

We now give a procedure for getting certain clutters. First, define an augmented matrix to be a matrix $[M \mid b]$ whose last column is distinguished and called the right-hand side. Assume that M is m×n and b is m×1. For our purposes, M and b are assumed to have 0-1 entries and are arbitrary except for the condition:

feasibility condition: $b = Mt^*$ (modulo 2) for some 0-1 vector t^*.

The reason for this condition will be made clear.

The row space of $[M \mid b]$ is formed by taking the sums (modulo 2) of all 0-1 combinations of rows of $[M \mid b]$. We can partition the rows in the row space by their entry in the $n+1^{\underline{st}}$ column, corresponding to b, so that the row space can be written

$$\begin{bmatrix} R_1 & 1 \\ \hline R_0 & 0 \end{bmatrix}$$

The feasibility condition is equivalent to not having the row

$[0 \ldots 0 \mid 1]$

in the row space. Of course, the row of all zeros,

$[0 \ldots 0 \mid 0]$

is in the row space as can be seen by multiplying each row of $[M \mid b]$ by zero and adding. Hence, the row of all zeros is in R_0 so is not in row R_1. Thus, the matrix Q whose rows are the minimal rows of R_1 represents a clutter. Define any clutter obtained in this way to be a binary clutter.

For a binary matrix M, the associated binary matroid is the matroid whose independent sets correspond to independent columns of M, with addition taken modulo 2. The minimal non-zero rows of the row space are called the cocircuits of M.

Define an elementary row operation on a binary matrix to be one of the following two operations:

(i) replace any row by the sum, modulo 2, of that row with any different row;

(ii) Remove a row that is all zero.

It is easily verified that an elementary row operation does not change the independent columns or the row space of a matrix. A pivot step on a binary matrix $A = (a_{ij})$ consists of picking an entry of 1, say $a_{ij} = 1$ and adding (modulo 2) row i to every row $k \neq i$ having $a_{kj} = 1$. The entry a_{ij} is called the pivot element. The pivot step brings column j to a unit column having a 1 in row i and 0's elsewhere. Bring the augmented matrix $[M \mid b]$ to standard form $[I\ N \mid \tilde{b}]$ by pivoting successively on columns in M with the pivot element chosen in different rows, deleting rows of all zeros, and possibly reordering the columns and remaining rows. Because of the feasibility assumption, $[0\ldots0 \mid 1]$ cannot occur so that we never have to pivot on the right-hand side column b. The columns in M corresponding to I in the standard form are a basis of M and are called basic columns. The others, including b, are called non-basic columns. There are, in general, many bases of M, and the basis can be changed without changing the matroid or row space. Assume that some basis has been chosen, and the reduction to standard form has been performed so that $M = [I\ N]$. Now, drop the bar on $\tilde{b}$ and let us write $[M \mid b] = [I\ N \mid b]$.

Following matroid terminology, define a matrix $[M \mid b]$ to be separable if in standard form, for any basis, there is a reordering of the rows and columns so that the matrix is block diagonal:

$$[M \mid b] = \left[\begin{array}{c:c:c|c} I & N_1 & 0 & 0 \\ \hdashline & 0 & N_2 & b_2 \end{array}\right] ,$$

and the matrix is <u>non-separable</u> otherwise. Separable can be defined for any matrix, but when we define it for augmented matrices, we include the right-hand side column as a column of the matrix. It is straight-forward to show that an augmented matrix is separable if and only if its binary clutter Q has no column of all zeros.

We now show that if $[M \mid b]$ is non-separable, then it can be recovered from Q by simply forming $[Q \mid 1]$ (see (46) of [15]). That is, if $[Q \mid 1]$ is reduced to standard form, eliminating redundant rows, we will get back the same matrix $[M \mid b]$ if it was brought to standard form with respect to the same basis. Clearly, the row space generated by $[Q \mid 1]$ must be a subspace of the row space of $[M \mid b]$ so it suffices to show the other direction. Further, it suffices to show that any minimal row in the row space of $[M \mid b]$ can be generated by $[Q \mid 1]$ since the rows of $[M \mid b]$, when brought to standard form, are minimal. Since $[Q \mid 1]$ already has all minimal rows with a 1 in the right-hand side column, all of those rows can trivially be generated. Thus, suppose $[r \mid 0]$ is a minimal row in the row space of $[M \mid b]$ and cannot be generated by $[Q \mid 1]$. Since $[M \mid b]$ is non-separable, there is a row of $[Q \mid 1]$, say $[q \mid 1]$ that properly intersects with $[r \mid 0]$. Let $[q \mid 1]$ also be chosen so that $r \vee q$ is minimal. Let $[p \mid 1]$ be defined by

$$[p \mid 1] = [q \mid 1] \hat{+} [r \mid 0] ,$$

where $\hat{+}$ means + (modulo 2), as defined earlier. Either $[p \mid 1]$ is a row of $[Q \mid 1]$ and we are done since $[p \mid 1] \hat{+} [q \mid 1] = [r \mid 0]$, or there is a row $[\bar{p} \mid 1]$ with $\bar{p} \leq p$ such that $[\bar{p} \mid 1]$ is minimal in the row space of $[M \mid b]$ and hence is in $[Q \mid 1]$. Since q was chosen with $r \vee q$ minimal and since $\bar{p} \leq r \vee q$, it must be that $\bar{p} \backslash s = q \backslash s$. Hence $\bar{p}+q$ (modulo 2) gives a row in the row space of $[M \mid b]$ that is a subvector of s, contradicting s being minimal. Thus we have proven the following result.

<u>Theorem 2.1</u> (Lehman) The binary group constraints

$$Mt^* \equiv b,\ t^* \geq 0 \text{ and integer, and}$$

$$Qt^* \equiv 1,\ t^* \geq 0 \text{ and integer}$$

have the same solutions t^* provided $[M \mid b]$ is non-separable, where Q is the binary clutter generated from $[M \mid b]$.

We now show another result of Lehman [15]. Actually, he defines binary clutter in the second form.

<u>Theorem 2.2</u> (Lehman) A clutter Q is a binary clutter if and only if every odd sum (modulo 2) of rows of Q is either a row of Q or has a subvector that is a row of Q.

Proof: If Q is a binary clutter, then every odd sum of rows of Q is in R_1 so must either be in Q or represent a super set of a row of Q.

For the other direction, for any clutter Q we can form the augmented matrix $[Q \mid 1]$ and its row space

$$\begin{bmatrix} R_1 & 1 \\ R_0 & 0 \end{bmatrix}$$

The rows of R_1 are odd sums of rows of Q. If Q satisfies the condition that its odd sums are either in Q or supervectors of a row of Q, then the clutter $\bar{Q}$ of minimal rows of R_1 will be exactly Q completing the proof since $\bar{Q}$ is obviously a binary clutter.

The dual matrix of an augmented matrix $[M \mid b] = [I\ N \mid b]$, in standard form, is the augmented matrix

$$[M^* \mid b^*] = \begin{bmatrix} N^T & I & 0 \\ b^T & 0 & 1 \end{bmatrix}$$

where I is the $n\text{-}m \times n\text{-}m$ identity matrix and 0 represents a row or column vector of all zeros. In this form, the right-hand side b^* is in the basis, and leaving it in the basis we say that the matrix $[M^* \mid b^*]$ as above is in <u>right-hand side form</u> as opposed to standard form. We could, of course, bring it to standard form by pivoting on a non-zero in the row b^T. However, it will be convenient for many discussions to leave the dual matrix in right-hand side form.

The dual row space can be formed, in the same way as for $[M \mid b]$, and partitioned to give

$$\begin{bmatrix} R_1^* & 1 \\ R_0^* & 0 \end{bmatrix}$$

and let Q^* denote the clutter of minimal rows of R_1^*.

The row spaces of $[M \mid b]$ and $[M^* \mid b^*]$ are dual, or orthogonal, spaces in the sense that the inner product, taken modulo 2, (i.e. $q \wedge q^*$) of any two row vectors from the respective spaces is zero. The dual row spaces can properly intersect because the inner product does not generally satisfy the condition $q \wedge q = 0$ implies $q = 0$. In fact, the two spaces can be equal. However, the dimensions of the two spaces add up to n+1, the dimension of the entire space, and it is easily shown that the dimensions could not add up to a larger number.

The <u>blocking clutter</u> $\mathcal{B}$ of a given clutter Q is defined by

$$\mathcal{B} = \{A^* \subseteq J \mid \text{(i) } A^* \cap A \neq \emptyset \text{ for all } A \in Q, \text{ and (ii) } A^* \text{ is minimal with respect to (i)}\}.$$

The matrix B corresponding to $\mathcal{B}$ has rows q^* satisfying $q^* \cdot q \geq 1$, i.e. for every $q \in Q$ and $q^* \in B$ there is a $j \in J$ such that $q_j = q_j^* = 1$. Furthermore, if r^* is a 0-1 vector such that $r^* < q^*$ for any $q^* \in B$ then there is a $q \in Q$ such that $r^* \cdot q = 0$. Clearly the blocking clutter of $\mathcal{B}$ is Q itself; a fact that depends only on Q being a clutter to begin with.

An example of a clutter is {{1,2}, {2,3}, {3,4}}, and its blocking clutter is {{1,3}, {2,3}, {2,4}}. The corresponding matrices are

$$Q = \begin{bmatrix} 1 & 1 & 0 & 0 \\ 0 & 1 & 1 & 0 \\ 0 & 0 & 1 & 1 \end{bmatrix} \quad \text{and} \quad B = \begin{bmatrix} 1 & 0 & 1 & 0 \\ 0 & 1 & 1 & 0 \\ 0 & 1 & 0 & 1 \end{bmatrix}$$

Let Q be a binary clutter, i.e. the minimal rows of R_1 from the row spaces of $[M \mid b]$, and let Q^* be the clutter from its dual row space. Because $[q \mid 1]$ and $[q^* \mid 1]$, for $q \in Q$ and $q^* \in Q^*$, are in dual row spaces, it is easily seen that $q \wedge q^* = 1$ and hence $q \cdot q^* \geq 1$. In order to complete the proof that Q^* is the blocking clutter of Q [15], we must show two things:

(1) every $q^* \epsilon Q^*$ is minimal with respect to $q^* \cdot q \geq 1$ for all $q \epsilon Q$; and

(2) any 0-1 r^* satisfying $r^* \cdot q \geq 1$ for all $q \epsilon Q$ and minimal with respect to this property is in Q^*.

To prove (1), note that $q^* \epsilon Q^*$ must be a solution to

$Qq^* \equiv 1 \pmod 2$, and hence to

$Mq^* \equiv b \pmod 2$,

and the columns j with $q_j^* = 1$ must be linearly independent in M or else a 0-1 solution s^* to $Ms^* \equiv 0$ could be added, modulo 2, to q^* to get a vector r^* less than or equal to q^* still satisfying $Mr^* \equiv b$. Let J_1 be the set of those j for which $qj^* = 1$. The matrix M can be brought to standard form with J_1 being a subset of the basic columns. In such a standard form,

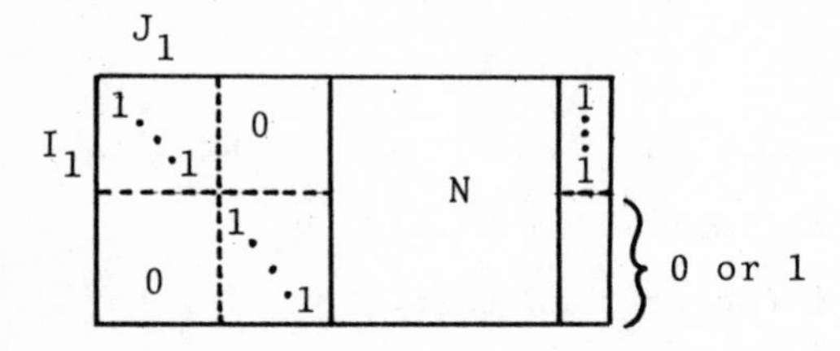

the rows I_1 corresponding to the identity columns with 1's in the columns J_1 must have a 1 in the right-hand side of the standard form because $q_j^* = 1$, $j \epsilon J_1$ is a solution to $Mt^* \equiv b$. Hence those rows q_i for $i \epsilon I_1$ are rows of Q, and any 0-1 vector t^* satisfying $t^* \leq q^*$ and $Qt^* \geq 1$ must have $t_j^* = 1$ for all $j \epsilon J_1$. Therefore $t^* = q^*$.

To prove (2), suppose r^* is such a 0-1 blocking vector and is not a row of Q^*. Let J_1 be the indices j such that $r_j^* = 1$. Since r^* is a blocking vector for Q, every row of Q has at least one 1 among the columns J_1. Since r^* is minimal with respect to $q \cdot r^* \geq 1$, for each $j \epsilon J_1$ there must be a row $q_{i(j)}$ such that $q_{i(j)}$ has a 1 in column j and a 0 in all other columns of J_1. Thus, we can reorder the rows and columns of Q so that it has the form

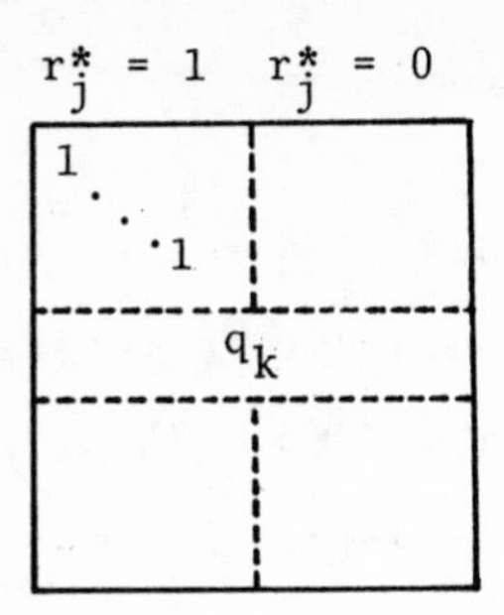

that is, an identity in the upper left-hand corner corresponding to the columns J_1 where $r_j^* = 1$. Suppose there is a row, say q_k, of Q that has an even intersection with r^*. Since $r^* \cdot q_k \geq 1$, there must be a positive, even number of 1's in columns J_1 for the row q_k. Now, adding (modulo 2) q_k to the sum of the (even number) of rows of Q from among the first $|J_1|$ rows having a 1 in columns j where $q_{kj} = 1$ gives a row r having all 0's in the columns J_1. But r is either in Q or a supervector of a row q of Q (by Theorem 2.2). Hence, $q \cdot r^* = 0$ for a row q of Q, giving a contradiction. Thus, we have proven the theorem below.

<u>Theorem 2.3</u> (Lehman) The blocking clutter of a binary clutter Q from $[M \mid b]$ is the same as the binary clutter obtained from the dual matrix $[M^* \mid b^*]$ to $[M \mid b]$.

We now state as a separate lemma a result shown in the above proof.

3. Binary Group Problems

The group problem was introduced by Gomory [9], and he gave some results specific to binary and ternary groups. Here, we only consider binary groups: finite Abelian groups each element of which has order two. It can be shown that such groups are isomorphic to the direct product of cyclic groups of order two, denoted in $\mathbb{C}_2$. For a subset $\mathcal{M}$ of elements of $m\mathbb{C}_2$, we can form a matrix M having m rows and a column for each element in $\mathcal{M}$ with 0-1 entries. Addition can be taken to be componentwise addition modulo 2. For any column m-vector b, the <u>binary group problem</u> is to minimize ct subject to

$$Mt^* \equiv b \pmod 2$$

$$t^* \geq 0 \text{ and integer.}$$

Note that t^* is not required to be 0-1 but can be any non-negative integer. Thus, c_j can be assumed to be a non-negative real since otherwise the objective function would be unbounded. Then, the optimum t^* can be chosen from among the 0-1 solutions. We assume that there is an integer solution t^*, which is the feasibility assumption imposed in the previous section. Thus, for each non-zero and feasible b, there is a binary group problem associated with $[M \mid b]$ once an objective function is specified. We refer to the binary group problem $[M \mid b]$ to mean the associated group problem with any objective function.

Gomory defined the group polyhedron [9]:

$$P(M,b) = \text{conv } \{t^* \in \mathbb{Z}_+^n \mid Mt^* \equiv b \pmod 2)\},$$

where $\mathbb{Z}_+^n$ denotes the non-negative integer n-vectors. He gave several results, one of which was that for this binary case the vertices of $P(M,b)$ correspond to the solutions with independent columns in M. When b is adjoined, we get a circuit in the matroid over $[M \mid b]$ so the vertices are exactly the rows of Q^*. Thus, we have the result below.

<u>Theorem 3.1</u> (Gomory) For the binary group problem $[M \mid b]$,

$$P(M,b) = \text{conv } \{Q^*\} + \mathbb{R}_+^n .$$

Here, $\mathbb{R}_+^n$ means the non-negative real n-vectors. By conv $\{Q^*\}$ we mean the convex hull of the rows of Q^*.

Gomory gave two other main types of results for this problem. One was a subadditive characterization of facets. However, this result was for master problems, those where M has all 2^m-1 non-zero columns of 0's and 1's. For subproblems, it still had some interest. In section 5, we give an extension of his results applicable to binary group subproblems.

To explain further Gomory's subadditive characterization, first of all, facets are defining inequalities of a polyhedron and, here, are either of the form $t_j^* \geq 0$ or

$$\sum_j \pi_j t_j^* \geq \pi_0$$

where $\pi_j \geq 0$ and $\pi_0 > 0$, and the inequality is valid, i.e. satisfied by every group solution t^*. In order for the latter to be a facet, it is necessary and sufficient that there be K solutions t^k of the group problem, where K is the number of $\pi_j > 0$, such that

$$\sum_j \pi_j t_j^k = \pi_0, \quad k = 1,\ldots,K,$$

and such that the K×K matrix T, formed by taking a row for each t^k and a column for each j with $\pi_j > 0$, is non-singular.

What Gomory showed can be interpreted for subproblems as follows. For a facet $(\pi_0;\pi_1,\ldots,\pi_n)$,

$$\text{if } M^j + M^k = M^\ell, \quad \text{then } \pi_j + \pi_k \geq \pi_\ell, \quad \text{and}$$

$$\text{if } M^j + M^k = b, \quad \text{then } \pi_j + \pi_k = \pi_0 .$$

However, the more significant part of his result was the converse, in a certain sense, namely that the cone of subadditive inequalities on the π's given above together with $\pi \geq 0$ has extreme rays that are exactly all facets. But the result holds only for master problems. We know that all facets for subproblems come from master problems (see section 8), but here we are interested in knowing for which subproblems a given class of facets suffices (see section 8), and theorems about facets of master problems are not of much help. Section 5 extends Gomory's result to give a more interesting result for subproblems which allows us to give a sufficient condition for the Fulkerson property to hold. Seymour's work [19] on matroids having the "sums of circuits" property is used.

Gomory also gave two ways to get facets for problems from facets for smaller problems. The first of these we call facet extension corresponding to extension from a subproblem to a larger subproblem or to a master problem eventually. This method is explained in Theorem 8.3 and Theorem 8.4. The second lifting method was given in terms of homomorphisms and explained all facets having coefficients of zero. Gaston and Johnson [8] generalized it to say exactly when a facet for a subproblem lifts to give a facet for the preimage. We explain this result in sections 8 and 9 in terms of contractions and expansions.

One thing Gomory failed to do was to relate a group problem to any type of blocking problem. In the next section, an introduction to Fulkerson's blocking theory [7] for polyhedra is given. It was noted in [12] that Gomory's group polyhedra fitted into Fulkerson's framework but that while Fulkerson's framework is completely symmetric, Gomory's is not. That asymmetry has been partially alleviated by the development of the blocking group problem, first for the binary case [8] and then the general group problem [4]. For the binary case, there are pairs of binary clutters that may or may not have a certain property that Lehman [16] showed equivalent to several others and that he called the max-flow, min-cut property. What Fulkerson showed [7] is that when this property holds, then there is a pair of polyhedra with a one-to-one correspondence between vertices of one and non-zero facets (see section 4) of the other and vice versa. What it means for binary group problems is that the blocking polyhedron of one binary group problem is itself the polyhedron of another binary group problem, and we say that the polyhedra have the Fulkerson property (see section 6).

Given a binary group problem $[M \mid b]$, define its <u>blocking binary group problem</u> to be the binary group problem over the dual matrix $[M^* \mid b^*]$:

$$\begin{array}{lcl} t^* \geq 0, \text{ integer} & & t \geq 0, \text{ integer} \\ Mt^* \equiv b \pmod 2 & \text{and} & M^*t \equiv b^* \pmod 2 \\ ct^* = \min & & c^*t = \min \end{array}$$

There is no correspondence here between c and c^*; both are arbitrary vectors in $\mathbb{R}^n_+$. For M in standard form, the constraints, other than non-negativity, of the two problems are

$$[I\ N]\ t^* \equiv b \pmod 2, \quad \text{and} \quad \begin{bmatrix} N^T & I \\ b^T & 0\ldots0 \end{bmatrix} t^* \equiv \begin{bmatrix} 0 \\ \vdots \\ 0 \\ 1 \end{bmatrix} \pmod 2.$$

Whether or not the Fulkerson property holds, we always have

$$\begin{aligned} P(M,b) &= \text{conv}\ \{t^* \in \mathbb{Z}_+^n \mid Mt^* \equiv b \pmod 2\} \\ &= \text{conv}\ \{Q^*\} + \mathbb{R}_+^n, \quad \text{and} \\ P(M^*,b^*) &= \text{conv}\ \{t \in \mathbb{Z}_+^n \mid M^*t \equiv b^* \pmod 2\} \\ &= \text{conv}\ \{Q\} + \mathbb{R}_+^n . \end{aligned}$$

The Fulkerson property holds for $P(M,b)$ if

$$P(M,b) = \{t^* \in \mathbb{R}_+^n \mid Qt^* \geq 1\}.$$

In the next section, we show that it holds for $P(M,b)$ if and only if it holds for $P(M^*,b^*)$.

4. Blocking Pairs of Polyhedra

Given a polyhedron P contained in $\mathbb{R}_+^n$, define the blocking polyhedron $B(P)$ of P [7] to be

$$B(P) = \{x^* \geq 0 \mid x^* \cdot x \geq 1, \text{ for all } x \in P\}.$$

In order for

$$B(B(P)) = P$$

to hold, it is necessary and sufficient that

$$P = P + \mathbb{R}_+^n \quad \text{and} \quad P \subseteq \mathbb{R}_+^n .$$

The development here is from Fulkerson [7]. For other discussion, see [12], where the initial relationship to Gomory's work was made.

If the polyhedron P is given by

$$P = \text{conv}\ \{V\} + \mathbb{R}_+^n ,$$

for a finite set V of vectors in $\mathbb{R}_+^n$, then it can be shown that

$$B(P) = \{x^* \geq 0 \mid x^* \cdot v \geq 1, v \in V\}.$$

A minimal such set V defining P are, of necessity, the vertices of P, and by the above are the facets, along with $x^* \geq 0$, of $B(P)$.

The blocking polyhedron $B(P)$ has a similar representation:

$$B(P) = \text{conv}\ \{V^*\} + \mathbb{R}_+^n,$$

and for a minimal such set V^*, the $v^* \in V^*$ are the vertices of $B(P)$ and the facets, along with $x \geq 0$, of $\mathcal{P}$, i.e.

$$\mathcal{P} = \{x \geq 0 \mid x \cdot v^* \geq 1,\ v^* \in V^*\}.$$

This theory is completely symmetric in $\mathcal{P}, \mathcal{B}(\mathcal{P})$ and in V and V^*, but finding one of V, V^* given the other is not necessarily easy.

Although Fulkerson's theory of blocking pairs of polyhedra can be carried through for arbitrary real vectors, we will specialize it to vectors V that are 0-1 vectors. A set of 0-1 vectors V is minimal in defining

$$\mathcal{P} = \text{conv}\ \{V\} + \mathbb{R}^n_+ ,$$

i.e. no subset of V defines the same polyhedron $\mathcal{P}$, if and only if the vectors in V are incidence vectors of a clutter. This result follows from the fact that if $v^1, v^2 \in V$ and $v^1 \geq v^2$, then v^1 can be deleted from V without changing conv $\{V\} + \mathbb{R}^n_+$. Thus, we can think of V as representing a clutter, but it need not be a binary clutter as defined in section 2.

For any clutter Q, binary or not, there is always a blocking clutter

$$Q^* = \{q^* \mid q^* \text{ is 0-1 and } q \cdot q^* \geq 1, \text{ for all } q \in Q\}.$$

However, it is not always true that

$$\mathcal{P} = \text{conv}\ \{Q\} + \mathbb{R}^n_+$$

has blocking polyhedron given by

$$\text{conv}\ \{Q^*\} + \mathbb{R}^n_+ .$$

However, $q^* \in Q^*$ is always in $\mathcal{B}(\mathcal{P})$ so

$$\mathcal{B}(\mathcal{P}) \supseteq \text{conv}\ \{Q^*\} + \mathbb{R}^n_+ .$$

The case where equality holds is characterized by

$$\mathcal{P} = \{x \geq 0 \mid Q^* x \geq 1\}.$$

By the previous discussion of Fulkerson's results [7], we then have

$$\mathcal{B}(\mathcal{P}) = \{x^* \geq 0 \mid Q x^* \geq 1\}.$$

For this case, consider the linear program:

$$\text{minimize } c^* x$$

$$\text{subject to } x \geq 0 \text{ and } Q^* x \geq 1,$$

for $c^* \in \mathbb{R}^n_+$. For such a non-negative objective function, the optimum solution can be taken to be a vector in Q, and we have by linear programming duality:

$$\min \{c^*x \mid x \in Q\} = \max \{ \sum_{i=1}^{m} \pi_i \mid \pi \geq 0 \text{ and } \pi Q^* \leq c^*\}.$$

This equality is called by Lehman [16] the max-flow, min-cut equality and we say that Q has the max-flow, min-cut property. He shows that Q has the max-flow, min-cut property if and only if Q* does. Seymour [18] refers to this property as the weak max-flow, min-cut property.

We now give an example of the type that motivates the terminology. Consider the graph (undirected)

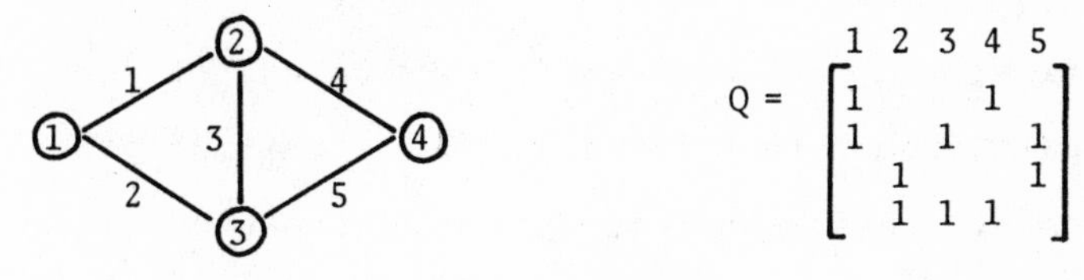

$$Q = \begin{bmatrix} 1 & & & 1 & \\ 1 & & 1 & & 1 \\ & 1 & & & 1 \\ & 1 & 1 & 1 & \end{bmatrix}$$

where Q is the clutter of undirected, simple paths from node 1 to node 4. The linear program mentioned before is

$$\text{minimize } c_1x_1 + c_2x_2 + c_3x_3 + c_4x_4 + c_5x_5$$

subject to $x_j \geq 0$ and

$$\begin{aligned} x_1 \qquad\qquad + x_4 \qquad &\geq 1 \\ x_1 \qquad + x_3 \qquad + x_5 &\geq 1 \\ x_2 \qquad\qquad\qquad + x_5 &\geq 1 \\ x_2 + x_3 + x_4 \qquad &\geq 1 . \end{aligned}$$

The optimum solution will be a cut separating nodes 1 and 4. The dual linear program assigns a variable π_i to each row, i.e. to paths, such that on any edge e the sum of the π_i's over paths containing edge e is no more than c_e, the capacity of edge e. Thus, the π_i represent a path decomposition of a flow from 1 to 4 subject to edge capacities and maximizing the sum of the π_i, i.e. the total flow. In this undirected version of the flow problem, flows on edges add up regardless of direction in a path. In this case, the problem we have posed is the binary group problem with constraints:

$$\begin{bmatrix} 1 & 1 & & & \\ 1 & & 1 & 1 & \\ & 1 & 1 & & 1 \\ & & & 1 & 1 \end{bmatrix} t^* \equiv \begin{bmatrix} 1 \\ 0 \\ 0 \\ 1 \end{bmatrix}, \text{ or in standard form}$$

$$\begin{bmatrix} 1 & & 1 & & 1 \\ & 1 & 1 & & 1 \\ & & & 1 & 1 \end{bmatrix} t^* \equiv \begin{bmatrix} 1 \\ 0 \\ 1 \end{bmatrix}, \text{ or reordered } \left[\begin{array}{ccccc|c} 1 & & & 1 & 1 & 1 \\ & 1 & & 1 & 1 & 0 \\ & & 1 & & 1 & 1 \end{array}\right].$$

The blocking problem, thus, has constraints

$$\begin{bmatrix} 1 & 1 & 1 & 0 & 0 \\ 1 & 1 & 0 & 1 & 1 \\ 1 & 0 & 0 & 1 & 0 \end{bmatrix} t \equiv \begin{bmatrix} 0 \\ 0 \\ 1 \end{bmatrix} \quad \text{or reordered} \quad \left[\begin{array}{ccccc|c} 1 & 1 & 0 & 1 & & 0 \\ 1 & 1 & 1 & & 1 & 0 \\ 1 & 0 & 1 & 0 & 0 & 1 \end{array}\right].$$

The blocking clutter is

$$Q^* = \begin{bmatrix} 1 & 1 & & & \\ 1 & & 1 & & 1 \\ & 1 & 1 & 1 & \\ & & & 1 & 1 \end{bmatrix}.$$

There is a max-flow, min-cut theorem for this Q^* and it says that the minimum length path from 1 to 4 is equal to the maximum packing of cuts separating 1 from 4. This result is less well-known but is in a sense a twin to the more well known max-flow, min-cut theorem.

For the example given, the dual problem has an optimum integer answer whenever c is integer. Lehman [16] refers to this property as the strong max-flow, min-cut property for Q, and Seymour [18] calls it the max-flow, min-cut property. This property may hold for Q but not Q^*, but where it holds for one of Q, Q^*, then the polyhedra are a blocking pair and the weaker version of the max-flow, min-cut property holds for both Q and Q^*. We do not discuss the strong version in this paper, and will focus on binary clutters and the Fulkerson property for the associated group polyhedra.

Fulkerson's terminology is that the pair of polyhedra $P(M,b)$ and $P(M^*,b^*)$ are a <u>blocking pair of polyhedra</u> whenever the Fulkerson property holds. There is some potential for confusion in that we have a blocking pair of binary group problems whose polyhedra may or may not be a blocking pair of polyhedra. However, this confusion is already present in the notion of blocking pairs of clutters whose polyhedra (the convex hull of the clutters plus the non-negative orthant) may or may not be blocking pairs of polyhedra.

We now show a result that holds in any case, whether or not the Fulkerson property holds.

Theorem 4.1 Each inequality in the system $Qt^* \geq 1$ is a facet of $P(M,b)$ and every facet with 0-1 coefficients π of the form $\pi t^* \geq 1$ is among the facets in the system $Qt^* \geq 1$.

The proof of this result requires the following lemma.

Lemma 4.2 An inequality $\pi t^* \geq 1$ is a facet of $P(M,b)$ if and only if $\pi \geq 0$, $\pi \cdot q^* \geq 1$ for all $q^* \epsilon Q^*$, and there is a non-singular submatrix T of Q^* with columns corresponding to columns j where $\pi_j > 0$ and rows from among the rows of Q^* for which $\pi \cdot q^* = 1$.

The proof of the lemma is direct from the shape of $P(M,b)$: it is a proper subset of $\mathbb{R}^n_+$ with recession cone equal to $\mathbb{R}^n_+$.

The first half of the theorem is easily proven from the lemma by the fact that every row q of Q is a (minimal) blocker of the rows of Q^*. Thus, for every column j where $q_j > 0$ we can find a row of Q^* with a 1 in column j and no other 1 among the columns k where $q_k = 1$. Therefore, the matrix T required by the lemma can be found and is in fact an identity I.

The converse requires showing that a facet $\pi t^* \geq 1$ having 0-1 coefficients must satisfy $\pi \cdot q^* = 1$ (modulo 2) for all rows q^* of Q^*. To show this fact, suppose not. The contradiction hypothesis is that there is a row r^* of Q^* for which $\pi \cdot r^*$ is even, i.e. r^* has an even number of 1's where π is equal to 1. Since $\pi t^* \geq 1$ is a facet with 0-1 coefficients π, the non-singular matrix T in Lemma 4.2 must be an identity matrix I. If we add (modulo 2) to r^* the rows of Q^* corresponding to the rows of T having a 1 in the columns where r^* is 1, then we get a row q^* which is all 0 in the places where π is 1. Thus, $\pi \cdot q^* \geq 1$ is not satisfied. However this row q^* is a sum of an odd number of rows of Q^* so is either a row of Q^* or dominates a row of Q^*. Thus, we reach the desired contradiction because π should satisfy $\pi Q^* \geq 1$.

To complete the proof of the theorem, we must show that a facet $\pi t^* \geq 1$ with 0-1 coefficients π not only satisfies the congruences $\pi \cdot q^* = 1$ (modulo 2) for all rows q^* of Q^*, but π is minimal among all 0-1 vectors satisfying this system of congruences. If π were a facet and some 0-1 vector $\rho \geq \pi$, $\rho \neq \pi$, were to satisfy the system of congruences, then $\rho t^* \geq 1$ would be a valid inequality for $P(M,b)$, and $\pi t^* \geq 1$ would be the sum of it and the non-negativity restrictions

$$t^*_j \geq 0, \text{ for } j \text{ with } \pi_j = 1 \text{ and } \rho_j = 0,$$

contradicting $\pi t^* \geq 1$ being a facet. The proof of the theorem is thus completed.

Let Π denote the real matrix whose rows are facets $\pi t^* \geq 1$ of $P(M,b)$ and let Π^* denote the matrix whose rows are facets of $P(M^*,b^*)$. By Theorem 4.1 we know that $Q(Q^*)$ is a submatrix of $\Pi(\Pi^*)$. We use $\Pi \backslash Q$ to denote the matrix whose rows are in Π but not in Q.

<u>Theorem 4.3</u> [7] The rows of $\Pi \backslash Q$ are the fractional vertices of

$$\pi \geq 0,$$
$$\pi q^* \geq 1, \quad \text{for all } q^* \in Q^*.$$

Proof: This result also follows from lemma 4.2 because the facets of $P(M,b)$ are, in fact, the vertices of the inequality system indicated there and repeated in the statement of the theorem. The fractional (not all 0-1) vertices there are the rows of $\Pi \backslash Q$.

The Fulkerson property holds whenever $\Pi = Q$ and that holds if and only if $\Pi^* = Q^*$. When it does not hold we are really interested in two pairs of blocking polyhedra:

$$P(M,b) \ \& \ \text{conv}(\Pi) + \mathbb{R}^n_+ \text{ and } P(M^*,b^*) \ \& \ \text{conv}(\Pi^*) + \mathbb{R}^n_+$$

or, equivalently

$$\text{conv}(Q^*) + \mathbb{R}^n_+ \ \& \ \text{conv}(\Pi) + \mathbb{R}^n_+ \text{ and } \text{conv}(Q) + \mathbb{R}^n_+ \ \& \ \text{conv}(\Pi^*) + \mathbb{R}^n_+$$

Even when the Fulkerson property does not hold, we have, by theorem 4.1 that the vertices of $P(M,b)$ are the 0-1 vectors π^* for which $\pi^* t \geq 1$ is a facet of $P(M^*,b^*)$ and the vertices of $P(M^*,b^*)$ are all of the 0-1 vectors π for which $\pi t^* \geq 1$ is a facet of $P(M,b)$. When the Fulkerson property holds, there are no other facets of either polyhdron.

5. Subadditive Characterization of Facets

This section generalizes a result of Gomory (Theorem 18 of [9]) and can be found in Chopra and Johnson (section 7 of [4]) for general group problems. Here, we only consider binary group problems. First, we state Gomory's result. It applies when M has all 2^m-1 non-zero columns of 0's and 1's. Define the subadditive cone in that case to be

$$S = \{\pi \in \mathbb{R}^n_+ \mid \pi_k \le \pi_i + \pi_j \text{ if } M^k = M^i + M^j, \text{ and}$$
$$\pi_k = \pi_i + \pi_j \text{ if } b = M^k = M^i + M^j\}.$$

The inequality $\pi_k \le \pi_i + \pi_j$ need only be imposed if none of M^k, M^i, or M^j is equal to b or to 0. The stronger constraint, namely equality, is imposed when $M^k = b$.

Gomory's result is that the extreme rays of this cone give precisely all facets of $P(M,b)$, other than $t_j \ge 0$, and the facets are the inequalities

$$\sum_{j=1}^{n} \pi_j t_j \ge \pi_k, \quad \text{where } b = M^k,$$

for $(\pi_1,\ldots,\pi_n)$, where $n = 2^m-1$, on an extreme ray of the cone.

For a given binary group problem $[M \mid b]$, where M is now any m×n 0-1 matrix and b is not all 0's, and such that the problem is feasible, define the subadditive cone by

$$S(M,b) = \{\pi \in \mathbb{R}^{n+1}_+ \mid \pi_k \le \sum_{j=1}^{n} \pi_j t^*_j \quad \text{if } M^k = Mt^*,\ t^* \in \mathbb{Z}^n_+,$$
$$\pi_0 \le \sum_{j=1}^{n} \pi_j t^*_j \quad \text{if } b = Mt^*,\ t^* \in \mathbb{Z}^n_+\}.$$

Since b may not be a column of M we have to expand the definition so that $\pi = (\pi_0,\pi_1,\ldots,\pi_n)$ and π_0 corresponds to b.

We have defined $S(M,b)$ for many more t^* than necessary. We give two lemmas to cut down on the number of inequalities defining $S(M,b)$ by eliminating some redundant ones.

<u>Lemma 5.1</u> The subadditive cone can be defined by

$$S(M,b) = \{\pi \in \mathbb{R}^{n+1}_+ \mid \pi_k \le \sum_{\substack{j=1 \\ j \ne k}}^{n} \pi_j q^*_j, \quad \text{if } q^* \in Q^*_0 \text{ and } q^*_k = 1,$$

$$\pi_0 \le \sum_{j=1}^{n} \pi_j q_j^*, \text{ if } q^* \in Q^*\}.$$

Here Q_0^* is the clutter of R_0^* so that

$$\begin{bmatrix} Q_0^* & 0 \\ Q^* & 1 \end{bmatrix}$$

have rows that are incidence vectors of circuits of $[M \mid b]$.

Proof: First, if $t_k^* \ge 1$ where $M^k = Mt^*$, then the inequality $\pi^k \le \pi \cdot t^*$ is redundant using only $\pi \ge 0$.

Next, if any $t_j^* \ge 2$, then we can reduce it by 2 to get another inequality that, together with $\pi_j \ge 0$, implies the original one. Thus, the t^*, when augmented by $t_k^* = 1$ in the first type of inequality and by $t_0^* = 1$ in the second, is in the dual row space of $[M \mid b]$. To reduce them to the clutter of that dual row space only requires, again, using $\pi \ge 0$; completing the proof of the lemma.

We now need a definition. Given a circuit (q^*, q_0^*) of $[M \mid b]$ define a <u>chord</u> of the circuit to be an index $j \in \{0,1,\ldots,n\}$ such that $q_j^* = 0$ and such that there is another circuit (r^*, r_0^*) of $[M \mid b]$ satisfying

$$r_j^* = 1, \text{ and}$$

$$r_k^* \le q_k^*, \; k \ne j.$$

This definition corresponds to that of a chord of a circuit in a graph, i.e. a chord is an edge, corresponding to an index, such that if the chord is adjoined to the circuit there is another circuit in the union. The importance of chords in similar problems has been shown by Barahona and Groetschel [2].

By considering circuits of $[M \mid b]$ to be of the form (q^*, q_0^*) we can restate lemma 5.1 and add the restriction that there be no chord.

<u>Lemma 5.2</u> The subadditive cone can be defined by

$$S(M,b) = \{\pi \in \mathbb{R}_+^{n+1} \mid \pi_k \le \sum_{\substack{j=1 \\ j \ne k}}^{n} \pi_j q_j^*, \text{ for all circuits}$$

(q^*, q_0^*) of $[M \mid b]$ containing no chord

and for all $k \in \{0, \ldots, n\}$ with $q_k^* = 1\}$.

Proof: We must show that if (q^*, q_0^*) is a circuit of $[M \mid b]$ containing a chord, say ℓ, then

$$\pi_k \leq \sum_{\substack{j=0 \\ j \neq k}}^{n} \pi_j q_j^* , \quad \text{where } q_k^* = 1,$$

is redundant in defining the cone $S(M,b)$. Let (r^*, r_0^*) be the circuit containing the chord ℓ so that

$$r_\ell^* = 1 \quad \text{and} \quad q_\ell^* = 0,$$

$$r_j^* \leq q_j^*, \; j \neq \ell, \; j = 0, 1, \ldots, n.$$

If $r_k^* = 0$, then consider $(r^*, r_0^*) + (q^*, q_0^*)$ (modulo 2). This sum, modulo 2, has $k^{\underline{th}}$ component equal to 1 and is either a circuit or partitions into circuits (because the matroid is binary). However, it must be a circuit because if it partitioned into two or more circuits, then one of them would be a subset of q^*. Now, we can change (r^*, r_0^*) to be this sum so that we can assume, without loss of generality, that $r_k^* = 1$. Now, (r^*, r_0^*) gives the inequality

$$\pi_k \leq \sum_{\substack{j=0 \\ j \neq k}}^{n} \pi_j r_j^* = \sum_{\substack{j=0 \\ j \neq k,\ell}}^{n} \pi_j r_j^* + \pi_\ell .$$

As mentioned before, $(r^*, r_0^*) + (q^*, q_0^*)$ (modulo 2) is a circuit (s^*, s_0^*), and it satisfies $(s^*, s_0^*) \leq (r^*, r_0^*) + (q^*, q_0^*)$ (modulo 2) and $s_\ell^* = 1$. Consider the subadditive inequality given by s^*:

$$\pi_\ell \leq \sum_{\substack{j=0 \\ j \neq \ell}}^{n} \pi_j s_j^* = \sum_{\substack{j=0 \\ j \neq \ell,k}}^{n} \pi_j s_j^*, \quad \text{by } s_k^* = 0 .$$

Adding to it the previously derived inequality:

$$\pi_k \leq \sum_{\substack{j=0 \\ j \neq \ell,k}}^{n} \pi_j r_j^* + \pi_\ell$$

gives the inequality

$$\pi_k \le \sum_{\substack{j=0 \\ j\ne k,\ell}}^{n} \pi_j(s_j^* + r_j^*) .$$

But in real addition, we have

$$s_j^* + r_j^* = q_j^*, \; j \ne k,\ell,$$

so the above inequality is the same as the inequality for (q^*,q_0^*).

The proof is completed except in the case where (s^*,s_0^*) has a chord. However, we can eliminate any inequality shown to be implied by others when $S(M,b)$ is full dimensional. This is the case unless $[M \mid b]$ has circuits of length 1 or 2 since then $(1,1,\ldots,1)$ is interior to the cone. Since $[M \mid b]$ has no 0 columns, it has no length 1 circuits. A length 2 circuit means $[M \mid b]$ has duplicate columns, and then the π_j's must be equal for those two columns. We can remove duplicate columns, prove the theorem, and then add back the duplicate columns to complete the proof.

Define a $\pi \in S(M,b)$ to be <u>b-minimal</u> if there does not exist a $\rho \in S(M,b)$ satisfying

$$\rho_0 \ge \pi_0, \quad \text{and}$$
$$\rho_j \le \pi_j, \; j = 1,\ldots,n .$$

We are now in a position to state the main result of this section.

<u>Theorem 5.3</u> The facets of $P(M,b)$, other than $t_j^* \ge 0$, are the inequalities

$$\sum_{j=1}^{n} \pi_j t_j^* \ge \pi_0 ,$$

where:

(i) π is on an extreme ray of $S(M,b)$;

(ii) $\pi_0 > 0$; and

(iii) π is b-minimal in $S(M,b)$.

We remark that this theorem generalizes Gomory's result in that for the master problem every circuit of length four or more has a chord. Thus, only inequalities of the form $\pi_j \le \pi_k + \pi_\ell$ need be imposed in defining $S(M,b)$.

Define (π_0,π) to be a <u>valid inequality</u> for $P(M,b)$ if

$$\sum_{j=1}^{n} \pi_j t_j \geq \pi_0 ,$$

for all $t \in P(M,b)$. It should be clear that $(\pi_0,\pi) \in S(M,b)$ implies that it is a valid inequality by the definition of $S(M,b)$. The $(\pi_0,\pi) \in S(M,b)$ are referred to as <u>subadditive valid inequalities</u>. For any valid inequality (π_0,π), define it to be a <u>minimal valid inequality</u> if there does not exist another valid inequality (ρ_0,ρ) with

$$\rho_0 \geq \pi_0, \quad \text{and} \quad \rho_j \leq \pi_j, \; j = 1,\ldots,n .$$

Not all valid inequalities are in $S(M,b)$, but the next lemma says that all minimal valid inequalities are in $S(M,b)$.

<u>Lemma 5.4</u> If (π_0,π) is a minimal valid inequality, then it is in $S(M,b)$.

Proof: Suppose (π,π_0) is minimal valid inequality, but $(\pi_0,\pi) \notin S(M,b)$. Then there is some circuit (q^*,q_0^*) of $[M \mid b]$ containing no chord with $q_k^* = 1$ and

$$\pi_k > \sum_{\substack{j=0 \\ j \neq k}}^{n} \pi_j q_j^* .$$

This $k \neq 0$ since that would contradict (π,π_0) being a valid inequality. Also, $q_0^* = 0$ because if $q_0^* = 1$, then $\pi_k > \pi_0$, and clearly $\pi_k \leq \pi_0$ in any minimal valid inequality. Define (ρ,ρ_0) by

$$\rho_j = \pi_j , \quad j \neq k$$

$$\rho_k = \sum_{\substack{j=1 \\ j \neq k}}^{n} \pi_j q_j^* .$$

A contradiction will be reached if (ρ,ρ_0) is shown to be a valid inequality. If a solution $t^* \in \mathbb{Z}_+^n$, $Mt^* = b$ has $t_k^* = 0$ then

$$\sum_{j=1}^{n} \rho_j t_j^* = \sum_{j=1}^{n} \pi_j t_j^* \geq \pi_0 = \rho_0 .$$

On the other hand, if $t_k^* \geq 1$, then

$$\sum_{j=1}^{n} \rho_j t_j^* = \sum_{\substack{j=1 \\ j \neq k}}^{n} \pi_j t_j^* + \Big(\sum_{\substack{j=1 \\ j \neq k}}^{n} \pi_j q_j^*\Big) t_k^*$$

$$= \sum_{\substack{j=1 \\ j \neq k}}^{n} \pi_j (t_j^* + q_j^* t_k^*) \geq \pi_0 = \rho_0$$

because $(t_j^* + q_j^* t_k^*)$, $j = 1,\ldots,n$, $j \neq k$, is a solution by

$$\sum_{\substack{j=1 \\ j \neq k}}^{n} M^j (t_j^* + q_j^* t_k^*) = \sum_{\substack{j=1 \\ \neq k}}^{n} M^j t_j^* + t_k^* \sum_{\substack{j=1 \\ \neq k}}^{n} M^j q_j^*$$

$$= \sum_{\substack{j=1 \\ \neq k}}^{n} M^j t_j^* + t_k^* M^k = b ,$$

completing the proof.

Proof of Theorem 5.3: Suppose first that

$$\sum_{j=1}^{n} \pi_j t_j^* \geq 0$$

is a facet, other than $t_j^* \geq 0$, of $P(M,b)$. Then, clearly, $\pi_0 > 0$ and (π,π_0) is a minimal valid inequality. Since it is a minimal valid inequality, it is in $S(M,b)$ by lemma 5.4. Hence, it is b-minimal in $S(M,b)$. That it is on an extreme ray of $S(M,b)$ follows from the fact that if not then it would be the sum of two other vectors in $S(M,b)$, neither of which is a multiple of (π,π_0) and both being subadditive valid inequalities, contradicting (π,π_0) being a facet.

Conversely, if (π,π_0) satisfies (i), (ii), and (iii) of Theorem 5.3, then it is a minimal valid inequality with $\pi_0 > 0$ so can be scaled so that $\pi_0 = 1$. If it is not a facet, then there are two other valid inequalities that can also be scaled so that $\pi_0^1 = \pi_0^2 = 1$ and where

$$(\pi,1) = \tfrac{1}{2}(\pi^1,1) + \tfrac{1}{2}(\pi^2,1) .$$

If either of $(\pi^1,1)$ or $(\pi^2,1)$ are not minimal valid inequalities, then π^1, say, can be lowered to be a minimal valid inequality. Here, this result follows from the fact that $\pi \geq 0$ in any valid inequality, and there are only a finite number of 0-1 solutions, which are clearly the ones limiting how small any π_j can be made in keeping π a valid inequality.

Since minimal valid inequalities are subadditive, we have a contradiction to (π, π_0) being on an extreme ray of $S(M,b)$, completing the proof.

6. The Fulkerson Property and Sums of Circuits

The problem $[M \mid b]$ satisfies the Fulkerson property if

$$P(M,b) = \{t^* \in \mathbb{R}^n_+ \mid Qt^* \geq 1\},$$

and holds if and only if it holds for $P(M^*,b^*)$ and then $P(M,b)$ and $P(M^*,b^*)$ are a blocking pair of polyhedra. In section 4 we showed that each of the inequalities $q \cdot t^* \geq 1$, $q \in Q$, is a facet of $P(M,b)$. From the previous section, we know that the facets of $P(M,b)$ are among the vectors on extreme rays of $S(M,b)$. Thus, we know that $[q \mid 1]$ is on an extreme ray of $S(M,b)$. There may be other extreme rays of $S(M,b)$ even when the Fulkerson property holds because not all contain facets.

Seymour [19] has given a very interesting result characterizing matroids M having the sums of circuits property. He defines a matroid M to have the <u>sums of circuits property</u> if the cone defined by the inequality system

$$t \geq 0, \quad \text{and}$$

$$\sum_{\substack{j=1 \\ j \neq k}}^{n} r_j^* t_j \geq t_k, \quad \text{for all cocircuits } r^* \text{ of M with } r_k^* = 1,$$

has extreme rays that are the circuits of M. His result is:

<u>Theorem 6.1</u> (Seymour) A matroid M has the sums of circuits property if and only if M is binary and has no F_7^*, R_{10}, or K_5^* minor.

The forbidden minors F_7^*, R_{10}, and K_5^* are, respectively, the dual Fano, the famous non-regular R_{10} that is neither graphic nor cographic, and the dual to the complete graph on five nodes. When we say that a binary matroid, whose independent sets are independent columns of the binary matrix M, does not have an F_7^*, R_{10}, or K_5^* minor, we mean that there is no sequence of contractions and deletions of M bringing it to, respectively,

$$\begin{bmatrix} 1 & & & & 1 & 1 & \\ & 1 & & & 1 & & 1 \\ & & 1 & & & 1 & 1 \\ & & & 1 & 1 & 1 & 1 \end{bmatrix} \qquad F_7^*$$

$$\begin{bmatrix} 1 & & & & & 1 & & & 1 & 1 \\ & 1 & & & & & & 1 & 1 & 1 \\ & & 1 & & & & 1 & 1 & & 1 \\ & & & 1 & & 1 & 1 & & & 1 \\ & & & & 1 & 1 & 1 & 1 & 1 & 1 \end{bmatrix} \qquad R_{10}$$

$$\begin{bmatrix} 1 & & & & & & 1 & 1 & & \\ & 1 & & & & & 1 & & 1 & \\ & & 1 & & & & 1 & & & 1 \\ & & & 1 & & & & 1 & 1 & \\ & & & & 1 & & & 1 & & 1 \\ & & & & & 1 & & & 1 & 1 \end{bmatrix} \qquad K_5^* .$$

A counter-example for the sums of circuits property is provided by a non-negative vector r that is not a non-negative combination of circuits but does satisfy the specified inequalities for every cocircuit. For F_7^*, a counter-example is given by $(111\tfrac{1}{2}\tfrac{1}{2}\tfrac{1}{2}\tfrac{1}{2})$, which cannot be a non-negative sum of circuits because

$$-t_1^* - t_2^* - t_3^* + t_4^* + t_5^* + t_6^* + t_7^* \geq 0$$

is satisfied by every circuit but strictly violated by $(111\tfrac{1}{2}\tfrac{1}{2}\tfrac{1}{2}\tfrac{1}{2})$.

Consider now a binary group problem $[M \mid b]$. The cocircuits of the augmented matrix are the rows of

$$\begin{bmatrix} Q & 1 \\ Q_0 & 0 \end{bmatrix} .$$

The inequality system used in defining the sums of circuits property is exactly the subadditive cone, but for $[M^* \mid b^*]$. Thus, we can define the sums of circuits property to hold for the augmented matrix $[M \mid b]$ if and only if $S(M^*,b^*)$ has extreme rays that are circuits of $[M \mid b]$. Thus, a restatement of Seymour's result for augmented matrices is below.

<u>Corollary 6.2</u> The subadditive cone $S(M,b)$ for the binary group problem

[M | b] has extreme rays that are incidence vectors of cocircuits of [M | b] if and only if [M | b] has no F_7, R_{10}, or K_5 minor.

The duals of the minors in Theorem 6.1 are in Corollary 6.2 since the inequalities defining $S(M,b)$ are given by circuits, i.e. cocircuit duals. We can now state the theorem below. Its proof is immediate from Theorem 5.3 and Corollary 6.2.

<u>Theorem 6.3</u> The binary group problem [M | b] has the Fulkerson property whenever [M | b] has no F_7, R_{10}, or K_5 minor.

While the sums of circuits property is not dual symmetric, the Fulkerson property is. Thus, we have a dual result.

<u>Theorem 6.4</u> The binary group problem [M | b] has the Fulkerson property whenever [M | b] has no F_7^*, R_{10}, or K_5^* minor.

Proof: If [M | b] has the Fulkerson property, then so does [M* | b*].

Theorem 6.3 and 6.4 are blocking, or dual, results. However, they apply now to both [M | b] and [M* | b*]. In this way, the result is strengthened. The stronger result says that [M | b] has the Fulkerson property whenever it has no R_{10} minor and is missing either F_7 and K_5 or F_7^* and K_5^*. However, an even stronger result is given in section 11, using the same type of duality.

There is one other characterization known [6]: [M | b] has the Fulkerson property whenever it is regular, i.e. has no F_7 or F_7^* minor (see [3] for an extension to problems not binary). This result is not implied by the above theorems. Together, they give the result that, for the Fulkerson property to hold, it suffices to have either both no F_7 and no F_7^* or to have one of them (say no F_7) together with no R_{10} and no K_5 with star if F_7 did not.

7. <u>Chinese Postman and Related Problems in Graphs</u>

Let G = (V,E) be an undirected graph that may have duplicate edges or loops and may be disconnected. That is, an edge is an unordered pair of nodes, and there are no restrictions on what pairs of nodes make up the edge set. Let $c: E \to \mathbb{R}_+$ be a non-negative cost function on the edges. Define $c(S)$, the cost of a set $S \subseteq E$, to be

$$c(S) = \sum_{e \in S} c(e).$$

This section draws upon results from [14].

Chinese Postman Problem: In order to state the problem, define the degree $d_i(S)$ of a node i for a subset S of edges to be the total number of edges $e \in S$ meeting node i. However, if e is a loop meeting node i, then it counts twice in determining $d_i(S)$. The Chinese postman problem is to find a minimum cost subset S of edges such that $d_i(S)$ is an odd integer for $i \in U$ and an even integer for $i \notin U$, where U is a subset of V called the odd nodes. In order that there exist a postman solution, every connected component of G must be even, that is, contain an even number of odd nodes.

The original version of this problem came from the problem of finding a minimum cost postman tour in a graph. A tour of a graph is a path, not necessarily simple, that returns to its origin. A postman tour is a tour that uses every edge at least once. The problem of finding a minimum cost postman tour [11] is equivalent to the special case of the above described problem when the graph G is connected and odd nodes U are those nodes having odd degree over the entire edge set E. Then, the edges in a postman solution are the edges that are traversed twice in a corresponding postman tour. When the edges in a postman solution are duplicated in the graph, the resulting graph has an Euler tour because it has even degree and is connected, and that Euler tour is the corresponding postman tour of the original graph. There is a good algorithm for solving this problem [5].

Odd Cut Problem: A cut in a graph is a set of edges whose deletion from the graph increases the number of connected components of the graph. A minimal cut, or cocircuit of the graph is a cut that is minimal with respect to being a cut, i.e. no proper subset of edges in the cut is a cut. For a subset U of V designated as odd nodes such that every connected component of G is even, as in the Chinese postman problem, define an odd cut to be a minimal cut of G such that deleting the edges in the cut from G leaves a subgraph with an odd connected component. In this case, there will be two odd components. The odd cut problem is to find the minimum cost odd cut [1].

The family of odd cuts are, in fact, the blocking clutter of the clutter of postman solutions. To see this result, observe that a postman solution exists if and only if each connected component contains an even number of odd nodes. Thus, for a set C of edges to be a blocker of Chinese postman solutions, its deletion from G must leave an odd connected component, and the set C must be minimal with respect to this property. Clearly, C must be a cut, but it remains to show that C is a minimal cut, or cocircuit, of G. This fact has already been shown in section 2 for arbitrary binary clutters.

Padberg and Rao [17] gave a good algorithm for solving the odd cut problem.

To pose the Chinese postman problem as a binary group problem is easy: M can be taken to be the node-edge incidence matrix of G, i.e. M has exactly two 1's per column. Alternatively, the problem can be brought to standard form $[I \quad N \mid b]$ where the basic columns correspond to a spanning tree (when the graph is connected), and column j of N has 1's corresponding to edges in the tree forming a path between the two ends of edge e_j. The right-hand side b in standard form has a 1 in rows corresponding to the (unique) Chinese postman solution that is a subset of the spanning tree. The ith row of $[I \quad N]$ corresponds to a cocircuit, and the set of cocircuits for all rows of $[I \quad N]$ is a fundamental set [21] of cocircuits. The cocircuit for row i is an odd cut if and only if $b_i = 1$. An alternative way to define the Chinese postman problem in terms of the standard form $[I \quad N \mid b]$ is that, for this particular fundamental set of cocircuits, the sum of t_j^* over e_j in a cocircuit must be odd if the cocircuit is odd and must be even otherwise. The more usual definition for degree constraints on t^* corresponds to M being a node-edge incidence matrix and is, in fact, in terms of cuts (but not a fundamental set of cocircuits) since the edges meeting a node do form a cut.

When we put the blocking problem in right-hand side form

$$\left[\begin{array}{cc|c} N^T & I & 0 \\ b^T & 0 & 1 \end{array}\right]$$

we see that the constraints say that the sum of t_j over edges e_j in a fundamental set of circuits must be even and the sum of t_j over a particular postman solution must be odd. Then, the sum of t_j over every postman solution will be odd, and minimal 0-1 solutions will be odd cuts.

Co-postman Problem: Given a subset D of the edges E, called odd edges, the co-postman problem is to find a minimum cost subset S of edges such that in the remaining graph having edge set E\S there are no odd circuits, i.e. a circuit containing an odd number of odd edges.

When every edge of G is odd, then the co-postman problem is to remove a minimum cost subset of edges so that the remaining graph is bi-partite (has no odd length circuit). This problem is equivalent to finding a maximum weight bipartite subgraph of a graph and is known to be an NP-complete problem [1]. Thus, the co-postman problem is, in general, NP-complete whereas the Chinese postman problem is polynomially solvable.

Odd Circuit Problem: Finding a minimum cost odd circuit, as defined for the co-postman problem, is called the odd circuit problem. By contrast with the co-postman problem, it has a good algorithm [10]. It is clearly the blocking problem to the co-postman problem.

In a similar way to that described for the Chinese postman problem, if we take a spanning tree (for a connected graph), then the out-of-tree edges form a circuit when adjoined to the tree. Some of these circuits are odd (if they contain an odd number of odd edges) and the rest are even. The out-of-tree edges that form odd circuits are the unique co-postman solution that is a subset of the out-of-tree edges.

In order to state the co-postman problem as a binary group problem, if

$$M = [I \quad N]$$

is a graphic matrix, then its dual matrix

$$[N^T \quad I]$$

is co-graphic, and the binary group problem

$$[N^T \; I]\, t \equiv b^*$$

is a co-postman problem having constraints that say that the sum of t_j over a fundamental set of circuits must be odd if b_i^* is 1 and even otherwise. The b_i^* can be used to designate odd and even circuits among a fundamental set of circuits and thereby designate all circuits as being odd or even. Every circuit is a sum (modulo 2) of circuits in the fundamental circuits, and a circuit is odd if its sum has an odd number of odd circuits in it. Equivalently, we can designate out-of-tree edges as odd if the corresponding $b_i^* = 1$ and even otherwise. Then the same set of circuits are odd when they are defined as odd by having an odd number of odd edges. The same remarks hold in a dual fashion for cocircuits and odd cuts when we designate odd edges to be a particular Chinese postman solution.

In right-hand side form, the odd circuit problem is

$$\left[\begin{array}{cc|c} I & N & 0 \\ 0 & b^{*T} & 1 \end{array}\right]$$

where $[I \quad N]$ is graphic and $[0 \; b^*]$ is a designation of even and odd ($b_i^* = 1$) edges. The constraints say that the sums of t_j over e_j in a fundamental set of cocircuits must be even and the sum of t_j over out-of-tree edges having $b_i^* = 1$ ($j = m^* + i$) must be odd. The minimal 0-1 solutions t will be odd circuits.

There is a duality in that the Chinese postman problem requires an even or odd intersection with a fundamental set of cocircuits whereas the co-postman problem requires edge sets to have an even or odd intersection with a fundamental set of circuits. Somehow, intersections with co-circuits gives an easier problem than intersections with circuits. However, both problems have polynomially solvable blocking problems.

8. Minors and Majors of Binary Matrices

Given a binary matrix M, a minor $\overline{M}$ of M is another binary matrix obtained by sequentially performing two operations:

> deletion of a column of M means simply leaving it out;
>
> contraction of a column of M is performed by pivoting on a column and then deleting the row and column pivoted on.

In case we try to contract a column of all zeros, we cannot pivot on it, so contraction in that case means just deleting it. On the other hand, if we delete a column that has the only non-zero in some row, then we would delete the resulting row of zeros, and deletion is the same as contraction.

To be precise and without loss of generality, let us delete, but not contract, columns of all zeros and contract, but not delete, any column whose deletion would give a zero row. Having adopted this convention, M can be brought to standard form with columns to be contracted in the basis and columns to be deleted outside of the basis.

The dual $\overline{M}^*$ of a minor $\overline{M}$ is the minor of the dual M* obtained by interchanging contraction and deletion. This fact [20] is easily seen and shows an interesting duality between the two operations.

For an augmented matrix $[M \mid b]$, define a feasible minor $[\overline{M} \mid \overline{b}]$ to be a minor of $[M \mid b]$ such that

(i) b is not contracted or deleted;

(ii) the right-hand side $\overline{b}$ of the minor is not all zeros; and

(iii) there is a 0-1 solution $\overline{t}^*$ to $\overline{M}^* \; \overline{t}^* \equiv \overline{b}$ (modulo 2).

For an augmented matrix in standard form $[I\ N \mid b]$, a feasible minor is formed by contracting columns of I and deleting columns of N (but not the right-hand side) so that not all rows i having $b_i = 1$ are contracted away, i.e. corresponding basic columns are not contracted. Condition (iii) is then always satisfied. The dual matrix of $[I\ N \mid b]$ in right-hand side form is

$$\left[\begin{array}{cc|c} N^T & I & 0 \\ b^T & 0 & 1 \end{array}\right]$$

and a feasible minor of it, dual to a feasible minor just discussed for $[I\ N \mid b]$, is formed by deleting non-basic columns so that not all columns having $b_i^T = 1$ are deleted and contracting basic columns (but not the right-hand side). Condition (ii) is, then, always satisfied. Thus, the dual of a feasible minor is a feasible minor of the dual augmented matrix [18].

There is a close connection between contraction and homomorphic images. Let the dual matrices $[M \mid b]$ and $[M^* \mid b^*]$ be imbedded in the respective groups mC_2 and m^*C_2. That is, we think of the group mC_2, as being all m-columns of 0's and 1's, so that $[M \mid b]$ has columns which are a subset of mC_2. Now, a homomorphism of mC_2 is a mapping onto another binary group such that

$$\phi(g + h) = \phi(g) + \phi(h), \quad \text{for all } g,h \in mC_2 .$$

The kernel K of the homomorphism is the set of $g \in C_2$ such that $\phi(g) = 0$. Let K_0 denote a minimal (and, therefore, linearly independent) generating set for the kernel. If we picture the group mC_2 as a matrix whose columns are all possible 0-1 m-columns, then the mapping ϕ is given by first pivoting K_0 into the basis and then leaving out the rows pivoted on. The resulting matrix gives the mapping ϕ. However, in the image, duplicate elements are identified as being the corresponding image group element. By the image $[\phi(M) \mid \phi(b)]$ of $[M \mid b]$ we mean the following: first, replace each column of $[M \mid b]$ by its homomorphic image; and, then, remove duplicate columns and 0-columns among the image of M. It should be clear that the resulting $[\phi(M) \mid \phi(b)]$ is a feasible minor of $[M \mid b]$ provided two conditions hold: $\phi(b) \neq 0$; and a generator K_0 of the kernel K of ϕ is a subset of the columns of M. The importance of this latter condition in the context of lifting facets was shown by Gastou and Johnson [8].

When we perform such a homomorphism on $[M \mid b]$, the effect on the dual matrix is to delete some non-basic columns followed by contracting basic columns (not the right-hand side b*) which are basic in a row that is either 0 except for the 1 in the basis or that is equal to some other row except for the basic columns.

When a column of $[M \mid b]$ is contracted or deleted, one can see directly the effect on the clutter Q. When a column is contracted, the only row containing a non-zero in that column is left out. Thus, in forming row sums, any row sum including that row should be left out. Therefore, the resulting $\overline{Q}$ is formed by deleting all rows with a 1 in the column to be contracted and then deleting that column [7]. What remains will be exactly the new clutter because all rows will remain minimal. This new clutter $\overline{Q}$ is called a <u>deletion</u> of Q.

When a column of $[M \mid b]$ is deleted, we just leave out that column in the clutter. However, the resulting rows may not be minimal, so non-minimal rows must be deleted to form the new clutter $\overline{Q}$ called a <u>contraction</u> of Q [7]. Thus, a contraction of $[M \mid b]$ leads to a deletion of Q, and a deletion of $[M \mid b]$ leads to a contraction of Q. This usage is conventional [7] and, despite its appearance, is convenient in defining the geometrical notions below because a contraction of $[M \mid b]$ leads to a contraction of Q*.

Since the vertices of $P(M,b)$ are the rows of Q*, a contraction of $[M \mid b]$ to give $[\overline{M} \mid \overline{b}]$ results in a polyhedron $P(\overline{M},\overline{b})$ whose vertices are obtained by contraction of Q*. That is, we take every vertex and leave out the column contracted and then check for minimality among all of the potential vertices so formed. We refer to this operation on the set of vertices as contraction of the vertices of $P(M,b)$. Thus, contraction of the vertices is accomplished by a contraction of Q^*. Geometrically, $P(\bar{M},\bar{b})$ is formed from $P(M,b)$ by projecting onto the face $\{t \mid t_j = 0\}$ where j is the column contracted. We see this projection from the fact that the vertices are moved onto the face perpendicularly by just changing t_j to 0 and then leaving it out.

On the other hand, a deletion on $[M \mid b]$ results in a polyhedron $P(\bar{M},\bar{b})$ whose vertices are obtained by deletion of Q^*. That is, we take every vertex of $P(M,b)$ having $t_j = 0$ for j the column to be deleted to get the vertices of $P(\bar{M},\bar{b})$. We refer to this operation on the set of vertices as deletion of the vertices of $P(M,b)$. Thus, deletion of the vertices is accomplished by a deletion of Q^*. Geometrically, $P(\bar{M},\bar{b})$ is formed from $P(M,b)$ by intersecting with the face $\{t^* \mid t_j^* = 0\}$. The vertices of $P(\bar{M},\bar{b})$ are simply all of the vertices of $P(M,b)$ which already lay in that face.

The two geometric notions, projection onto a face and intersection with a face are dual to each other [8] in that when we do one to $P(M,b)$ the other gets done to the polyhedron $P(M^*,b^*)$ of the blocking problem. When the Fulkerson property holds, we already know what happens to facets as well as vertices: for a contraction of $[M \mid b]$ the facets of $P(\bar{M},\bar{b})$ are obtained by a deletion of Q and the vertices by a contraction of Q^*; and for a deletion of $[M \mid b]$ the facets are obtained by a deletion of Q and the vertices by a contraction of Q^*.

In general, the matrix of facets contains rows other than those of Q. Let Π be a matrix such that every row π_i gives a facet $\Sigma \pi_{ij} t_j^* \geq 1$ of $P(M,b)$. A deletion of column j in $[M \mid b]$ results in $\bar{\Pi}$ given from Π by deleting the column j from Π and then checking which of the resulting valid inequalities are facets. Gomory [9] refers to $[\bar{M} \mid \bar{b}]$ as a subproblem of $[M \mid b]$ and shows this way of getting facets. Checking which are facets is more complicated than simply checking minimality but can be accomplished by eliminating any row i whose non-zeros are a super-set of the non-zeros of another row k and for those same rows i and k, the inequality from row i holds with equality for vertices which are a subset of the vertices for which the inequality from row k holds with equality. We refer to the $\bar{\Pi}$ obtained from Π in this way as a <u>facet deletion</u>. We summarize these results in the theorem below.

<u>Theorem 8.1</u> Deletion of column j from $[M \mid b]$ gives a polyhedron $P(\bar{\mathbf{M}},\bar{b})$ whose vertices are obtained by taking those rows of Q^* having a 0 in column j (vertex deletion = deletion of Q^*) and whose facets are obtained by deleting column j from the facets Π and then selecting from among the resulting valid inequalities those which are facets (facet deletion is similar to contraction of a clutter but the selection procedure is more complicated than checking minimality).

A contraction of column j in $[M \mid b]$ results in facets $\bar{\Pi}$ given from Π by leaving out those rows having non-zero entries in column j from Π and then deleting column j. In this case, as for deletion of a clutter, no checking is necessary. However, we must prove the result that exactly all of the facets result from this simple procedure. The proof is part of the proof of Gastou and Johnson [8] which generalized Gomory's result [9] on lifting of facets. Here, we have broken a homomorphism into several mappings, of which this one is the first.

Before giving the proof let us recall the geometry. Contraction gives a projection of the polyhedron onto a face $\{t^* \mid t^*_j = 0\}$. The result we wish to establish is that all of the resulting facets come from the facets of the original polyhedron which were perpendicular to that face, i.e. had $\pi_j = 0$.

To see that the inequalities are valid is clear from the fact that vertices are obtained by contraction and our facets had a 0 in the column deleted. To prove that they are all facets is easy, too, if we slightly extend Lemma 4.2 to allow solutions which may not be vertices in forming T there. Since each new facet has the same non-zero set as before and all of the old vertices give solutions for which the inequality still holds with equality where the old facet did, we can use the same T of Lemma 4.2 as for the old facet.

To prove that there are no other facets for $P(M,b)$, a new concept is useful. A contraction of a column of $[M \mid b]$ gives a minor $[\bar{M} \mid \bar{b}]$. The inverse operation is to perform an expansion of an augmented matrix. It is convenient to consider the problem in right-hand side form, and let us denote it:

$$[M \mid b] = \left[\begin{array}{cc|c} N & I & 0 \\ b & 0 & 1 \end{array}\right]$$

We <u>expand</u> $[M \mid b]$ by adjoining a new row and a new column which is equal to 0 except in the new row. The new row is 0 for the old basic columns including the right-hand side column and is arbitrary for the non-basic columns. Let us suppose the new row is inserted just above the bottom row so that the expanded matrix is:

$$\left[\begin{array}{ccc|c} N & I & 0 & 0 \\ N_{n+1} & 0 & 1 & 0 \\ b & 0 & 0 & 1 \end{array}\right] = [\hat{M} \mid \hat{b}]$$

where N_{m+1} is then $m+1^{\underline{st}}$ row of the expanded matrix. We refer to the new augmented matrix as an <u>expansion</u> of $[M \mid b]$ and we get solutions to the expanded binary group problem from old solutions t^* by inserting a new entry for the new column with entry equal to $-N_{m+1}\, t^*_N$, where t^*_N denotes the subvector of non-basic columns of t^*. When the solution t^* is also a vertex of $P(M,b)$, it should be clear that we get a vertex of $P(\hat{M},\hat{b})$ by this procedure, which we call <u>vertex expansion</u>. Thus, expansion of $[M \mid b]$ is the inverse of contraction of $[M \mid b]$.

<u>Facet expansion</u> involves simply putting a 0 coefficient in π to give a new inequality. We must show that facet expansion gives facets to the new $P(M,b)$. Again, Lemma 4.2 is useful because all of the old vertices used to form T for a given facet can be used for its facet expansion, which has the same set of non-zeros. We have, thus,

completed the proof of the following theorem.

Theorem 8.2 Contraction of column j from $[M \mid b]$ gives a polyhedron $P(\bar{M},\bar{b})$ whose vertices are obtained by contracting column j from the clutter Q* and then selecting from among the rows those which are minimal (vertex contraction = contraction of Q*) and whose facets are obtained by taking the rows of Π having a 0 in column j and then deleting column j.

A deletion of a column of $[M \mid b]$ gives a minor $[\bar{M} \mid \bar{b}]$. The inverse operation is called an extension of an augmented matrix. Let $[I\ N \mid b]$ be the original augmented matrix in standard form. We extend $[I\ N \mid b]$ by adjoining a new column to N, and we refer to the new augmented matrix as an extension of $[I\ N \mid b]$. To get some solutions to the extended binary group problem from old solutions t* insert a new entry for the new column with that entry equal to 0.

On a clutter Q, an expansion is as described above in defining vertex expansion. On the other hand, extension of a clutter Q is simply adjoining a column of 0's.

Facet extension involves a procedure similar to expansion of a clutter. Gomory uses such a procedure to generate some facets of group problems [9]. When we extend the facet $\pi t \geq 1$, the coefficient in the new column is

$$1 - \min \{\pi t^* \mid t^* \geq 0 \text{ and integer, } Mt^* = b - N_{n+1} \pmod 2\}.$$

The next two theorems summarize expansion and extension of facets and vertices.

Theorem 8.3 Expansion of $[\bar{M} \mid \bar{b}]$ gives a polyhedron $P(M,b)$ some of whose vertices are obtained by expanding the clutter $\bar{Q}^*$ and some of whose facets are obtained by expanding the facets of $P(\bar{M},\bar{b})$. The vertices of $P(M,b)$ which come from expansion of column j are those which are minimal after contracting column j of Q*. The facets of $P(M,b)$ which come from expansion of column j are those facets with a 0 in column j.

Theorem 8.4 Extension of $[\overline{M} \mid \overline{b}]$ gives a polyhedron $P(M,b)$ some of whose vertices are obtained by extending the clutter $\overline{Q}^*$ and some of whose facets are obtained by extending the facets of $P(\overline{M},\overline{b})$. The vertices of $P(M,b)$ which come from extension of column j are those which have 0 entries in the column j. The facets π of $P(M,b)$ which come from extension of column j are those facets such that either:

(i) $\pi_j > 0$ and there exists a matrix T as required by Lemma 4.2 with a singleton column for column j; or

(ii) $\pi_j = 0$ and there exists a matrix T as required by Lemma 4.2 such that the solutions corresponding to rows of T all have $t_j^* = 0$, i.e. the column j of the matrix Q* restricted to rows of T is all zero.

We note that, by comparison with facet expansion, facet extension is more difficult to perform, and it is more difficult to determine which facets come from extension. Similarly, by comparison with vertex extension, vertex expansion is more difficult to perform, and it is more difficult to determine which vertices come from expansion. Facet expansion is practically the same as vertex extension, but facet extension is more complicated than, although similar to, vertex expansion. When the Fulkerson property holds, they are exactly the same.

9. Homomorphic Lifting of Facets

The main purpose of this section is to prove Theorem 9.1 below. This theorem extends a similar result of Gomory [9] for master problems and is due to Gastou and Johnson [8]. Its main use here is to be able to say that if there is a fractional facet (one not coming from Q), then there is a fractional facet for a minor such that the facet has only positive coefficients. Sections 10 and 11 use this result. See

[4] for a similar line of proof for the general group problem.

Theorem 9.1 Let $[M \mid b]$ be a binary group problem and let $[\phi(M) \mid \phi(b)]$ be its homomorphic image under the homomorphism ϕ such that the kernel K of ϕ is generated by a subset K_0 of the column of M. If $\bar{\pi}\bar{t}^* \geq 1$ is a facet of $P(\phi(M),\phi(b))$, then $\pi t^* \geq 1$ is a facet of $P(M,b)$ where

$$\pi(M^j) = \begin{cases} 0 \text{ if } M^j \in K, \\ \bar{\pi}(\phi(M^j)) \text{ otherwise.} \end{cases}$$

Furthermore, for every facet $\pi t^* \geq 1$ of $P(M,b)$ with some zero coefficients, there is such a homomorphism ϕ so that π is obtained by lifting a facet $\bar{\pi}$ of the image problem $[\phi(M) \mid \phi(b)]$.

Proof: We must begin by discussing two dual notions. The first is very simple and is to extend $[M \mid b]$ by a column of all zeros. The new variable can take on any value in any solution but will be zero in any vertex. New facets will be zero in the new column. The extended polyhedron is obtained by adding a ray in the new axis direction to every point of the old polyhedron. In this case, vertex and facet extension give the entire set of vertices and facets. The dual notion is to expand by a row which is zero except in its basic column. Define a row to be an essentially zero row of $[M \mid b]$ if it is zero except in one column of M and is zero in the right-hand side. This variable corresponding to the non-zero entry in the row must be equal to zero (modulo 2) in every solution so can take on any even value. As before, all vertices and facets have zero entries in the new column, and vertex and facet expansion give the full set of vertices and facets of the new polyhedron.

The other dual operations are column and row replication. Column replication means extending by adjoining a column already present. Facet extension amounts in this case to giving the same coefficient to the new column as the one already there. In fact, facet extension gives the full set of facets also in this case. In order to see that result, let M have duplicate columns, say columns n-1 and n. For any facet of the binary group problem [M | b], every facet has equal coefficients in columns n-1 and n. To see that the facet comes from extension of one of the columns, the matrix T in Lemma 4.2 for the facet in question must be taken as a subset of the rows of the submatrix of vertices (i.e. Q*) for which equality holds in the facet. There are no vertices with 1's in both columns, and the rows with 1's in either duplicate column come in pairs:

$$t^*_{i_1} = (\ldots 1\ 0)$$

$$t^*_{i_2} = (\ldots 0\ 1)$$

where the rows $t^*_{i_1}$ and $t^*_{i_2}$ are equal except in columns n-1 and n. Picking any one of these pairs, we can reduce every other such pair to:

$$(\ \ldots\ \ 1\ 0)$$

$$(0\ldots 0\ 0\ 0)$$

by adding $t^*_{i_1}$, subtracting $t^*_{i_2}$, and subtracting one of the two of this pair from the other. These elementary row operations do not change the column rank, so we must be able to still find a non-singular submatrix T of these columns. This T will have a unit column for one of the two duplicate columns. We now use Theorem 8.4 to complete the proof.

The dual operation to column replication we call row replication but what is meant is expansion where the non-basic part of the adjoined row is equal to the non-basic part of another row. The new variable must be equal to the other basic variable for the row whose non-basic part is equal to the non-basic part of the new row. Every facet of the resulting expansion must have a 0 coefficient in one of the two columns because otherwise we would have two equal columns in the non-singular matrix T required by Lemma 4.2. Hence, every facet comes from facet expansion of at least one of the two columns.

Vertex expansion gives all of the vertices of the expanded

problem. In fact, vertex expansion here is clutter expansion by adjoining a duplicate column, which clearly does not affect minimality.

Two columns which are equal to each other will be called duplicate columns, two rows which are equal to each other except in their basic columns will be called essentially duplicate rows.

To complete the proof of Theorem 9.1, the homomorphism ϕ with kernel generated by a subset K_0 of the columns of M amounts to a contraction of some columns (the columns in K_0) of $[M \mid b]$ followed by deletion of zero columns and duplicate columns. For a facet of the minor obtained, we can extend it using the above discussion to a facet for the problem with zero columns and duplicate columns. In fact, the extension up to this point is complete in the sense that every facet of the extended problem comes from such a construction beginning with a facet of the minor. We can now expand this facet to a facet of the original $[M^* \mid b^*]$ using Theorem 8.3. This facet expansion amounts to just putting a coefficient of zero for the columns contracted. Furthermore, all facets with any zero coefficients come from lifting a facet from a minor in this way.

We can now specify a dual construction on the blocking problem. Begin by deleting some columns and then contracting away any row which is an essentially zero row and contracting away any row which is an essentially duplicate row. For a facet of the minor obtained, we can expand it using the above discussion to a facet for the problem with zero rows and essentially duplicate rows. As before, the expansion up to this point is complete in the sense that every facet of the expanded problem comes from such a construction beginning with a facet of the minor. We can now extend this facet to a facet of the original $[M^* \mid b^*]$ using Theorem 8.4. This facet extension is more complicated than the corresponding facet expansion, and it is more difficult to say which facets come from such an extension (see Theorem 8.4). Thus, the dual of a homomorphism is to delete some columns and then contract essentially zero rows and essentially duplicate columns. There is then an analogous version of Theorem 9.1 except that the algorithm for lifting π from $\bar{\pi}$ is not so simple and the characterization of facets obtained in this way is not so simple.

We can give another dual construction but on the original problem by discussing homomorphic lifting of vertices. Begin by contracting in $[M \mid b]$ a subset K_0 of M followed by deletion of zero columns and duplicate columns. For a vertex of the minor obtained, we can extend it using the above discussion to a vertex for the problem with zero columns and duplicate columns. As before, the extension up to this point is complete in the sense that every vertex of the extended problem comes from such a construction beginning with a vertex of the minor. We can now expand this vertex to a vertex of the original $P(M,b)$ using Theorem 8.3. This vertex expansion amounts to putting a uniquely determined value for each column contracted so that the resulting vector is a solution. The vertices which come from such a lifting are those which are still minimal among all vertices when the columns expanded are deleted (see Theorem 8.3).

Moreover, we can also discuss extension of vertices of the blocking problem. As in the discussion of extension of facets, begin by deleting some columns and then contracting away any row which is an essentially zero row and contracting away any row which is an essentially duplicate row. For a vertex of the minor obtained, we can expand it to a vertex for the problem with zero rows and duplicate rows. As before, the expansion up to this point is complete in the sense that every vertex of the expanded problem comes from such a construction beginning with a vertex of the minor. We can now extend this vertex to a vertex of the original $[M^* \mid b^*]$ using Theorem 8.4. This vertex extension is simply giving value zero to the coefficients of the vertex for the columns deleted. Every vertex with a zero coefficient comes from such a lifting.

It should be clear that except for the trivial problem:

$$t_1^* \equiv 1 \pmod 2,\ t_1^* \geq 0 \text{ and integer},$$

every vertex of every problem without essentially zero rows or essentially duplicate rows comes from the extension procedure just described. The dual statement is that when the Fulkerson property holds for a problem without zero columns or duplicate columns then every facet comes from the homomorphic lifting of facets of Gomory, Gastou, and Johnson unless the problem is the trivial one-variable problem above. In fact, we can say that the Fulkerson property holds if and only if every facet can be lifted from a minor of the form

$$[\bar{M} \mid \bar{b}] = [1 \mid 1]$$

by the homomorphic lifting procedure.

10. Critically Non-Fulkerson Problems and the Chinese Postman Polyhedron

A problem (or, simply, $[M \mid b]$) is critically non-Fulkerson if $P(M,b)$ does not have the Fulkerson property, but $P(\bar{M},\bar{b})$ does for every minor $[\bar{M} \mid \bar{b}]$. Some preliminary results will first be given. Theorem 10.1 below is due to Seymour [18]. Our proof of it is given by Theorem 10.2. By a fractional facet, we mean a facet $\pi t^* \geq 1$ such that π includes some non-integer entries. (See Theorem 4.3.)

Theorem 10.1 If $[M \mid b]$ has the Fulkerson property, then so does every feasible minor.

Theorem 10.2 If $[\bar{M} \mid \bar{b}]$ does not have the Fulkerson property, then for any major $[M \mid b]$ of $[\bar{M} \mid \bar{b}]$ there is at least one facet of $P(M,b)$ which is fractional and comes from lifting (expansion and extension) a fractional facet of $P(\bar{M},\bar{b})$.

Theorem 10.3 $[M \mid b]$ is critically non-Fulkerson if and only if the blocking problem $[M^* \mid b^*]$ is also critically non-Fulkerson.

The above three theorems appear in Gastou and Johnson [8]. Let us remark that every binary group problem with three or fewer rows satisfies the Fulkerson property (see the tables of facets given by

Gomory [9] for a proof). Thus, $m > 3$ for any critically non-Fulkerson problem $[M \mid b]$, and, by the above theorem, $n-m > 3$.

<u>Theorem 10.4</u> If $[M \mid b]$ is critically non-Fulkerson, then it contains no zero column, no duplicate columns, no essentially zero row, and no essentially duplicate rows.

Proof: The proof is straight-forward from the results in section 9. It is somewhat easier to give the proof for essentially duplicate rows since every facet will have a zero coefficient in one of the two columns which are basic with 1's in the two rows. Then the result for duplicate columns can be derived from the duality between duplicate columns and essentially duplicate rows.

<u>Theorem 10.5</u> If $[M \mid b]$ is critically non-Fulkerson, then M does not contain a column equal to b and $[M \mid b]$ does not contain a constraint saying that a variable is equal to 1 (modulo 2).

Proof: The two conditions are dual in that if M contains the column b, then the blocking problem contains the two rows:

$$(b^T\ 0\ldots0\ 1\ 0\ldots0 \mid 0), \quad \text{and}$$

$$(b^T\ 0\ldots0\ 0\ 0\ldots0 \mid 1)$$

and adding them together gives the constraint that a variable is equal to 1 (modulo 2). The proof is somewhat easier for the second of the two conditions and will be given only for that case.

Suppose that a variable, say t_n^*, is equal to 1 (modulo 2) in every solution. Then, $t_n^* \geq 1$ is a valid inequality and, trivially, a facet of $P(M,b)$ by Lemma 4.2. If some other facet $\pi t^* \geq 1$ has $\pi_n > 0$, then

$$\pi_1 t^*_1 + \ldots + \pi_{n-1} t^*_{n-1} + (\pi_n - \varepsilon) t^*_n \geq 1-\varepsilon, \quad \text{and}$$

$$\varepsilon t^*_n \geq \varepsilon$$

are both valid inequalities for any ε satisfying $0 < \varepsilon < \pi_n$. The original facet $\pi t^* \geq 1$ is the sum of these two valid inequalities, giving a contradiction.

A consequence of the above two theorems is that for any column j of a critically non-Fulkerson problem [M | b], there is at least one basis containing column j and at least one basis not containing column j. Thus, there is at least one vertex t* with $t_j^* = 0$ by just taking the solution given by a basis not containing column j. Also, there is at least one vertex with $t_j^* = 1$. This assertion is slightly harder to prove and requires a result of Lehman (see page 721 of [15]) which implies that if a column of Q* was all zeros then the same column of Q_0^* would be all zeros. Thus, the column in M would have to be all zeros.

Theorem 10.6 If [M | b] is critically non-Fulkerson, then M does not contain two columns which add up to b (modulo 2), and [M | b] does not contain a constraint saying that two variables add up to 1 (modulo 2).

Proof: The two conditions are again dual, and the proof is similar to that of Theorem 10.5 since if two variables must add to 1 (modulo 2) in every solution, then one of the two variables must be 1 and the other 0 in every solution. The result does not extend to three variables because then all three variables could be equal to 1 in a solution.

We can view the above results in terms of the clutters Q and Q*. Let us also consider the clutters Q_0 and Q_0^* where the right-hand sides are all equal to 0. What we have shown is that when [M | b] is critically non-Fulkerson, then there are no rows of any of the four clutters Q, Q*, Q_0, Q_0^* having fewer than three entries equal to 1.

We now present two results generalizing the proof of Gastou and Johnson [8] that the Chinese postman polyhedra have the Fulkerson property (a result of Edmonds and Johnson [5]). However, we are not successful in using the method to characterize binary group problems having the Fulkerson property. Such a characterization remains a major unresolved problem. The following two results do shed more light on the structure of critically non-Fulkerson problems.

Recall the notation $\hat{+}$, \, ∨, and ∧ from section 2. We continue to let s + t and s - t mean ordinary (real) plus and minus.

Define δ^j for $0 \le j \le n$ by

$$\delta^j_k = \begin{cases} 1 & \text{if } k = j \\ 0 & \text{otherwise} \end{cases}$$

Given $[M \mid b]$, define a <u>chord</u> of a solution t^* to be a 0-1 vector s^* such that $Ms^* \equiv 0$ (modulo 2) and s^* has exactly one non-zero entry s^*_j where $t^*_j = 0$. That is, s^* is a row of Q^*_0 such that except for one entry s^* would be less than or equal to t^*. This terminology is consistent with the definition of a chord of a row of Q^* (section 5).

For a facet $\pi t^* \ge 1$ of $P(M,b)$, define an equality solution t^* to be a 0-1 solution of $Mt^* \equiv b$ (modulo 2) such that $\pi t^* = 1$.

<u>Theorem 10.7</u> For a critically non-Fulkerson facet π of $P(M,b)$, any equality solution t^* has no chord.

Proof: Suppose the contrary. Then for a critically non-Fulkerson facet π and an equality solution t^*, there is a chord s^*. Clearly, $u^* = s^* + t^*$ (modulo 2) is another solution with $u^*_j = 1$ and $t^*_j = 0$. Since π is a critically non-Fulkerson facet, $\pi_j > 0$ by Theorem 8.3, and there must be at least two equality solutions with entry of 1 in column j by Theorem 8.4. Let v^* be such a solution not equal to t^* or u^*. We now have:

$$\pi t^* = 1,$$

$$\pi u^* \ge 1,$$

$$\pi v^* = 1.$$

From the first two we derive

$$\pi_j \ge 1 - \pi(u^* - \delta^j) = \pi t - \pi(u^* - \delta^j) = \pi(t^* \setminus u^*).$$

Therefore,

$$1 = \pi v^* = \pi(v^* - \delta^j) + \pi_j \ge \pi(v^* = \delta^j) + \pi(t^* \setminus u^*).$$

However, adding (modulo 2) all three of t^*, u^*, and v^* gives another solution $w^* = t^* \hat{+} u^* \hat{+} v^*$ which satisfies $\pi w^* \ge 1$ and $w^* \le (v^* - \delta^j) \vee (t^* \setminus u^*)$. Thus, the above inequalities must hold with equality, u^* and w^* must also be equality solutions, and v^* must be disjoint from

$t^* - u^*$. The solution w^* is the sum of t^* and v^* minus u^*, $w^* = (t^* + v^*) - u^*$ (in ordinary arithmetic). In the matrix T required by Lemma 4.2, we can substitute w^* in place of v^*, and thus eliminate from T all other (than u^*) solutions with a 1 in column j. Thus, the facet π came from extension, contradicting its being critical.

Theorem 10.7 implies that for the rows of Q^* which are equality solutions for a critical facet, the row cannot miss by only one column from being non-minimal in Q^*. The result was stated for s^* such that $Ms^* = 0$ (for example, a row of Q_0^*); however, that condition implies the same result for s^* a solution (or, equivalently, a vertex, i.e. row of Q^*).

<u>Theorem 10.8</u> For a critically non-Fulkerson facet π of $P(M.b)$ and any two equality solutions t^* and u^* with a non-zero intersection, there is no 0-1 vector s^* contained in the union $t^* \vee u^*$, intersecting the intersection $t^* \wedge u^*$, and satisfying $Ms^* \equiv 0$ (modulo 2).

Proof: Suppose not. Then both $v^* = t^* \hat{+} s^*$ and $w^* = u^* \hat{+} s^*$ are solutions. But, $\pi v^* + \pi w^*$ is strictly smaller than $\pi t^* + \pi u^*$, because both v^* and w^* are zero in the columns where s^* intersects $t^* \wedge u^*$. Thus $\pi v^* + \pi w^* < \pi t^* + \pi u^*$. We use the fact that π must be positive because it is critical. A contradiction is thus reached because $\pi v^* \geq 1$, $\pi w^* \geq 1$, and $\pi t^* = \pi u^* = 1$.

This theorem, as for Theorem 10.7, would also be true if the s^* is required to be a solution. Therefore, among the equality solutions for a critical facet there are not two that intersect and a third contained in their union which does not include all of their intersection.

<u>Corollary 10.9</u> For a critically non-Fulkerson facet π of $P(M,b)$ and any two equality solutions t^* and u^*, if s^* is a 0-1 vector satisfying $Ms^* \equiv 0$ (modulo 2) and if $s^* \leq t^* \vee u^*$, then the two solutions $t^* \hat{+} s^*$ and $u^* \hat{+} s^*$ are also equality solutions and

$$\pi \cdot (t^* \wedge s^*) = \pi \cdot (s^* \setminus t^*).$$

Proof: In the proof of Theorem 10.8, even if s^* does not intersect $t^* \wedge u^*$, then still $\pi v^* + \pi w^* \leq \pi t^* + \pi u^*$ where $v^* = t^* \hat{+} s^*$ and $w^* = u^* \hat{+} s^*$. As is shown there, the reverse inequality must also hold so $\pi v^* = \pi w^* = 1$, showing that both $t^* \hat{+} s^*$ and $u^* \hat{+} s^*$ are equality solutions. But

$$\pi v^* = \pi t^* + \pi\cdot(s^* \setminus t^*(- \pi\cdot(t^* \wedge s^*),$$

so the equation $\pi\cdot(t^* \wedge s^*) = \pi\cdot(s^* \setminus t^*)$ follows.

In the proof of Gastou and Johnson [8] for the Chinese postman polyhedra, the critical lemma in their development says that a circuit in the symmetric difference of two equality solutions for a positive facet is component simple with respect to either solution. The proof there used the result in Corollary 10.9 (see (9) of [8]). The proof here is simpler and is valid for any binary group problem, not just Chinese postman problems. However, pushing Corollary 10.9 further in order to characterize those binary group problems having the Fulkerson property seems difficult.

11. Binary Group Problems Having the Fulkerson Property

The main result of this section is the theorem below.

Theorem 11.1 (Seymour) The binary group problem $[M \mid b]$ has the Fulkerson property whenever M contains no F_7^*, R_{10}, or K_5^* minor. Thus, for this matrix M and any feasible right-hand side b, $[M \mid b]$ has the Fulkerson property.

There is a blocking result immediate from the fact that when $[M \mid b]$ has the Fulkerson property then so does $[M^* \mid b^*]$. Deleting b from $[M \mid b]$ is dual to contracting b^* in $[M^* \mid b^*]$. Theorem 11.2 is stated in terms of $[M \mid b]$.

Theorem 11.2 The binary group problem $[M \mid b]$ has the Fulkerson property whenever the matrix $\tilde{M}$, obtained from $[M \mid b]$ by contracting b, contains no F_7, R_{10}, or K_5 minor.

Theorem 11.1 follows from (8.3) and (10.2) of [19]. Here, we show how to prove it using Theorem 6.1 ((8.3) of [19]) and a construction of Barahona and Groetschel [2]. Theorem 11.1 can be proven [13] directly from the characterization [2] of the facets of the polytope of 0-1 solutions using the fact that a facet $\pi t^* \geq 1$ of $P(M,b)$ having all $\pi_j > 0$ is also a facet of the polytope of 0-1 solutions. For a facet with some $\pi_j = 0$, we can contract $[M \mid b]$ to a problem $[\overline{M} \mid \overline{b}]$ such that $P(\overline{M},\overline{b})$ has a facet $\overline{\pi}\overline{t}^* \geq 1$ with all $\overline{\pi}_j > 0$ and with the same set of non-zero entries in $\overline{\pi}$ as in π. The reverse direction can also be proven using the mapping given following the proof of Theorem 11.1.

To prove Theorem 11.1, suppose it is false. Then, there is a fractional facet π of $P(M,b)$ where M has no F_7^*, R_{10}, or K_5^* minor. By taking the homomorphic image of $[M \mid b]$ using the homomorphism ϕ mapping to zero M^j having $\pi_j = 0$, a problem $[\phi(M) \mid \phi(b)]$ is obtained still having a fractional facet $\overline{\pi}$ and where $\phi(M)$ has no F_7^*, R_{10}, or K_5^* minor since $\phi(M)$ is a minor of M. Hence, let the image problem be the problem violating Theorem 11.1, or in other words, assume that π has only positive coefficients. By Lemma 5.2, there is a subset of n rows of Q^* forming the matrix T^* such that T^* is $n \times n$ non-singular, and $\pi t^* = 1$ for all rows t^* of T^*.

Thus, π is a positive, fractional facet of $P(M,b)$ and T^* is an $n \times n$ non-singular matrix of equality solutions. Recall that $\pi r^* \geq 1$ for all rows r^* of R_1^* since every row of R_1^* is a solution of $Mt^* \equiv b$, (modulo 2), $t^* \in \mathbb{Z}_+^n$.

Now, the transformation of Barahona and Groetschel will be introduced. Let p^* be a particular solution, $p^* \in Q^*$, such that $\pi \cdot p^* = 1$, i.e. let p^* be a row of T^*. Define σ by

$$\sigma_j = \begin{cases} \pi_j & \text{if } p_j^* = 0 \\ -\pi_j & \text{if } p_j^* = 1 . \end{cases}$$

We claim that σ is a facet of the cone of circuits of M. To show that $\sigma \cdot s^* \geq 0$ is a valid inequality is equivalent to showing

$$\sigma \cdot s^* \geq 0, \quad s^* \in R_0^* ,$$

where R_0^* is the row space of M taken modulo 2. In order to show these facts about σ, consider the transformation $\mu: \mathbb{R}^n \to \mathbb{R}^n$ given by

$$\mu(x) = \begin{cases} x_j & \text{if } p_j^* = 0 \\ 1-x_j & \text{if } p_j^* = 1. \end{cases}$$

For 0-1 vectors t^*, it is easily seen that

$$\mu(t^*) = t^* \hat{+} p^* .$$

Thus μ is a 1-1 mapping between the rows of R_0^* and the rows of R_1^* Further, $\mu(\mu(t^*)) = t^*$. Hence, for $s^* \epsilon R_0^*$,

$$\begin{aligned} \sigma \cdot s^* &= \sigma \cdot \mu(\mu(s^*)) \\ &= \sum_{p_j^*=0} \pi_j t_j^* + \sum_{p_j^*=1} (-\pi_j)(1-t_j^*), \text{ where } t^* = \mu(s^*), \\ &= \sum_{j=1}^{n} \pi_j t_j^* - \sum_{j=1}^{n} \pi_j p_j^* \\ &\geq 0 , \end{aligned}$$

because

$$\sum_{j=1}^{n} \pi_j t_j^* \geq 1 , \quad \text{by } t^* \epsilon R_1^*, \text{ and}$$

$$\sum_{j=1}^{n} \pi_j p_j^* = 1 , \quad \text{by definition of } p^* .$$

Hence, $\sigma \cdot s^* \geq 0$ is a valid inequality for the cone of circuits of M, and furthermore

$$\sigma \cdot s^* = 0 \quad \text{for } s^* \epsilon R_0^*$$

if and only if

$$\pi \cdot \mu(s^*) = 1, \text{ where } \mu(s^*) \epsilon R_1^* .$$

Let S^* be the $n \times n$ submatrix of R_0^* such that the rows s_i^* of S^* are defined from $t_i^* \epsilon T^*$ by

$$s_i^* = \mu(t_i^*) .$$

The proof that $\sigma \cdot s^* \geq 0$ is a facet of the cone of R_0 is completed once we show that the rows of S^* are affinely independent. But to get S^* from T^* only involves changing the sign of a subset of columns and then subtracting the row p^* of T^* from all rows, including itself. This transformation introduces a zero row, in place of p^*, and the other n-1 rows remain linearly independent in $\mathbb{R}^n_+$. Thus $\sigma \cdot s^* \geq 0$ is a facet of the cone of circuits.

Seymour's result is that when M has no F^*_7, R_{10}, or K^*_5 then the facets of the cone of circuits are all of the form

$$\sum_{j \neq k} r_j s^*_j \geq s^*_k ,$$

where r_j is 0-1. Hence, σ having proper fractions could not be a facet of the cone, giving a contradiction and completing the proof.

We remark here that a similar mapping to the one used in the above proof allows one to map problems over the 0-1 polytope of solutions [2] to binary group problems. For a problem:

$$\min ct^*$$
$$Mt^* \equiv b \pmod 2,$$
$$t^*_j = 0 \text{ or } 1, \ j = 1,\ldots,n,$$

where c_j is real and may be negative, form the binary group problem

$$\min \tilde{c}\tilde{t}^*, \text{ where } \tilde{c}_j = | c_j | ,$$
$$M\tilde{t}^* \equiv b - \sum_{j \in J_-} M^j$$
$$t^* \in \mathbb{Z}^n_+ ,$$

where $J_- = \{j \mid c_j < 0\}$.

Having an optimum solution $\tilde{t}$ to the binary group problem, a solution t to the original 0-1 problem is obtained by letting

$$t_j = \begin{cases} 1-\tilde{t}_j , & j \in J_- , \\ \tilde{t}_j , & \text{otherwise.} \end{cases}$$

In making this observation, it should be pointed out that neither problem is necessarily any simpler than the other. However, the binary group approach does have some appeal, e.g. the blocking pairs structure that is missing from the 0-1 problem.

We know two pairs of critically non-Fulkerson binary group

problems [18].

$$\left[\begin{array}{ccccccc|c} 1 & & & & 1 & 1 & & 1 \\ & 1 & & & 1 & & 1 & 1 \\ & & 1 & & & 1 & 1 & 1 \\ & & & 1 & 1 & 1 & 1 & 0 \end{array}\right]$$

which is self-blocking and for which

$$t_1^* + \cdots + t_7^* \geq 3$$

is a facet; and

$$\left[\begin{array}{cccccccccc|c} 1 & & & & & & 1 & 1 & & & 1 \\ & 1 & & & & & 1 & & 1 & & 1 \\ & & 1 & & & & 1 & & & 1 & 1 \\ & & & 1 & & & & 1 & 1 & & 1 \\ & & & & 1 & & & 1 & & 1 & 1 \\ & & & & & 1 & & & 1 & 1 & 1 \end{array}\right] \quad \text{and} \quad \left[\begin{array}{cccccccccc|c} 1 & 1 & 1 & & & & 1 & & & & 0 \\ 1 & & & 1 & 1 & & & 1 & & & 0 \\ & 1 & & 1 & & 1 & & & 1 & & 0 \\ & & 1 & & 1 & 1 & & & & 1 & 0 \\ 1 & 1 & 1 & 1 & 1 & 1 & & & & & 1 \end{array}\right]$$

which are a blocking pair of binary group problems for which

$$t_1^* + \cdots + t_{10}^* \geq 4 \quad \text{and} \quad t_1^* + \cdots + t_{10}^* \geq 3 ,$$

respectively, are facets. The matrix M in the first example is the dual Fano matrix F_7^*. Note that contracting the right-hand side in that example by pivoting on the first element of b gives the matrix

$$\left[\begin{array}{ccccccc} 1 & 1 & & & & 1 & 1 \\ 1 & & 1 & & 1 & & 1 \\ & & & 1 & 1 & 1 & 1 \end{array}\right]$$

which is the Fano matrix F_7. Thus, we have an instance of the Theorem 11.1 where $[M \mid b]$ contains F_7^* after deleting b and F_7 after contracting b and does not have the Fulkerson property.

In the second example, the first of the blocking pair has M that is K_5^*. Contracting b gives R_{10}. Thus, $[M \mid b]$ contains K_5^* after deleting b and contains R_{10} after contracting b and does not have the Fulkerson property. The second of the blocking pair, say $[M^* \mid b^*]$, is clearly K_5 after contracting b* and is R_{10} after deleting b* by pivoting on the bottom row to take b* out of the basis.

The conjecture is that these are the only critically non-Fulkerson binary group problems (see conjecture 9.2 of [19]).

We know a class of non-Fulkerson problems that has facets with arbitrarily large right-hand sides and, therefore, coefficients. It generalizes the first example above but contains it as a minor. Let k be an odd number greater than 1 and let N below be $k+3$ by $k+2$:

$$[I\ N \mid b] = \left[\begin{array}{ccccccc|c}
1 & & & & & & 011\cdots111 & 1 \\
 & 1 & & & & & 101\cdots111 & 1 \\
 & & 1 & & & & 110\cdots111 & 1 \\
 & & & \ddots & & & \vdots\ \ddots\ \vdots & \vdots \\
 & & & & 1 & & 111\quad 011 & 1 \\
 & & & & & 1 & 111\quad 101 & 1 \\
 & & & & & & 1\ 111\quad 110 & 1 \\
 & & & & & & 1111\cdots111 & 0
\end{array}\right]$$

That is, $[N \mid b] = E-I$, where E is a $k+3$ by $k+3$ matrix of 1's. Then,

$$t_1^* + t_2^* + \cdots + t_{k+1}^* + t_{k+2}^* + kt_{k+3}^* + t_{k+4}^* + \cdots + t_{2k+5}^* \ge k+2$$

is a facet. That it is a facet follows from the fact that it is valid, and the matrix below gives the T required from Lemma 4.2.

$$\left[\begin{array}{cccccccccc}
111\cdots111 & & & & & & & & & \\
1 & & & & & & 011\cdots111 & & & \\
 & 1 & & & & & 101\cdots111 & & & \\
 & & 1 & & & & 110\cdots111 & & & \\
 & & & \ddots & & & \vdots\ \ddots\ \vdots & & & \\
 & & & & 1 & & 111\cdots011 & & & \\
 & & & & & 1 & 111\cdots101 & & & \\
 & & & & & & 1\ 111\cdots110 & & & \\
1 & & & & & & 11 & & & \\
 & 1 & & & & & 1\ 1 & & & \\
 & & 1 & & & & 1 & 1 & & \\
 & & & \ddots & & & 1 & & \ddots & \\
 & & & & & & 1 & & & \\
 & & & & 1 & & 1 & & 1 & \\
 & & & & & 1 & 1 & & & 1 \\
 & & & & & & 11 & & & \quad 1
\end{array}\right]$$

Any characterization of facets by giving critical problems (that is, $[M \mid b]$ for which the facet does not remain a facet for any minor) must include this class of facets, because it is critical in this sense. The fact that the right-hand sides get arbitrarily large answers negatively the question as to whether there is a bound on how big the numbers may get in a facet. It also casts some doubt on the possibility that one can characterize all critical problems and, thereby, all facets. However, the fact that all of these $[M \mid b]$ contain as a minor the dual Fano with right-hand side $(1,1,1,0)$ means that none of these problems, for $k \ge 3$, is critically non-Fulkerson. So hope remains that the three critically non-Fulkerson problems are the only ones there are.

REFERENCES

1. F. Barahona, "The Max Cut Problem in Graphs not Contractible to K_5," Operations Research Letters 2 (1983), 107-111.

2. F. Barahona and M. Groetschel, "On the Cycle Polytope of a Binary Matroid," to appear in Journal of Combinatorial Theory.

3. S. Chopra, D.L. Jenson, and E.L. Johnson, "Polyhedra of Regular Binary Group Problems," IBM RC, IBM Watson Research Center, Yorktown Heights, NY 1986

4. S. Chopra and E.L. Johnson, "Dual Row Modules and Polyhedra of Blocking Group Problems," to appear in Mathematical Programming.

5. J. Edmonds and E.L. Johnson, "Matching, Euler Tours, and the Chinese Postman Problem," Mathematical Programming 5 (1973), 88-124.

6. D.R. Fulkerson, "Networks, Frames, and Blocking Systems," in: G.B. Dantzig and A.F. Veinott, Jr. (eds.), Mathematics of the Decision Sciences, Part 1. AMS (1968), 303-334.

7. D.R. Fulkerson, "Blocking Polyhedra," in B. Harris (ed.), Graph Theory and its Applications, Academic Press, NY, 1970, 93-112.

8. G. Gastou and E.L. Johnson, "Binary Group and Chinese Postman Polyhedra," Mathematical Programming 34 (1986), 1-33.

9. R.E. Gomory, "Some Polyhedra Related to Combinatorial Problems," Linear Algebra and Applications 2 (1969), 451-558.

10. M. Groetschel and W.R. Pulleyblank, "Weakly Bipartite Graphs," Operations Research Letters 1 (1981), 23-27.

11. M. Guan, "Graphic Programming Using Odd or Even Points," Chinese Mathematics 1 (1962), 273-277.

12. E.L. Johnson, "Integer Programming: Facets, Subadditivity, and Duality for Group and Semigroup Problems," CBMS-NSF Regional Conferences Series in Applied Mathematics 32 (Society for Industrial and Applied Mathematics) (1980), Philadelphia, PA.

13. E.L. Johnson and O. Marcotte, manuscript (1984).

14. E.L. Johnson and S. Mosterts, "Oh Four Problems in Graph Theory," to appear in SIAM Journal on Algebraic and Discrete Methods.

15. A. Lehman, "A Solution of the Shannon Switching Game," SIAM Journal of Applied Mathematics 12 (1964), 687-725.

16. A. Lehman, "On the Width-Length Inequality," Mathematical Programming 16 (1979), 245-259.

17. M. Padberg and M.R. Rao, "Odd Minimum Cuts and b-matching," Mathematics of Operations Research 7 (1982), 67-80.

18. P.D. Seymour, "Matroids with the Max-flow Min-cut Property," Journal of Combinatorial Theory Series B 23 (1977), 189-222.

19. P.D. Seymour, "Matroids and Multicommodity Flows," European Journal of Combinatorics 2 (1981), 257-290.

20. W.T. Tutte, "Lectures on Matroids," Journal of Research of the National Bureau of Standards Section B 69 (1965), 1-47.

21. H. Whitney, "On the Abstract Properties of Linear Dependence," American Journal of Mathematics 57 (1935), 509-533.

EXAMPLES AND ALGORITHMIC PROPERTIES OF GREEDOIDS*

O. Goecke, Bonn
B. Korte, Bonn
L. Lovász, Budapest

0. Introduction

Greedoids have been introduced in KORTE and LOVÁSZ [1981] as relaxations of matroids. Since then several other authors have studied structural, algorithmic and other properties of greedoids. The interested reader is referred to the bibliography at the end of this paper.

Many algorithmic approaches in continuous optimization as well as in discrete optimization are based on the principle of "greediness". In continuous optimization all steepest descent or gradient methods are of greedy–type. In discrete optimization certain combinatorial structures can be defined by the optimality of the greedy algorithm. Matroids may be characterized axiomatically as those independence systems for which the greedy solution is optimal for certain objective functions (e.g. linear or bottleneck functions). Greedoids can be also characterized by the optimality of the greedy algorithm for a broad class of (nonlinear) objective functions (e.g. generalized bottleneck functions).

On the other hand many algorithmic approaches in different areas of combinatorics and other fields of numerical mathematics give rise to the definition of special greedoids. Indeed, greedoids can be found very often in practice, i.e. with many different algorithms, e.g. scheduling under precedence constraints, breadth first search, shortest path, Gaussian elimination, shellings of trees, chordal graphs, convex sets and oriented matroids, line and point search, series–parallel decomposition, ear–decomposition, retracting and dismantling of posets and graphs, bisimplicial elimination, blossom algorithm and others.

Thus, in this paper we will focuss on these two aspects: We will give an extensive list of examples of greedoids which can be derived from different algorithmic and structural properties in combinatorics and other fields and we will report on certain algorithmic properties of greedoids, especially on algorithmic characterizations of greedoids and certain subclasses of them.

In chapter 1 we give some definitions and basic facts about greedoids. This is a very brief sketch of notations, definitions and results which are necessary for the subsequent chapters. For

* Supported by the joint research project "Algorithmic Aspects of Combinatorial Optimization" of the Hungarian Academy of Sciences (Magyar Tudományos Akadémia) and the German Research Association (Deutsche Forschungsgemeinschaft, SFB 303)

those readers which are not familiar with greedoid theory we recommend to study some of the previous papers about structural and algorithmic properties of greedoids (e.g. KORTE, LOVÁSZ [1983], [1984a], [1984c], [1985d]).

Chapter 2 gives an extensive list of different examples and subclasses of greedoids. Some of them have already been introduced in previous papers but they are discussed here in greater detail. Other examples have not appeared in the literature. We also give additional structural properties of these examples.

In the last chapter we report algorithmic properties of greedoids. We start with the algorithmic characterization of general greedoids via the optimality of the greedy algorithm for a certain class of objective functions. We continue with a more specific algorithmic characterization of antimatroids and we conclude the chapter with algorithmic properties of strong matroid systems. This provides a framework for different elimination greedoids.

1. Definitions and Basic Facts about Greedoids

We assume that the reader is familiar with the basic fact of matroid theory. In general our notation is in accordance with the standard matroid terminology (cf. WELSH [1976]).

A *set system* over a finite ground set E is a pair $(E, \mathcal{F})$ with $\mathcal{F} \subseteq 2^E$. Sets belonging to $\mathcal{F}$ are called *feasible* otherwise *infeasible.* A set system $(E, \mathcal{F})$ is a *matroid* if the following hold:

(M1) $\emptyset \in \mathcal{F}$,

(M2) if $X \in \mathcal{F}, X \neq \emptyset$ then $X - x \in \mathcal{F}$ for all $x \in X$,

(M3) if $X, Y \in \mathcal{F},\ |X| > |Y|$ then there exists $x \in X \backslash Y$ such that $Y \cup x \in \mathcal{F}$.

We call $(E, \mathcal{F})$ an *independence system* if (M1) and (M2) are satisfied. We call $(E, \mathcal{F})$ a *greedoid* if (M1) and (M3) are satisfied. We observe that $(E, \mathcal{F})$ is a greedoid if and only if

(M1) $\emptyset \in \mathcal{F}$,

(M2') $X \in \mathcal{F}, X \neq \emptyset$ then there exists $x \in X$ such that $X - x \in \mathcal{F}$

(M3') if $X, Y \in \mathcal{F},\ |X| = |Y| + 1$ then there exists $x \in X \backslash Y$ such that $Y \cup x \in \mathcal{F}$.

A set system satisfying (M1) and (M2') is called an *accessible set system.* We refer to (M3) as the *augmentation property.* For an arbitrary set system $(E, \mathcal{S})$ we define its *accessible kernel* $\mathcal{K}$ as

$$\mathcal{K} := \{X \in \mathcal{S} : X = \{x_1, \dots, x_k\} \text{ and } \{x_1, \dots, x_i\} \in \mathcal{S} \text{ for all } 1 \leq i \leq k\}.$$

Let $(E, \mathcal{F})$ be an accessible set system then any $F \in \mathcal{F}$ has at least one *feasible ordering* of its elements, i.e. the elements of F can be ordered, say $F := \{x_1, \dots, x_k\}$, such that $\{x_1, \dots, x_i\} \in \mathcal{F}$ for $1 \leq i \leq k$. This leads us to the definition of greedoids as a collection of ordered sets:

A *(simple) language* over a finite ground set E is a pair $(E, \mathcal{L})$ where $\mathcal{L}$ is a collection of finite sequences $x_1 \dots x_k$ of (distinct) elements $x_i \in E$. We call these sequences *strings* or *words,* the elements of E are also called *letters.* Words will be abbreviated by small greek letters, $\emptyset$ will denote the empty word; maximal words in $\mathcal{L}$ are called *basic words.* If $\alpha \in \mathcal{L}$ the $|\alpha|$ denotes the *length* of string α and $\tilde{\alpha} \subseteq E$ denotes the set of letters in α. A simple language $(E, \mathcal{L})$ is called a *hereditary language* if

(G1) $\emptyset \in \mathcal{L}$,

(G2) if $x_1 \ldots x_k \in \mathcal{L}$ then $x_1 \ldots x_i \in \mathcal{L}$ for all $1 \leq i \leq k$.

$(E, \mathcal{L})$ is called a *greedoid* if in addition to (G1) and (G2) the following axiom holds:

(G3) if $\alpha, \beta \in \mathcal{L}$, $|\alpha| < |\beta|$ then there exists $x \in \tilde{\beta} \backslash \tilde{\alpha}$ such that $\alpha x \in \mathcal{L}$.

We observed that (M2') is a consequence of (M3) and (M1), however, dispite the resemblence (G2) is not a consequence of (G3) and (G1).

By now we have two definitions of greedoid: the unordered version $(E, \mathcal{F})$ satisfying (M1), (M3) or equivalently (M1), (M2'), (M3') and the ordered version $(E, \mathcal{L})$ satisfying (G1), (G2),(G3). The following theorem shows that there is a one to one correspondence between these definitions. Thus according to convenience we will use in this paper the ordered or the unordered version.

Theorem 1.1: *If $(E, \mathcal{L})$ is a language satisfying (G1),(G2), and (G3) then $(E, \mathcal{F}(\mathcal{L}))$, where*

$$\mathcal{F}(\mathcal{L}) := \{\tilde{\alpha} : \alpha \in \mathcal{L}\},$$

is a set system satisfying (M1), (M3). Conversely, if $(E, \mathcal{F})$ is a set system with properties (M1), (M3), or equivalently (M1), (M2'), (M3'), then $(E, \mathcal{L}(\mathcal{F}))$, where

$$\mathcal{L}(\mathcal{F}) := \{x_1 \ldots x_k : \{x_1, \ldots, x_i\} \in \mathcal{F} \text{ for } 1 \leq i \leq k\}$$

satisfies (G1), (G2) and (G3).

Proof. (KORTE, LOVÁSZ [1984a])

Remark: If $(E, \mathcal{L})$ is a hereditary language and $(E, \mathcal{F})$ an accessible set system then $\mathcal{F}(\mathcal{L}(\mathcal{F})) = \mathcal{F}$ and $\mathcal{L}(\mathcal{F}(\mathcal{L})) \supseteq \mathcal{L}$. Equality holds, i.e. $\mathcal{L}(\mathcal{F}(\mathcal{L})) = \mathcal{L}$ if and only if for all $\alpha, \beta \in \mathcal{L}$ such that $\tilde{\alpha} \cup \{x\} = \tilde{\beta}$ and $x \notin \tilde{\alpha}$ we have $\alpha x \in \mathcal{L}$.

For a set system $(E, \mathcal{F})$ we define

$$r(X) := max\{|F| : F \subseteq X,\ F \in \mathcal{F}\}$$

as its *rank function*.

Theorem 1.2: *A mapping $r : 2^E \to \mathbb{Z}$ is a rank function of some greedoid $(E, \mathcal{F})$ if and only if for all $X, Y \subseteq E$ and $x, y \in E$*

(R1) $r(\emptyset) = 0$,

(R2) $r(X) \leq r(X \cup \{x\}) \leq |X| + 1$

(R3) *if $r(X) = r(X \cup \{x\}) = r(X \cup \{y\})$ then $r(X) = r(X \cup \{x, y\})$.*

Furthermore, the rank function determines the greedoid uniquely and we have $\mathcal{F} = \{X \subseteq E : r(X) = |X|\}$.

Proof. (KORTE, LOVÁSZ [1983])

Remark: If we replace (R2) by

(R2') $r(X) \leq r(X \cup \{x\}) \leq r(X) + 1$

then (R1), (R2'), (R3) are the axioms for a matroid rank function.

For completeness we mention a third characterization of greedoids. If $(E, \mathcal{F})$ is a set system and $r : 2^E \to \mathbb{Z}$ the rank function of $(E, \mathcal{F})$ then

$$\sigma(X) := X \cup \{x \in E \backslash X : r(X \cup x) = r(X)\}$$

is called the *closure operator* of $(E, \mathcal{F})$.

Theorem 1.3: *A mapping $\sigma : 2^E \to 2^E$ is the closure operator of some greedoid $(E, \mathcal{F})$ if and only if*

(S1) *$X \subseteq \sigma(X)$ for all $X \subseteq E$;*

(S2) *$\sigma(X) = \sigma(Y)$ for all $X, Y \subseteq E$ with $X \subseteq Y \subseteq \sigma(X)$*

(S3) *if $z \notin \sigma(X \cup x - z)$ for all $z \in X \cup x$ and if $x \in \sigma(X \cup y)$ then $y \in \sigma(X \cup x)$.*

If $\sigma : 2^E \to 2^E$ satisfies (S1), (S2), (S3) then

$$\mathcal{F} := \{F \subseteq E : x \notin \sigma(F - x) \text{ for all } x \in F\}.$$

Proof. (KORTE, LOVÁSZ [1983]).

Similar to matroid minors we can define *greedoid minors.* If $(E, \mathcal{F})$ is greedoid and $E' \subseteq E$ then

$$(E', \mathcal{F} \mid E') \text{ where } \mathcal{F} \mid E' := \{F \subseteq E' : F \in \mathcal{F}\}$$

is called the *restriction* of $\mathcal{F}$ to E'. If $F \in \mathcal{F}$ then

$$(E \backslash F,\ \mathcal{F}/F) \text{ where } \mathcal{F}/F := \{A \subseteq E \backslash F : A \cup F \in \mathcal{F}\}$$

is called the *contraction* of $\mathcal{F}$ with respect to $F \in \mathcal{F}$.

Finally, for $k \in \mathbb{Z},\ k \geq 0$

$$\mathcal{F}^{(k)} := \{F \in \mathcal{F} : |F| \leq k\}$$

is called the *k–truncation* of $\mathcal{F}$.

While the restriction, contraction and truncation of a greedoid always results in a greedoid the following minor operation preserves the greedoid property only in special cases.

If $(E, \mathcal{F})$ is a greedoid and $E' \subseteq E$, then we call the set system

$$(E', \mathcal{F} : E') \text{ where } \mathcal{F} : E' = \{F \cap E' : F \in \mathcal{F}\}$$

the *trace* of $\mathcal{F}$ to E'. Note that for an independence system restriction and trace coincide.

Antimatroids, also called *shelling structures*, form an important subclass of greedoids. A greedoid $(E, \mathcal{A})$ is an *antimatroid* if $\mathcal{A}$ is closed under union, i.e. if $A, B \in \mathcal{A}$ then $A \cup B \in \mathcal{A}$. An easy observation is the following:

A set system $(E, \mathcal{A})$ is an antimatroid if and only if

(A1) $\emptyset \in \mathcal{A}$

(A2) if $A, B \in \mathcal{A}$ and $B \backslash A \neq \emptyset$ then there exists $x \in B \backslash A$ such that $A \cup x \in \mathcal{A}$.

We say that a language $(E, \mathcal{L})$ is an antimatroid if $(E, \mathcal{A})$ with $\mathcal{A} := \mathcal{F}(\mathcal{L})$ is an antimatroid. In the sequel we will always assume that antimatroids are *full* greedoids, i.e. greedoids where the ground set E is feasible. Thus by definition antimatroids do not contain *dummy elements*, i.e. elements which do not occur in any feasible set.

If $(E, \mathcal{A})$ is an antimatroid then the restriction and contraction is an antimatroid, furthermore the trace of $\mathcal{A}$ to an arbitrary set is again an antimatroid.

A *rooted subset* of E is a pair (C, r) with $r \in C \subseteq E$, r is called the *root* of C. Let $(E, \mathcal{A})$ be an antimatroid, a rooted subset (C, a) of E is called a *circuit with root a* or, shorter , a circuit if

$$\mathcal{A} : C = \{A \subseteq C : A \neq \{a\}\}.$$

Theorem 1.4: (DIETRICH [1985]) *Let E be a finite set and $\mathcal{C}$ a collection of rooted subsets of E. Then $\mathcal{C}$ is the set of circuits of some antimatroid $(E, \mathcal{A})$ if $\mathcal{C}$ satisfies the following axioms:*

(C1) *if $(C_1, a), (C_2, a) \in \mathcal{C}$ then $C_1 \not\subset C_2$;*

(C2) *if $(C_1, a_1), (C_2, a_2) \in \mathcal{C}$ and $a_1 \in C_2 - a_2$ then there exists a rooted subset $(C_3, a_2) \in \mathcal{C}$ such that $C_3 \subseteq (C_1 \cup C_2) - a_1$.*

If (C1) and (C2) are satisfied then

$$\mathcal{A} = \{A \subseteq E : A \cap C \neq \{a\} \text{ for all } (C, a) \in \mathcal{C}\}.$$

Recall that $C \subseteq E$ is a circuit of a matroid $(E, \mathcal{M})$ if

$$\mathcal{M} \mid C = \{A \subseteq C : A \neq C\}.$$

To demonstrate the similarities between matroids and antimatroids we restate the circuit characterization of matroids.

Theorem 1.5: *Let E be a finite set and $\mathcal{C} \subseteq 2^E$, then $\mathcal{C}$ is the set of circuits of some matroid $(E, \mathcal{M})$ if $\mathcal{C}$ satisfies the following axioms:*

(C1') *if $C_1, C_2 \in \mathcal{C}$ then $C_1 \not\subset C_2$;*

(C2') *if $C_1, C_2 \in \mathcal{C}$, $C_1 \neq C_2$ and $z \in C_1 \cap C_2$, then there exists $C_3 \in \mathcal{C}$ such that $C_3 \subseteq (C_1 \cup C_2) - z$.*

If $\mathcal{C}$ satisfies (C1') and (C2') then $\mathcal{M} = \{A \subseteq E : C \not\subseteq A$ for all $C \in \mathcal{C}\}$.

We saw above that for an antimatroid $(E, \mathcal{A})$ with $\mathcal{C}$ as the collection of its circuits

$$\mathcal{A} = \{A \subseteq E \quad A \cap C \neq \{a\} \text{ for all } (C, a) \in \mathcal{C}\}.$$

The term circuit is quite intuitive for matroids since circuits in matroids are minimal dependent sets and for a graphic matroid circuits correspond to the circuits in the representing graph.

We want to give a similiar intuitive description for the circuits of an antimatroid. To this end let us rephrase the above statement about $\mathcal{A}$ in the ordered version. If $(E, \mathcal{L})$ is an antimatroid with $\mathcal{C}$ as the collection of its circuits then

$$\mathcal{L} := \{x_1 \dots x_k : \text{ if } (C, x_i) \in \mathcal{C} \text{ for some } 1 \leq i \leq k \text{ then } C \cap \{x_1, \dots, x_i\} \neq \{x_i\}\}.$$

This observation justifies the term "circuit" for antimatroids:

If one builds up a feasible word $x_1 \dots x_k \in \mathcal{L}$ letter by letter and if one tries to add a root $x_{k+1} \in E$ of some circuit (C, x_{k+1}) then one has to pick first at least one letter from $C - x_{k+1}$. One could say that the root $a \in E$ of a circuit (C, a) is "encircled" by $C - a$.

This somewhat vague intuitive description will become clearer in the case of convex–shelling antimatroids which will be discussed in the next chapter.

Consider an arbitrary collection $\mathcal{C} = \{(C, a) : a \in C \subseteq E\}$ of rooted subsets of E and let

$$\mathcal{L}_{\mathcal{C}} := \{x_1 \dots x_k : \text{ if } (C, x_i) \in \mathcal{C} \text{ for some } 1 \leq i \leq k \text{ then } C \cap \{x_1, \dots, x_i\} \neq \{x_i\}\}.$$

Then $(E, \mathcal{L}_{\mathcal{C}})$ is an antimatroid (with possibly dummy elements).

This observation leads us to the question whether there is a minimal system of rooted subsets of E representing a given antimatroid $(E, \mathcal{L})$. In fact, there is a unique minimal such system:

Theorem 1.6: *Let $(E, \mathcal{L})$ be an antimatroid then there is a unique minimal collection $\mathcal{C}^*$ of rooted subsets of E representing $\mathcal{L}$. We call the elements of $\mathcal{C}^*$ critical circuits of $(E, \mathcal{L})$.*

Proof. (KORTE, LOVÁSZ [1984 b]).

Thus $\mathcal{L} = \mathcal{L}_{\mathcal{C}^*}$ and whenever $\mathcal{L} = \mathcal{L}_{\mathcal{C}}$ for some $\mathcal{C} \subseteq \{(C, a) : a \in C \subseteq E\}$ then $\mathcal{C}^* \subseteq \mathcal{C}$. In particular, every critical circuit is a circuit.

Theorem 1.7: *If $(E, \mathcal{L})$ is an antimatroid then a rooted subset (C, a) of E is a critical circuit if and only if for a maximal feasible word $\alpha \in \mathcal{L} \mid (E \backslash C)$ we have*

$$\alpha a \notin \mathcal{L}$$
$$\text{and } \alpha x \in \mathcal{L} \text{ for all } x \in C - a.$$

Proof. (KORTE, LOVÁSZ [1984b])

Let $(E, \mathcal{F})$ be a greedoid. A rooted subset (P, a) of E is called a *path with endpoint a* or shorter an *a–path*

$$\text{if } P \in \mathcal{F}$$
$$\text{and } P - x \notin \mathcal{F} \text{ for all } x \in P - a.$$

Since $\mathcal{F}$ is hereditary for every a–path (P, a) we have $P - a \in \mathcal{F}$.

Theorem 1.8: *A finite collection $\mathcal{P} = \{(P_i, e_i) : i \in I\}$ of rooted subsets of E is the collection of paths of an antimatroid $(E, \mathcal{A})$ if and only if*

(P1) $\bigcup_{i \in I} P_i = E$.

(P2) *for all $(P, e) \in \mathcal{P}$, e is the unique element for which $P - e$ can be represented as $P - e = \bigcup_{j \in J} P_j$ for some $J \subseteq I$.*

If (P1) and (P2) are satisfied then

$$\mathcal{A} := \{\bigcup_{j \in J} P_j : J \subseteq I\}.$$

Proof. (GOECKE [1986b])

The next theorem shows that paths and circuits of an antimatroid are in a certain sense dual objects. To make this precise let $(E, \mathcal{A})$ be an antimatroid and denote by $\mathcal{P}$ and $\mathcal{C}$ the collection of paths and circuits, resp. For $x \in E$ let

$$\mathcal{P}_x := \{P - x : (P, x) \in \mathcal{P}\}$$
$$\text{and } \mathcal{C}_x := \{C - x : (C, x) \in \mathcal{C}\}.$$

Note that for every $x \in E$ the set systems $\mathcal{P}_x$ and $\mathcal{C}_x$ are *clutters*, i.e. set systems such that for any two elements of them none is included in the other. If $\mathcal{K} \subseteq 2^E$ is a clutter then

$$\mathcal{K}^b := \{B \subseteq E : B \cap K \neq \emptyset \text{ for all } K \in \mathcal{K} \text{ and there is no strict subset of } B \text{ with this property}\}$$

is called the *blocker* of $\mathcal{K}$. The blocker of a clutter is again a clutter, moreover for any clutter $\mathcal{K}$ it follows that

$$(\mathcal{K}^b)^b = \mathcal{K}$$

(cf. EDMONDS and FULKERSON [1970]).

Theorem 1.9: *Let $(E, \mathcal{F})$ be an antimatroid and let $\mathcal{P}_x, \mathcal{C}_x$ be defined as above then*

$$\mathcal{P}_x^b = \mathcal{C}_x \text{ and } \mathcal{C}_x^b = \mathcal{P}_x$$

for all $x \in E$.

Proof. (GOECKE [1986b])

Antimatroids were first introduced by EDELMAN [1980] and JAMISON [1982] as a combinatorial abstraction of convexity. The closure operator of matroids can be considered as a combinatorial abstraction of the linear hull operator in vector spaces. A slight modification of the Steinitz–McLane exchange property of this operator leads to an appropriate convex hull operator:

A mapping $\tau : 2^E \to 2^E$ is called the *convex hull operator* if

(K1) $X \subseteq \tau(X)$ for all $X \subseteq E$

(K2) $X \subseteq Y$ implies $\tau(X) \subseteq \tau(Y)$

(K3) $\tau(\tau(X)) = \tau(X)$

(K4) If $y, z \notin \tau(X)$ and $z \in \tau(X \cup y)$ then $y \notin \tau(X \cup z)$.

(K4) is called the *anti-exchange property*. A set $X \subseteq E$ is called *convex* if $X = \tau(X)$. It can be easily seen that the complement of convex sets form the feasible sets of an antimatroid, i.e.

$$\mathcal{A} = \{X \subseteq E : E - X = \tau(E - X)\}.$$

On the other hand given an antimatroid $(E, \mathcal{A})$ we have convex sets as complements of elements of $\mathcal{A}$ and we define the convex hull of a set X as the intersection of all convex supersets of X. It can be also easily seen that this convex hull satisfies (K1), (K2), (K3), (K4).

In the rest of this chapter we will introduce two other classes of greedoids interval greedoids and transposition greedoids. We will see that every antimatroid is an interval greedoid and every interval greedoid is a transposition greedoid.

A greedoid $(E, \mathcal{F})$ is called an *interval greedoid* if for all $A, B, C \in \mathcal{F}$, $A \subseteq B \subseteq C$ the following holds:

$$\text{if } x \in E \backslash C \text{ and } A \cup x \in \mathcal{F} \text{ and } C \cup x \in \mathcal{F} \text{ then } B \cup x \in \mathcal{F}.$$

For short we call this property the *interval property.* We consider also two modifications of the interval property, namely the *interval property without upper bounds,* i.e. for all $A, B \in \mathcal{F}$ $A \subseteq B$ and $x \in E \backslash B$ such that $A \cup x \in \mathcal{F}$ it follows $B \cup x \in \mathcal{F}$, and the *interval property without lower bounds,* i.e. for all $B, C \in \mathcal{F}$, $B \subseteq C$ and $x \in E \backslash C$ such that $C \cup x \in \mathcal{F}$ it follows $B \cup x \in \mathcal{F}$.

It can be shown that antimatroids are exactly those greedoids for which the interval property without upper bounds hold, while matroids are exactly greedoids with the interval property without lower bounds.

Theorem 1.10: *The following statements are equivalent:*

(1) $(E, \mathcal{F})$ *is an interval greedoid.*

(2) For all $A, C \in \mathcal{F}$, $A \subseteq C$ *the following holds:*

if $x \in E \backslash C, y \in C \backslash A$ *and* $A \cup x, A \cup y, C \cup x \in \mathcal{F}$ *then also* $A \cup \{x, y\} \in \mathcal{F}$.

(3) For all $F \in \mathcal{F}$ *the set system* $(F, \mathcal{F} \mid F)$ *is an antimatroid.*

(4) If F_1 *and* F_2 *are two maximal feasible subsets of* $E' \subseteq E$, *then* $\mathcal{F}/F_1 = \mathcal{F}/F_2$.

Proof. (KORTE, LOVÁSZ [1983], GOECKE [1986b])

A greedoid $(E, \mathcal{F})$ is a *transposition greedoid* if

(TP) for all $A, B \subseteq E$, $A \subseteq B$, and $x, y \in E \backslash B$ the following holds:

if $A, A \cup x, A \cup y, B \cup x \in \mathcal{F}$ and $A \cup \{x, y\} \in \mathcal{F}$ then $B \cup y \in \mathcal{F}$.

A hereditary language satisfying (TP) is already a greedoid (cf. KORTE, LOVÁSZ [1986b]).

The transposition property (TP) becomes more transparent if formulated in the ordered version:

Theorem 1.11: *Let* $(E, \mathcal{L})$ *be an hereditary language then* $(E, \mathcal{L})$ *is a transposition greedoid if and only if the following holds:*

(T1) $\alpha x y \beta \in \mathcal{L}, \alpha y \in \mathcal{L}$ *implies* $\alpha y x \beta \in \mathcal{L}$

(T2) if $\alpha x, \alpha y \in \mathcal{L}, \alpha x y \notin \mathcal{L}$ *then* $\alpha x \beta \in \mathcal{L}, y \notin \tilde{\beta}$ *implies* $\alpha y \beta \in \mathcal{L}$ *and* $\alpha x \beta y \gamma \in \mathcal{L}$ *implies* $\alpha y \beta x \gamma \in \mathcal{L}$.

Proof. (KORTE, LOVÁSZ [1986b])

Finally, we formulate an ordered version of the interval property:

Theorem 1.12: *Let* $(E, \mathcal{L})$ *be a hereditary language the* $(E, \mathcal{L})$ *is an interval greedoid if and only if*

(T1) $\alpha x y \beta \in \mathcal{L}, \alpha y \in \mathcal{L}$ *implies* $\alpha y x \beta \in \mathcal{L}$

(T2') if $\alpha x, \alpha y \in \mathcal{L}, \alpha xy \notin \mathcal{L}$ then $\alpha x\beta \in \mathcal{L}$ implies $\alpha y\beta \in \mathcal{L}$.

Proof. (KORTE, LOVÁSZ [1986b])

2. List of Examples

In this chapter we give an extensive list of examples of greedoids. Some of them have not yet appeared in the literature others which have been mentioned elsewhere are discussed in more detail.

2.1 Poset greedoids

Let $(E, \leq)$ be a partially ordered set. A subset $I \subseteq E$ is called a (lower) *ideal* if $x \leq y$ and $y \in I$ implies that $x \in I$. The set system

$$\mathcal{F} := \{I \subseteq E : I \text{ is an ideal in } (E, \leq)\}$$

defines a greedoid which we call *poset greedoid.* The maximal feasible words in the poset greedoid are the linear extensions of this partial order. Note that in a poset greedoid the union and the intersection of feasible sets are again feasible. Also the converse is true: If $(E, \mathcal{F})$ is an accessible set system which has the property that $E \in \mathcal{F}$ and that $\mathcal{F}$ is closed under union and intersection then E can be endowed with a partial order $\leq$, such that $\mathcal{F}$ is the poset greedoid with respect to $(E, \leq)$.

Let us mention two other characterizations of poset greedoid.

Theorem 2.1: *Let $(E, \mathcal{F})$ be an antimatroid then the following statements are equivalent:*

(i) $(E, \mathcal{F})$ is a poset greedoid;

(ii) if (C, a) is a circuit then $|C| = 2$;

(iii) for every $x \in E$, there is a unique x-path.

Proof.

(i) $\Rightarrow$ (ii) obvious.

(ii) $\Rightarrow$(iii) By Theorem 1.9 we know that the blocker of the circuits with root x are just the paths with head x. By (ii) the set of circuits of a given element $x \in E, \{x\} \notin \mathcal{F}$ consists of singletons. The blocker of such a set system consists of a single set.

(iii) $\Rightarrow$ (i) Define $y \leq x$ if *the* x-path P_x contains y. This defines a partial order the lower ideals of this order are the paths $P_x, x \in E$. From this follows that the poset greedoid with respect to $(E, \leq)$ is just $(E, \mathcal{F})$.

□

We call a greedoid $(E, \mathcal{F})$ a *local poset greedoid,* if for $A, B, C \in \mathcal{F}$ with $A, B \subseteq C$ we have $A \cup B$, $A \cap B \in \mathcal{F}$. This means that the restriction of a local poset greedoid to any feasible set results in a poset greedoid.

The poset greedoid can be considered as a shelling of a poset from the bottom. One can generalize this by allowing to pick elements which are "near" to the bottom (cf. EDELMAN, JAMISON [1984]).

Let $(E, \leq)$ be a poset and $k \geq 1$ an integer. Call $F \subseteq E$ feasible if for every $y \in F$ and every chain $x_1 < x_2 < \ldots < x_k < y$ we have $F \cap \{x_1, \ldots, x_k\} \neq \emptyset$. The feasible sets form an antimatroid which is a poset greedoid if $k = 1$. Note that circuits are chains of length $k+1$, with the top element being the root.

2.2 Double poset shelling

Instead of shelling a poset $(E, \leq)$ from the bottom we could shell it from the bottom and the top. This again defines a greedoid, the *double poset greedoid.* Every feasible set in this greedoid can be represented as the union of a lower ideal and an upper ideal (filter). $U \subseteq E$ is an *upper ideal* if $x \leq y$ and $x \in U$ implies $y \in U$. The circuits of the double poset greedoids are rooted sets $(\{u, v\}, x)$ such that $u < x < v$ and for each $x \in E, \{x\} \notin \mathcal{F}$ there are exactly two x–paths, namely the ideal and the filter generated by x. The following example, however, shows that these two properties together do not characterize double poset greedoids.

Let $E = \{u, v, w, x, y\}$ and let $(\{u, v\}, x), (\{u, w\}, y)$ and $(\{w, x\}, y)$ be the circuits of the antimatroid $(E, \mathcal{A})$. There are exactly two x–paths, namely $\{u, x\}, \{v, x\}$ and two y–paths, namely $\{u, x, y\}$ and $\{w, y\}$. This antimatroid is not a double poset greedoid. To see this let us mention another property of double poset greedoids. Denote for every $x \in E, \{x\}$ not feasible, the two x–paths by P_x and P'_x. If then $y \in P_x$ and $P_y \subseteq P_x$ then $P'_x \subseteq P'_y$. The example just mentioned does not have this property.

2.3 Chordal graph shelling

An undirected graph G is called *chordal* or *triangulated* if every cycle of length greater than three has a chord. Call a vertex $v \in V(G)$ *simplicial* if the neighbors of v form a complete subgraph of G. Chordal graphs can be characterized by the property that every induced subgraph has a simplicial vertex (cf. GOLUMBIC [1980]). Thus chordal graphs can be pruned by simplicial vertices. This defines in a natural way an antimatroid $(E, \mathcal{L})$, where $E = V(G)$ and

$$\mathcal{L} := \{v_1 \ldots v_k : \text{ for } i = 1, \ldots, k \;\; v_i \text{ is a simplicial vertex of } G - \{v_1, \ldots, v_{i-1}\}\}.$$

The unordered version $(E, \mathcal{A})$ can be described by characterizing the convex sets: A set K is called convex if for every $x, y \in K$ every chordless (x, y)–path with respect to G lies in K. Then $\mathcal{A} = \{E - K : K \text{ convex}\}$. Like double poset greedoids the circuits of chordal shelling greedoids have cardinality 3. $(\{u, v\}, x)$ is a circuit with root x if and only if there is a chordless path from u to v passing through x.

2.4 Shelling of a tree

Trees are special chordal graphs, the simplicial points are just the leafs of the tree. Thus the successive elimination of leafs gives rise to an antimatroid, the *tree shelling greedoid*. Here the convex sets are just the connected subtrees.

2.5 Edge-shelling of a tree

This example is very similar to the previous one. Let G be a tree with edge set E. If we reduce G by taking away terminal edges then the collection of elimination sequence defines an antimatroid.

The reader might convince himself that *edge-shelling of a tree* and vertex–shelling of a tree are overlapping classes of antimatroids and that both are special chordal graph shelling greedoids.

2.6 Point-line search

Let G be an undirected graph and $E = E(G) \cup V(G)$. Define

$$\begin{aligned}\mathcal{A} :=\{A \subseteq E;\ &\text{if } (u,v) \in A \cap E(G) \text{ then either}\\ &u \in A \cap V(G) \text{ or } v \in A \cap V(G)\}.\end{aligned}$$

Then $(E, \mathcal{A})$ is an antimatroid. The circuits of $(E, \mathcal{A})$ are just rooted sets $(\{u, v, e\}, e)$, where $u, v \in V(G), e \in E(G)$ and e is the edge between u and v.

2.7 Cover greedoid

Let $G = (V, E)$ be an undirected graph and let $x \notin V$ be an extra node. On the ground set $V \cup \{x\}$ we define an antimatroid $(V \cup \{x\}, \mathcal{F})$ where

$$\mathcal{F} = 2^V \cup \{C \cup \{x\}, C \subseteq V \text{ covers all edges in } G\}.$$

Note that the circuits of this antimatroid are of the form $(\{v_1, v_2\}, x)$ where v_1 and v_2 are linked by an edge.

Remark: Antimatroids whose circuits have cardinality 2 are well understood since they "are" partially ordered set. Examples 2.2 – 2.7 all share the property of having circuits of length 3. These examples are quite different in structure. It would be interesting to have a canonical example for the class of antimatroids with circuit length 3.

2.8 Point search

Given a directed graph $G = (V, A)$ and a distinguished node $r \in V$, called the root. Let $E := V - r$ and define the system of feasible sets $\mathcal{F} \subseteq 2^E$ recursively by $\emptyset \in \mathcal{F}$ and if $F \in \mathcal{F}$, then for $x \in E \backslash F$ $F \cup x \in \mathcal{F}$ if and only if there is a directed edge $(y, x) \in A$ such that $y \in F \cup r$. It is easy to see that $(E, \mathcal{F})$ is an antimatroid.

2.9 Line search

As in 2.8 let $G = (V, A)$ be a digraph with root $r \in V$. Here, however, we take $E := A$ as the ground set and define $\mathcal{F} \subseteq 2^E$ by $\emptyset \in \mathcal{F}$ and if $F \in \mathcal{F}$ and $(v,w) \in E \backslash F$ $F \cup (v,w) \in \mathcal{F}$ if and only if $v = r$ or $(u,v) \in F$ for some $u \in V$.

2.10 Undirected point search, undirected line search

If we replace the digraph in 2.8 and 2.9 by an undirected graph, we can similarly define *undirected point (line-)search greedoids.* The reader may verify the following inclusion relation between antimatroids:

- every undirected line search is an undirected point search,
- every undirected point search is a directed point search,
- every directed line search is a directed point search.

All inclusions are strict.

Remark: The antimatroids in Example 2.8 - 2.10 share the following property.

(P1) If P is a path with head x then $P - x$ is a path. This is the same as to say that every path has a unique feasible ordering.

This property is not sufficient to characterize directed point search greedoids among general antimatroids. It is easy to see that the set $\mathcal{P}$ of paths of a directed point search greedoid satisfies:

(P2) If $P_1, P_2, P_1 - x \in \mathcal{P}$ and if the head of P_2 equals the head of $P_1 - x$, then there exists a path $P_3 \in \mathcal{P}$ with head x such that $P_3 \subseteq P_2 \cup x$.

Theorem 2.2: *Let $(E, \mathcal{A})$ be an antimatroid, then $(E, \mathcal{A})$ is a directed point search greedoid if and only if the set $\mathcal{P}$ of paths in $(E, \mathcal{A})$ satisfies (P1) and (P2).*

Proof (GOECKE [1986a])

Similar characterizations can be found for the examples in 2.9 and 2.10 (cf. GOECKE [1986a]). Let us mention two further examples of antimatroids with property (P1).

2.11 Line search in matroids

Here we generalize the undirected line search for graphs (Example 2.10) to matroids : Let $(E', \mathcal{M})$ be a matroid, $B \subseteq E'$ a basis of $\mathcal{M}$ and $e_0 \in B$.

Define $E := E' \backslash B$ and

$$\mathcal{L} := \{x_1 \ldots x_k : \text{ for } i = 1, \ldots, k \;\; x_i \in E \text{ and there are circuits } C_1, \ldots, C_k \text{ of } (E', \mathcal{M}) \text{ such that } e_0, x_i \in C_i \text{ and } C_i \subseteq B \cup \{x_1, \ldots, x_i\}\}.$$

Then $(E, \mathcal{L})$ is an antimatroid. It may, however, contain dummy elements, for instance if e_0 is an isthmus of $(E', \mathcal{M})$, then $\mathcal{L} = \{\emptyset\}$. Let us see how this relates to the graphic line search: Let $G = (V, E)$ be an undirected graph with root $r \in V$. Now add a new vertex v_0 and connect

v_0 to every vertex $v \in V$ by an edge, the edge which links v_0 to the root r denote by e_0. These new edges together form a basis of the graphic matroid induced by the enlarged graph $G' = (V \cup v_0, E \cup \{(v, v_0) : v \in V\})$. Now the reader can easily convince himself that the line search on G corresponds to the line search with respect to the graphic matroid on G' with basis $B = \{(v, v_0) : v \in V\}$.

2.12 Capacitated point search

Let $G = (V, A)$ be a digraph with root $r \in V$ and let $c(v) \in \mathbb{Z}$ be the *capacity* of node $v \in V$. We call a dipath $(r, v_1, \ldots, v_k)$ from the root c–compatible if

$$\begin{aligned} c(r) &\geq k, \\ c(v_1) &\geq k-1, \\ &\vdots \\ c(v_{k-1}) &\geq 1, \\ c(v_k) &\geq 0. \end{aligned}$$

Define $E := V - r$ and

$$\begin{aligned} \mathcal{A} :=\{F \subseteq E :\ & \text{for every } v \in F \text{ there exists a } c\text{–compatible} \\ & \text{rooted dipath } (r, v_1, \ldots, v_k) \text{ such that } \{v_1, \ldots, v_k\} \subseteq F \text{ and } v_k = v\}. \end{aligned}$$

If $c(v) = |V|$ for all $v \in V(E, A)$ then $(E, \mathcal{A})$ is the point search greedoid with respect to G since every rooted dipath is c–compatible.

2.13 (a, b)–path shelling

Let $E = \{1, 2, \ldots, n\}$ and $0 \leq a, b \leq n$ integers. Define $\mathcal{A} \subseteq 2^E$ recursively by $\emptyset \in \mathcal{A}$ and if $F \in \mathcal{A}$, $j \in E \backslash F$ then $F \cup j \in \mathcal{A}$ if and only if there are less than a numbers smaller than j not in F or less than b numbers larger than j not in F. $(E, \mathcal{A})$ is an antimatroid which we denote by $G(a, b, n)$. Let us give some special examples of (a, b)–path shellings:

- $G(0, 1, n)$ is the poset greedoid of the linear order $1 < 2 < \ldots < n$.
- $G(1, 1, n)$ is the double poset greedoid.
- if $a + b \geq n$ then $G(a, b, n)$ is the free greedoid on E
- if $a + b = n - 1$ then $G(a, b, n)$ has exactly one circuit, namely $(E, a + 1)$.

If $\{j\}$ is not feasible in $G(a, b, n)$ then P is a path with head j if and only if

$$\begin{aligned} & P \subseteq \{1, \ldots, j\} \text{ and } |P| = j - a \\ \text{or } & P \subseteq \{j, j+1, \ldots, n\} \text{ and } |P| = n - j + 1 - b. \end{aligned}$$

From this observation we can deduce that (C, r) is a circuit with root r if $|C \cap \{1, \ldots, r-1\}| = a$ and $|C \cap \{r+1, \ldots, n\}| = b$.

2.14 Transivity antimatroid

Let $G = (V, E)$ be a transitively oriented graph. Call $F \subseteq E$ feasible if no arc in F is a chord of a dipath in $E \backslash F$. Then the feasible sets define an antimatroid $(E, \mathcal{F})$. The convex sets of this antimatroid are sets $K \subseteq E$ with the property that if $P \subseteq K$ is a path and $e \in E$ is a chord of P then $e \in K$. If $e \in E$ and $\{e\}$ not feasible then (P, e) is a circuit with root e if and only if P is a path from tail to head of e.

2.15 Shelling of an oriented matroid

This is a straightforward generalization of the last example to acyclic oriented matroids. Recall the definition of an oriented matroid: Let E be a finite set, a pair $X = (X^+, X^-)$ with $X^+, X^- \subseteq X$ and $X^+ \cap X^- = \emptyset$ is called a *signed set*. A system $\mathcal{O}$ of signed sets is called an oriented matroid if $\mathcal{O}$ satisfies the following axioms

(OM0) $(\emptyset, \emptyset) \notin \mathcal{O}$,

(OM1) $(X^+, X^-) \in \mathcal{O} \Rightarrow (X^-, X^+) \in \mathcal{O}$,

(OM2) $(X^+, X^-), (Y^+, Y^-) \in \mathcal{O}$ and $X^+ \cup X^- \subseteq Y^+ \cup Y^-$, then

$$X^+ = Y^+ \text{ and } X^- = Y^-$$
$$\text{or } X^+ = Y^- \text{ and } X^- = Y^+,$$

(OM3) if $(X^+, X^-), (Y^+, Y^-) \in \mathcal{O}$ and $e \in X^+ \cap Y^-, y \in (X^+ \cup X^-) \backslash (Y^+ \cup Y^-)$ then there exists $Z^+ \subseteq (X^+ \cup Y^+) - e$, $Z^- \subseteq (X^- \cup Y^-) - e$, such that $(Z^+, Z^-) \in \mathcal{O}$ and $y \in Z^+ \cup Z^-$.

$\mathcal{O}$ is called an acyclic antimatroid if in addition $X^+ \neq \emptyset$ for all $(X^+, X^-) \in \mathcal{O}$. For $X \subseteq E$ define the *convex hull* of X to be

$$\tau(X) = X \cup \{y \in E \text{ s.t. } (Y^+, \{y\}) \in \mathcal{O} \text{ for some } Y^+ \subseteq X\}.$$

Correspondingly we call $X \subseteq E$ *convex* if $X = \tau(X)$. It is not hard to prove that $\tau : 2^E \to 2^E$ is a closure operator satisfying the antiexchange property. Thus $\mathcal{F} := \{E - \tau(X) : X \subseteq E\}$ is a shelling structure. It is easy to see that if $\mathcal{O}$ stems from a transitively oriented graph then $\mathcal{F}$ is the transitivity greedoid of Example 2.6.

2.16 Convex shelling

Let E be a finite set of points in $\mathbb{R}^d$, call $C \subseteq E$ convex if $conv(C) \cap E = C$, where conv denotes the usual convex–hull operator in $\mathbb{R}^d$. This gives rise an antimatroid $(E, \mathcal{L})$. $\mathcal{L}$ can also be defined as

$$\mathcal{L} := \{x_1 \dots x_k : \text{ for } i = 1, \dots, k \ x_i \text{ is a vertex of } conv(\{x_1, \dots, x_i\})\}.$$

(K, x) is a circuit with root x if and only if the points in $K - x$ are the vertices of a $|K| - 1$ dimensional simplex such that x lies in the interior of $conv(K)$. If $x \in E$, then paths with head x are the (by inclusion) minimal sets in

$$\mathcal{H}_x = \{E \cap H : H \text{ is a halfspace with } x \text{ on its boundary}\}.$$

It is a very appealing and apparently very hard problem to characterize convex shelling greedoids even in dimension 2 among general shelling structures. One of these characterizing properties could be the following

(C4) If (C,x) and (C,y) are circuits, then there is a unique subset $C' \subseteq C$ such that $(C' \cup x, y)$ is a circuit.

It is easy to see that for example tree shelling greedoids share this property and that not every tree shelling can be represented as a convex shelling. It has also been proved that the antimatroids from shelling acyclic oriented matroids have property (C4) (cf. DIETRICHS [1985]). In fact, convex shelling is a special case of acyclic oriented matroid shelling. This follows from known results of oriented matroid (cf. BLAND [1977]). However, the problem of characterizing convex shelling among acyclic oriented matroid shelling is closely related to the representability problem for matroids which is still unsolved.

2.17 Semi convex shelling

This example is similar to 2.16 except that here we allow only shelling in the direction of a cone: Let E be a finite set of points in $\mathbb{R}^d$ and K a cone in $\mathbb{R}^d$. Define $C \subseteq E$ to be convex if $C = (conv(C) + K) \cap E$ and let $\mathcal{A} := \{E - C : C \textit{ convex}\}$. Then $(E, \mathcal{A})$ is an antimatroid. If the cone K consists only of the origin then $(E, \mathcal{A})$ will be the usual convex shelling greedoid. For $d = 2$ and $K = \{(x,y) \in \mathbb{R}^2 : y \geq 0\}$ we call $(E, \mathcal{A})$ the *lower convex shelling.*

2.18 The universal antimatroid

Let $G = (U, V, E)$ be a bipartite graph and suppose that the edge set E has been two–colored red and green. A vertex $u \in U$ is called *extreme* in G if u is not incident to any red edge. For $U' \subseteq U$ let $G : U'$ denote the bipartite graph we obtain from G by deleting U' and all neighbors of U'. Define

$$\mathcal{L}_G := \{u_1 \ldots u_k : \text{ for } i = 1, \ldots, k \;\; u_i \text{ is extreme in } G : \{u_1, \ldots, u_{i-1}\}\}.$$

Then $(U, \mathcal{L})$ is an antimatroid.

Surprisingly also the converse is true, i.e. every antimatroid can be represented in this way. To see this let $(U, \mathcal{L})$ be any antimatroid and denote by V the collection of critical circuits of $(U, \mathcal{L})$. Connect $u \in U$ to a critical circuit by a red edge if u is the root of this circuit and by a green edge if u is a non root element of this circuit. This defines a two–colored bipartite graph $G = (U, V, E)$ and $u \in U$ extreme in this graph if and only if u is not the root of any circuit of $(U, \mathcal{L})$. From this one can easily deduce that $\mathcal{L} = \mathcal{L}_G$.

So far we have studied examples of antimatroids which are interval greedoids without upper bound. We now turn to general interval greedoids.

2.19 Undirected branching greedoid

Let $G = (V, E)$ be an undirected graph with root $r \in V$. Define $\mathcal{F}$ to be the collection of trees in G covering the root r. $(E,\mathcal{F})$ is a greedoid, called the *undirected branching greedoid* with respect to G.

2.20 Directed branching greedoid

Let $D = (V, E)$ be a directed graph with root $r \in V$ and let $\mathcal{F}$ be the set of (directed) arborescenses rooted at r. We call $(E,\mathcal{F})$ the *undirected branching greedoid* with respect to D.

2.21 Matroid branching

This example is closely related to line search in matroids (Example 2.11). Again let $(E',\mathcal{M})$ be a matroid and $B \subseteq E'$ a basis with distinguished element $e \in B$. Let $E := E'\backslash B$ and define

$$\begin{aligned}\mathcal{L} :=\{x_1\dots x_k : \{x_1,\dots,x_k\} &\text{ is independent in } (E',\mathcal{M})\\ &\text{and for } i=1,\dots,k \;\; x_i \in E \text{ and there exists a circuit}\\ &C_i \text{ such that } e_0, x_i \in C_i \subseteq B \cup \{x_1,\dots,x_i\}\}.\end{aligned}$$

This example generalizes undirected branching greedoids as can be seen from the construction in Example 2.11.

2.22 Polymatroid greedoids

Let (E, f) be a polymatroid and associate with (E, f) a hereditary language $(E, \mathcal{L}_f)$, where

$$\mathcal{L}_f := \{x_1\dots x_k : f(\{x_1,\dots,x_i\}) = i \text{ for all } 1 \le i \le k\}.$$

We claim that $(E,\mathcal{L}_f)$ is an interval greedoid — $(E,\mathcal{L}_f)$ is called the *polymatroid greedoid* with respect to (E, f).
We verify the properties (cf. Theorem 1.12)

(T1) $\quad \alpha xy\beta \in \mathcal{L}_f,\ \alpha y \in \mathcal{L}_f$ implies $\alpha yx\beta \in \mathcal{L}_f$ and

(T2') $\quad \alpha x, \alpha y \in \mathcal{L}_f, \alpha xy \notin \mathcal{L}_f$ then $\alpha x\beta \in \mathcal{L}_f$ implies $\alpha y\beta \in \mathcal{L}_f$

(T1) is obvious from the definitions.

To verify property (T2') denote by A the underlying set of α and let $\beta = y_1\dots y_k$. We make use of the following characteristic property of polymatroid rank functions:

For every $X \subseteq Y$ and $z \notin Y$

$$f(Y \cup z) - f(Y) \le f(X \cup z) - f(X).$$

From this and the assumption that $\alpha xy \notin \mathcal{L}_f$ we deduce that

$$\begin{aligned} f(A\cup x) &= f(A \cup y) = f(A \cup \{x,y\})\\ \text{and } f(A \cup x \cup \{b_1,\dots,b_i\}) &= f(A\cup\{x,y\}\cup\{b_1,\dots,b_i\})\\ &= f(A\cup y\cup\{b_1,\dots,b_i\}).\end{aligned}$$

Thus $\alpha x\beta \in \mathcal{L}_f$ implies $\alpha y\beta \in \mathcal{L}_f$.

Undirected branching greedoids are special polymatroid greedoids. To see this, let $G = (E, V)$ be an undirected graph with root $r \in V$. For $X \subseteq E$ let $f(X)$ be the number of vertices in $V - r$ which are covered by an edge in X. It is easy to see that f is a polymatroid rank function and that $\mathcal{L}_f$ is just the branching greedoid with respect to G.

Another equivalent way to describe polymatroid greedoids is the following. Let $(E', \mathcal{M})$ be a matroid with rank function $r : 2^{E'} \to \mathbb{Z}$ and let $A_1, \ldots, A_n$ be a collection of subsets of E'. Define $E := \{1, \ldots, n\}$ and

$$\mathcal{L} := \{i_1 \ldots i_k : \text{ for } 1 \le \nu \le k \ i_\nu \in E \text{ and } r(A_{i_1} \cup \ldots \cup A_{i_\nu}) = \nu\}.$$

Then $(E, \mathcal{L})$ is a polymatroid greedoid and, moreover, every polymatroid greedoid arises from this construction. The equivalence of both descriptions follows from the fact that for every polymatroid (E, f) there is a matroid (E', r) and a collection $(A_e : e \in E)$ of subsets of E' such that

$$f(X) = r(\bigcup_{e \in X} A_e) \text{ for } X \subseteq E.$$

It should be mentioned that also poset greedoids are special polymatroid greedoids since for a poset $(E, \le)$

$$f(X) := |\{e \in E : e \le x \text{ for some } x \in X\}|$$

is a polymatroid rank function on E and $f(\{x_1, \ldots, x_i\}) = i$ for $1 \le i \le k$ if and only if $\{x_1, \ldots, x_i\}$ is an order ideal in $(E, \le)$.

2.23 Hypergraph branching

Let V be a finite set and $E = \{e_i : i \in I\}$ a family of nonempty subsets of V. Suppose $h : E \to V$ has the property that $h(e) \in e$ for every $e \in E$. Then

$$\mathcal{L} := \{e_1 \ldots e_k : \text{ for } i = 1, \ldots, k \ \ e_i \in E \text{ and } e_i \backslash (e_1 \cup \ldots \cup e_{i-1}) = \{h(e_i)\}\}$$

defines a greedoid, which we call the *hypergraph branching greedoid* with respect to (V, E). Hypergraph branching greedoids generalize directed branching greedoids. Let $D = (V, E)$ be a directed graph with root $r \in V$ and denote by $h(e)$ the head of arc $e \in E$. Assume that $h(e) \ne r$ for all $e \in E$. If we replace every arc e incident to the root by the single element set $\{h(e)\}$ and if we regard the other arcs as 2–element subsets of $V - r$, we obtain a hypergraph $(V - r, E)$. The directed branchings in D corresponds to the feasible set of the hypergraph branching greedoid with respect to $(V - r, E)$.

2.24 Polymatroid branching

Let (E_0, f) be a polymatroid, $E \subseteq E_0$, and $t : E \to E_0$ a mapping with the property that $f(\{e, t(e)\}) = f(\{e\})$ for every $e \in E$. Define

$$\begin{aligned}\mathcal{L} :=& \{e_1 \ldots e_r : \text{ for all } 1 \le i \le k \ \ f(\{e_1, \ldots, e_i\}) = i \text{ and} \\ & f(\{e_1, \ldots, e_{i-1}, t(e_i)\}) = i - 1\}.\end{aligned}$$

$(E, \mathcal{L})$ is a local poset greedoid and the restriction to a feasible set is identical with the restriction of the corresponding polymatroid greedoid (cf. KORTE, LOVÁSZ [1983b]). We call $(E, \mathcal{L})$ the *polymatroid branching greedoid* with respect to (E_0, f) and $t : E \to E_0$.

Let us mention special cases of this construction. First, let (E, f') be a polymatroid and let $E_0 := E \cup \{t\}, t \notin E$. For $X \subseteq E_0$ define

$$f(X) := \begin{cases} f'(X) & \text{if } t \notin X \\ f'(X \backslash t) & \text{if } t \in X \end{cases}$$

and for $e \in E$ let $t(e) := t$. It is easy to verify that (E, f) is a polymatroid. We claim that the polymatroid branching with respect to (E_0, f) and $t : E \to E_0$ is just the polymatroid greedoid with respect to (E, f'). First, suppose that

$$f'(\{x_1, \ldots, x_i\}) = i \text{ for } 1 \leq i \leq k.$$
$$\text{Then } f(\{x_1, \ldots, x_i\}) = i \text{ and}$$
$$f(\{x_1, \ldots, x_{i-1}, t(x_i)\}) = i \text{ for } 1 \leq i \leq k.$$

Thus every feasible word in the polymatroid greedoid w.r.t. (E, f') is feasible in the polymatroid branching greedoid w.r.t. (E_0, f) and $t : E \to E_0$. The reverse inclusion is trivial.

Let (V, E) and $h : E \to V$ be as in Example 2.23. Define $t(e) := e \backslash \{h(e)\}$ for $e \in E$, and $E_0 := E \cup \{t(e) : e \in E\}$. Then $f : 2^{E_0} \to \mathbb{Z}$, $f(X) := |\bigcup_{e \in X} e|$, is a polymatroid rank function and the polymatroid branching greedoid with respect to (E_0, f) and $t : E \to E_0$ is just the hypergraph branching greedoid described in Example 2.23. We conclude that directed and undirected branching greedoids are also hypergraph branching greedoids.

A further interesting instance of hypergraph branching greedoids is the following: Given a linear space L and let E be a finite collection of linear subspaces of L. For every subspace $e \in E$ choose a subspace $t(e)$ of e. Let $E_0 := E \cup \{t(e) : e \in E\}$, then

$$f(A) := dim(lim(\bigcup_{e \in A} e)), \; A \subseteq E_0,$$

is a polymatroid rank function on E_0. If $\mathcal{L}$ denotes the polymatroid branching greedoid with respect to (E_0, f) and $t : E \to E_0$ then we have $e_1 \ldots e_k \in \mathcal{L}$ if and only if for $i = 1, \ldots, k$

$$e_i \cap lim(e_1 \cup \ldots \cup e_{i-1}) = t(e_i).$$

2.25 Clique elimination greedoid

This example bears some resemblence with Example 2.3, where we pruned a chordal graph by picking successively simplical vertices. SEYMOUR [1985] suggested the following modification of chordal graph shelling: Given a chordal graph $G = (V, E)$, choose a simplical vertex $v_1 \in V$ and eliminate v_1 together with the unique maximal clique $V_1 \subseteq V$ containing v_1. Denote by $G : v_1$ the subgraph of G induced by $V \backslash V_1$. Now choose a simplical vertex v_2 in $G : v_1$ and eliminate v_2 together with the corresponding maximal clique. Continuing in this way we obtain a sequence $v_1 \ldots v_k$ of vertices.

The language $\mathcal{L}$ consisting of all these elimination sequences forms an interval greedoid with V as the ground set. We call $(V, \mathcal{L})$ the *clique elimination greedoid* with respect to G. Note, however, that $(V, \mathcal{L})$ will in general contain dummy elements. Since every chordal graph has a simplical vertex, the clique elimination process will eventually terminate with the empty graph. However, different from the situation in Example 2.3, this can also be true even if the graph is not chordal.

Let us mention one special property (the *partition property*) of clique elimination greedoids: There exists a (up to dummy elements) unique partition $V = V_1 \cup \ldots \cup V_k$, $V_i \cap V_j = \emptyset$ for $i \neq j$, of the ground set such that every basic word in $(V, \mathcal{L})$ is a transversal of $V_1, \ldots, V_k$. This partition can be determined by a greedy approach:

If $v_1 \ldots v_k$ is a basic word and if V_i is the unique maximal clique of $G : v_1 : v_2 \ldots : v_{i-1}$ containing V_i then $V_1, \ldots, V_k$ is such a partition.

2.26 Hypergraph elimination greedoids

This is a straightforward generalization of the preceding example. Let $(K, \mathcal{H})$ be a hypergraph. If $A \in \mathcal{H}$, $x \in A$ we say that A is maximal in $\mathcal{H}$ with respect to x if for all subsets $A' \subseteq A$ and $A' \in \mathcal{H}$ we have $x \in A'$. For $A \subseteq K$ let $\mathcal{H} \backslash A := \{F \backslash A : A \in \mathcal{H}\}$. Now suppose that $E \subseteq K$ then we define

$$\begin{aligned} \mathcal{L} := \mathcal{L}(\mathcal{H}, E) := \{ & x_1 \ldots x_k : x_i \in E \text{ and there are subsets } A_1, \ldots, A_k \subseteq K \\ & \text{with the property that } A_i \text{ is maximal in } \mathcal{H} \backslash (A_1 \cup \ldots \cup A_{i-1}) \\ & \text{with respect to } x_i (1 \leq i \leq k)\}. \end{aligned}$$

We call $(E, \mathcal{L})$ the *hypergraph elimination greedoid* with respect to $(K, \mathcal{H})$ and $E \subseteq K$. It is easy to show that $(E, \mathcal{L})$ is an interval greedoid with the partition property. However, also the converse is true: If $(E, \mathcal{L})$ is an interval greedoid with the partition property then there exists a hypergraph $(K, \mathcal{H})$ and $E \subseteq K$ such that $\mathcal{L} = \mathcal{L}(\mathcal{H}, E)$ (cf. GOECKE [1986b]). Obviously every clique simplical elimination greedoid is a special hypergraph elimination greedoid. Also the directed branching greedoid has the partition property and therefore is a hypergraph elimination greedoid. In this case the instars of the non–root vertices of the directed graph provide the partition of the ground set.

We now turn to examples of non interval greedoids $(E, \mathcal{L})$ with the transposition property

(T2) if $\alpha x, \alpha y \in \mathcal{L}$ and $\alpha x y \notin \mathcal{L}$ then $\alpha x \beta \in \mathcal{L}, y \notin \tilde{\beta}$ implies $\alpha y \beta \in \mathcal{L}$ and $\alpha x \beta y \gamma \in \mathcal{L}$ implies $\alpha y \beta x \gamma \in \mathcal{L}$.

Note that interval greedoids have the transposition property.

2.27 Perfect elimination greedoids

An edge (s, t) of a bipartite graph $G = (S, T, E)$ is called *bisimplical* if the set of neighbors of s and t induce a complete bipartite subgraph of G. If (s, t) is an edge in G then we write $G - (s, t)$

for the subgraph of G induced by $(S\backslash s) \cup (T\backslash t)$. We call $(s_1,t_1),\ldots,(s_k,t_k)$ an *elimination sequence* if (s_i,t_i) is a bisimplical edge of $G\backslash\{(s_1,t_1),\ldots,(s_{i-1},t_{i-1})\}$ for $1 \leq i \leq k$. Now we define a hereditary language $\mathcal{L}$ on the ground set S:

$$\mathcal{L} := \{s_1 \ldots s_k : \text{ there exist } t_1,\ldots,t_k \in T \text{ such that } (s_1,t_1),\ldots,(s_k,t_k) \text{ is an elimination sequence}\}.$$

The proof that $(S,\mathcal{L})$ is a transposition greedoid uses the following simple observation:

If (s,t_1) and (s,t_2) are bisimplical edges of G, then the map $\varphi : S \cup T \to S \cup T$ which maps t_1 onto t_2 and t_2 onto t_1 and leaves all the other elements unchanged is an automorphism of G. A detailed proof can be found in (KORTE, LOVÁSZ [1986b]). Perfect elimination greedoids are discussed in BRYLAWSKI, DIETER [1986] and in FAIGLE, GOECKE, SCHRADER [1986]. The latter paper contains a slight generalization of perfect elimination greedoids.

2.28 Series parallel reduction greedoids

Let G be a loopless graph. An edge of G is called *reducible* if it is either a pending edge, or is parallel with another edge or is in series with another edge (i.e. they have a common endpoint of degree 2). If e is reducible in G, we define $G \div e$ as the graph obtained from G by deleting e if e has a parallel edge and by contracting e if e is a pending edge or is in series with another edge.

A sequence $e_1 \ldots e_k$ of edges is called a *series-parallel reduction sequence* if e_i is reducible in $G \div e_1 \div \ldots \div e_{i-1}$ for $i = 1,\ldots,k$. Let $E = E(G)$ and let $\mathcal{L}$ denote the set of all series-parallel reduction sequences. Then $(E,\mathcal{L})$ is a greedoid, as it will be shown below. It does not necessarily have the interval property, as shown by the graph in Fig. 2.1, where there $acb \in \mathcal{L}$ and $b \in \mathcal{L}$ but $ab \notin \mathcal{L}$. If G is a tree, then the resulting greedoid is an edge-shelling of the tree.

One could slightly modify the definition of a reducible edge, by replacing "pending edge" by "coloop" and "two edges in series" by "two edges forming a cut". Then the definition of series-parallel reduction greedoids could be extended to matroids.

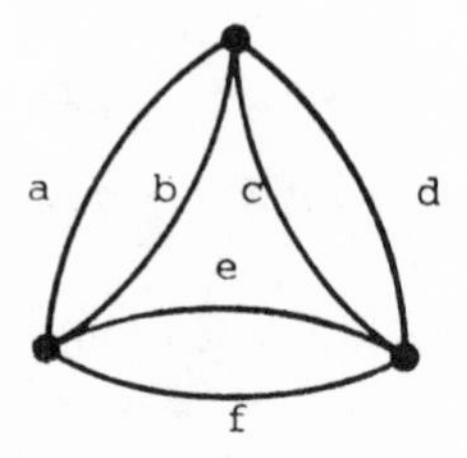

Figure 2.1

Claim 2.3: *$(E, \mathcal{L})$ is a transposition greedoid (cf. Example 2.22 and Example 2.26).*

Proof. First we verify (T1). Without loss of generality we may assume that $|\alpha| = 0$. Suppose that $xy \in \mathcal{L}$ and $y \in \mathcal{L}$. We are going to show that $yx \in \mathcal{L}$ and $G \div x \div y \simeq G \div y \div x$. This will imply that if $xy\beta \in \mathcal{L}$ for some word β then also $yx\beta \in \mathcal{L}$.

Unfortunately there are a number of cases to consider. Call two edges of G adjoint to each other if they are either in series or parallel.

Case 1. Suppose that both x and y have adjoints u and v and x, y, u, v are all distinct. Then v is in series (parallel) to y in $G \div x$ if it was in series (parallel) to y in G. Similar holds for x and u. Hence x is reducible in $G \div y$ and $G \div x \div y \simeq G \div y \div x$.

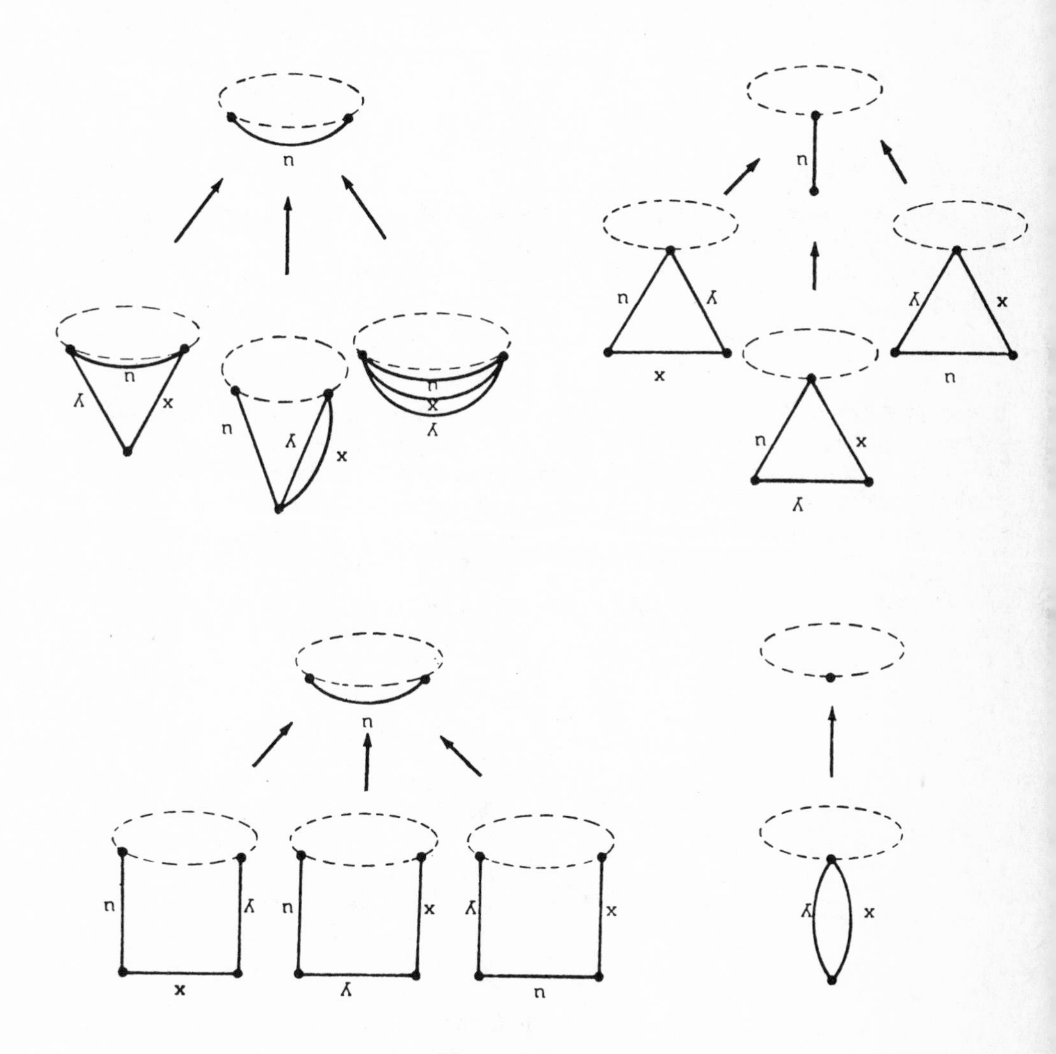

Figure 2.2

Case 2. Suppose that x and y have adjoints u, v but x, y, u, v are not all distinct. This results in 10 subcases as shown in Figure 2.2 . The lower rows show the graphs $G \div x \div y = G \div y \div x$.

Case 3. If one of x and y is a pending edge then the argument is trivial, since its contraction does not influence the reducibility of any other edge.

Thus, we have verified (T1).

To show (T2) we may again assume that $|\alpha| = 0$. Then from the hypothesis that $x \in \alpha$, $y \in \mathcal{L}$ but $xy \notin \mathcal{L}$, an easy argument shows that x and y have to be adjoint. Hence $G \div x \simeq G \div y$, and in fact there is an isomorphism which assigns x to y and any other edge to itself. Thus (T2) is satisfied.

□

2.29 Retract greedoids

Let G be a digraph, and $x, y \in V(G)$. We say that x is *retractable* to G if for any edge $xz \in E(G)$, $z \neq x$, we also have $yz \in E(G)$ and for any edge $zx \in E(G)$, $z \neq x$, we have $zy \in E(G)$, and if $xx \in E(G)$, then $yy \in E(G)$. In other words, the mapping $\rho_{xy} : V(G) \longrightarrow V(G) - x$ defined by

$$\rho_{xy}(z) = \begin{cases} y & \text{if } z = x, \\ z & \text{if } z \neq x \end{cases}$$

is edge-preserving. We say that x is *retractable* in G if it is retractable to some point $y \neq x$. A sequence $x_1 \dots x_k$ of points is called a retract sequence if x_i is retractable in $G - x_1 - \dots - x_{i-1}$ for $i = 1, \dots, k$. In KORTE, LOVÁSZ [1985a] we showed that retract sequences form a greedoid on the ground set $E = V(G)$. Again, this proof was rather complicated. Within the framework of transposition greedoids we shall obtain a shorter argument for this. This construction generalizes the retracts of partial orders, for which results were obtained by DUFFUS and RIVAL [1979].

The retract greedoid of the graph in Fig. 2.3 (which in fact corresponds to a poset) does not have the interval property.

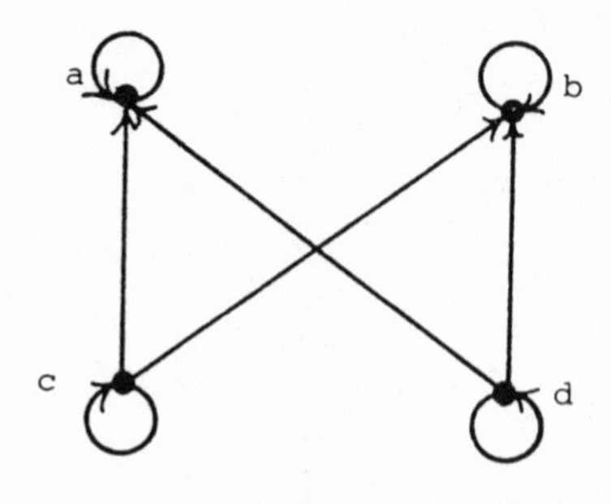

Figure 2.3

2.30 Monoid retract greedoids

The construction in Example 2.29 can be generalized to transformation monoids on a finite set E. A transformation monoid $\mathcal{S}$ is a collection of mappings $\varphi : E \longrightarrow E$ such that $id_E \in \mathcal{S}$ and $\varphi_1, \varphi_2 \in \mathcal{S}$ implies that the composition $\varphi_1 \circ \varphi_2 \in \mathcal{S}$. A subset $X \subseteq E$ is called a retract of E if there is an idempotent element $\varphi \in \mathcal{S}$ such that $X = \varphi(E)$. A sequence $x_1 \ldots x_k$ of distinct elements of E is called a *retract sequence* if $E - x_1 - \ldots - x_i$ is a retract of E for $i = 1, \ldots, k$.

Now let $\mathcal{L}$ be the set of all retract sequences in E. We are going to prove that $(E, \mathcal{L})$ is a transposition greedoid. Note that this greedoid can even easier be defined in the unordered version: its feasible sets are the members of the accessible kernel of the set-system

$$\{E - \varphi(E); \varphi \in \mathcal{S}, \varphi^2 = \varphi\}.$$

We also note that the class of greedoids in Example 2.29 is a proper subclass of the class of *monoid retract greedoids*. We use the unordered version of $(E, \mathcal{L})$ and define

$$\mathcal{F} := \{\{x_1, \ldots, x_k\} : x_1 \ldots x_k \in \mathcal{L}\}.$$

Claim 2.4: *$(E, \mathcal{F})$ satisfies*

(TP) *For every $A \subseteq E$, $x, y \in E \backslash A$ and $B \subseteq E \backslash (A \cup \{x, y\})$ such that $A \cup x, A \cup y, A \cup B \cup x \in \mathcal{F}$, and $A \cup \{x, y\} \notin \mathcal{F}$ it follows that $A \cup B \cup y \in \mathcal{F}$.*

Proof. Let $A \cup x, A \cup y \in \mathcal{F}, A \cup x \cup y \notin \mathcal{F}$ and $A \cup x \cup B \in \mathcal{F}$. Let $A \cup x = E - \eta(E), A \cup y = E - \mu(E)$ and $A \cup x \cup B = E = \varphi(E)$, where η, μ and φ are idempotent.

We claim that $\eta(x) = y$ and $\mu(y) = x$. Suppose not, and let, say, $\mu(y) \neq x$. Then consider $\rho = (\eta \circ \mu)^k$ for a k such that ρ is idempotent (it is an elementary result in semigroup theory that such a k exists). Then $\rho(E) = E - A - x - y$. In fact, both η and μ keep $E - A - x - y$ fixed, and hence so does ρ. Hence $\rho(E) \supseteq E - A - x - y$. Furthermore, $\rho(E) = \mu(\eta((\eta \circ \mu)^{k-1}(E))) \subseteq \mu(E) = E - A - y$. So it remains to show that $x \notin \rho(E)$.

Suppose $x \in \rho(E)$, then $x = \mu(z)$ for some $z \in \eta((\eta \circ \mu)^{k-1}(E)) \subseteq \eta(E) = E - A - x$. But if $z \in E - A - x - y$ then $\mu(z) = z \neq x$, and if $z = y$ then $\mu(z) \neq x$ by hypothesis. This contradiction proves that $\rho(E) = E - A - x - y$. However, this implies that $A \cup x \cup y \in \mathcal{F}$, contrary to the assumption. So we have proved that $\eta(x) = y$ and $\mu(y) = x$. Next we show that $\mu \circ \rho \circ \eta$ is idempotent and $(\mu \circ \rho \circ \eta)(E) = E - A - y - B$. Hence (TP) will follow immediately.

All three mappings η, ρ and μ are identical on $E - A - x - y - B$ hence so is $\eta \circ \rho \circ \mu$. Moreover

$$(\mu \circ \rho \circ \eta)(x) = (\mu \circ \rho)(y) = \mu(y) = x.$$

So $\mu \circ \rho \circ \eta$ is identical on $E - A - y - B$. Furthermore

$$(\mu \circ \rho \circ \eta)(E) \subseteq (\mu \circ \rho)(E) = \mu(E - A - x - B) = E - A - y - B.$$

Hence $\mu \circ \rho \circ \eta$ is indeed idempotent and hence $A \cup y \cup B \in \mathcal{F}$. So (TP) holds.

□

2.31 Dismantling greedoids

Let G be a digraph and $x \in V(G)$. We say that x is *dismantable* if there exists a $y \neq x$ such that x is retractable to y and x and y are adjacent in G. A sequence $x_1 \ldots x_k$ is called a *dismantling sequence* if x_i is dismantable in $G - x_1 - \ldots - x_{i-1}$. The collection of dismantling sequences forms a greedoid on $V(G)$.

This again generalizes a construction of DUFFUS and RIVAL [1976] for partially ordered sets. If P is a partial order and $x \in P$, then x is called *dismantable* if it has a unique upper cover or a unique lower cover. For the relation graph of the partial order, this coincides with the above definition. Dismantling greedoids are not necessarily interval greedoids even in the special case of dismantling partially ordered sets, as shown by the partial order in Fig. 2.4. There xyz and z are dismantling sequence but xz is not.

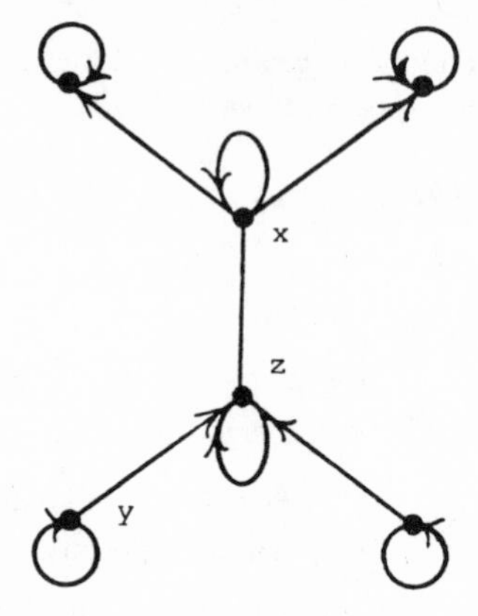

Figure 2.4

Claim 2.5: *The hereditary language $(E, \mathcal{L})$ formed by dismantling sequences of a digraph G satisfies (T1) and (T2). Hence $(E, \mathcal{L})$ is a transposition greedoid.*

Proof. Let $\alpha xy\beta \in \mathcal{L}$ and $\alpha y \in \mathcal{L}$. Without loss of generality, $|\alpha| = |\beta| = 0$. Since $xy \in \mathcal{L}$ there is a point u adjacent to x such that x is retractable to u in G and a point $v \neq x$ adjacent to y such that y is retractable to u in $G - x$. Furthermore, there exists a point v' such that y is adjacent to v' and y is retractable to v' in G. If $u \neq y$ then x is still retractable to u in $G - y$ and hence $yx \in \mathcal{L}$. So suppose that $y = u$. So x and y are adjacent, say $xy \in E(G)$. If $v' \neq x$ then since y is retractable to v' , we have $xv' \in E(G)$, and clearly z is retractable to v' in $G - y$. So $yx \in \mathcal{L}$. Finally, if $v' = x$ then v is adjacent to x since v is adjacent to y and xy is retractable to x. Since clearly x is retractable to v in $G - y$, we obtain again that $yx \in \mathcal{L}$. This proves (T1).

To prove (T2), we may assume again that $|\alpha| = 0$. The same argument as above implies that if $x \in \mathcal{L}, y \in \mathcal{L}$ but $xy \notin \mathcal{L}$ then x and y are adjacent and mutually retractable to each other. Hence $G - x \simeq G - y$ and an isomorphism is given by

$$\varphi(u) = \begin{cases} x & \text{if } u = y, \\ u & \text{if } u \neq y. \end{cases}$$

This proves (T2).

□

It appears from the above discussion that dismantling greedoids are strong-ly related to retract greedoids. The construction of dismantling greedoids can be generalized to transformation monoids as follows.

Let S be a transformation monoid on a set E and let R be a binary relation on E invariant under S. Call a sequence $x_1, \ldots, x_k$ of points a *dismantling sequence*, if there exist idempotent $\rho_1, \ldots, \rho_k$ of S such that maps $E - \rho_i(E) = \{x_1, \ldots, x_i\}$ for $i = 1, \ldots, k-1$ and moreover $(x_i, \rho(x_i)) \in R$ for each i.

Then the dismantling sequences form a transposition greedoid. This can be proved in the same way as above.

We close this section with a list of more or less sporadic examples of greedoids which do not have the transposition property, and thus in particular, they provide further examples of non-interval greedoids.

2.32 Bipartite matching greedoid

Let $G = (S, T, E)$ be a bipartite graph and let $t_1, \ldots, t_m$ an ordering of the vertices in T. On the ground set S we define an accessible set system

$$\mathcal{F} = \{F \subseteq S : \text{ the subgraph of } G \text{ induced by } F \cup \{t_1, \ldots, t_{|F|}\} \text{ has a perfect matching}\}$$

We call $(E, \mathcal{F})$ the *bipartite matching greedoid* with respect to G and the order $t_1, \ldots, t_m$. We will show below that $(E, \mathcal{F})$ is indeed a greedoid. The bases of the bipartite matching greedoid constitute the bases of a transversal matroid. $(E, \mathcal{F})$ does in general not have the transposition property as can be seen from Figure 2.5:

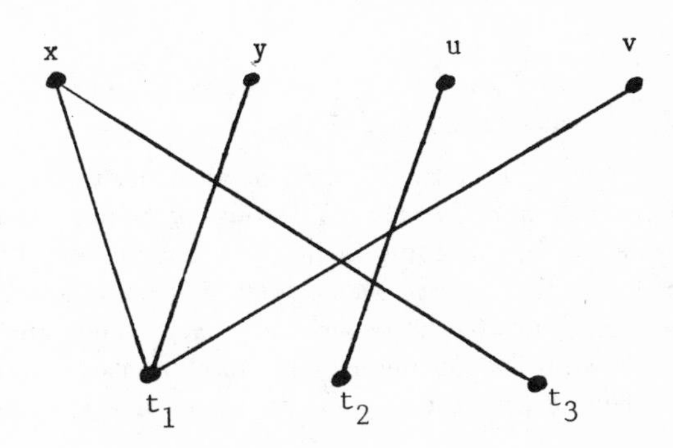

Figure 2.5

$\{x\}, \{y\} \in \mathcal{F}, \{x, y\} \notin \mathcal{F}$ and $\{x, u, v\} \in \mathcal{F}$, but $\{y, u, v\} \notin \mathcal{F}$.

2.33 Linking greedoids

As the bipartite matching greedoid relates to transversal matroid, the *linking greedoid* relates to linking matroids, introduced by WELSH [1976]:

Let $G = (V, E)$ be a directed graph and let $v_1, \ldots, v_m$ be an ordering of a subset B of nodes of G. Call $F \subseteq V$ feasible, if there exist $k := |F|$ node–disjoint path connecting each node in F with a different node in B.

Since we allow trivial path, in particular the sets $\emptyset, \{v_1\}, \{v_1, v_2\}, \ldots, \{v_1, \ldots, v_m\} = B$ are feasible. Later we will see that $\mathcal{F}$, the system of feasible set, is a greedoid on V.

2.34 Gaussian elimination greedoids

Let $A = (a(i,j) : 1 \leq i \leq m,\ 1 \leq j \leq n)$ be a matrix of a full row rank over some field K. Denote by E the set of column indices. To perform the Gaussian elimination process, one selects a column index $j_1 \in E$ such that $a(1, j_1) \neq 0$ and then replaces A by the $(m-1, n-1)$ matrix

$$\overline{A} = \{\overline{a}(i,j) : 2 \leq i \leq m,\ 1 \leq j \leq n,\ j \neq j_1\}$$

where

$$\overline{a}(i,j) := a(i,j) - a(i,j_1) \cdot a(1,j) \cdot [a(1,j_1)]^{-1}.$$

Proceding this way we can construct a sequence $j_1, \ldots, j_m (k \leq m)$ of column indices. Such a sequence we call an *elimination sequence.* Now let

$$\mathcal{F}_A := \{\{j_1, \ldots, j_k\} \subseteq E : j_1, \ldots, j_k \text{ is an elimination sequence}\}.$$

$(E, \mathcal{F}_A)$ is a greedoid (see below). We call it the *Gaussian elimination greedoid* with respect to the matrix A.

2.35 Graphic Gaussian elimination greedoid

Let $G = (V, E)$ be a graph and let $v_1, \ldots, v_m$ be an ordering of the vertex set V. Call $F \subseteq E$ feasible if F contains no cycle and if every connected component of F covers as many nodes of the set $\{v_1, \ldots, v_{|F|}\}$ as edges in this component. The feasible sets form a greedoid. In fact, this is a special instance of Example 2.34. To see this, let A' be the node-edge incidence matrix of G and order the rows in accordance to the order $v_1, \ldots, v_m$. Delete the last row of A', denote the resulting matrix by A. It is now a routine exercise to show that the feasible sets of the graphic Gaussian elimination greedoid are precisely the members of $\mathcal{F}_A$, as defined in 2.34.

2.36 Strong matroid systems

This example provides a common generalization of the last four examples. Let E be a finite set and let $\mathcal{M}_0, \ldots, \mathcal{M}_m$ be a sequence of matroids which are in strong map relation. Furthermore assume that rank $(\mathcal{M}_i) = i$ for $i = 0, \ldots, m$. Denote by $\mathcal{B}_i$ the collection of bases of $\mathcal{M}_i$.

Finally define

$$\mathcal{F} := \bigcup_{i=0}^{m} \mathcal{B}_i.$$

Claim 2.6: *$(E, \mathcal{F})$ is a greedoid.*

Proof. Trivially $(E, \mathcal{F})$ has the augmentation property.

To see that $(E, \mathcal{F})$ is an accessible set system let $F \in \mathcal{F}$ and we show that $F - x \in \mathcal{F}$ for some $x \in F$.

F is the basis of $\mathcal{M}_k$ where $k = |F|$. If we denote by r_k and r_{k-1} the rank functions of $\mathcal{M}_k$ and $\mathcal{M}_{k-1}$, we have that

$$0 \leq r_{k-1}(E) - r_{k-1}(F) \leq r_k(E) - r_k(F) = 0$$

by the strong map relation of $\mathcal{M}_k$ and $\mathcal{M}_{k-1}$. This implies that $r_{k-1}(F) = r_{k-1}(E) = k - 1$ and that there exists $x \in F$ such that $F - x$ is a basis of $\mathcal{M}_{k-1}$.

We can now easily prove that Examples 2.32 - 2.35 are indeed greedoids. In Example 2.32 observe that

$$\mathcal{M}_i = \{F \subseteq S;\ \text{the subgraph induced by } F \cup \{t_1, \ldots, t_i\} \text{ contains a matching covering } F\}$$

defines a matroid on S and that $\mathcal{M}_0, \ldots, \mathcal{M}_m$ are in strong map relation. Similarily, in Example 2.33

$$\mathcal{M}_i := \{F \subseteq S : \text{ there exist } |F| \text{ node disjoint paths connecting different node in } F \text{ to different nodes in } \{v_1, \ldots, v_i\}\}$$

is a matroid (cf. WELSH [1976]). Moreover $\mathcal{M}_0, \ldots, \mathcal{M}_m$ are in strong map relation. Finally, for Example 2.34, denote by $A_i (0 \leq i \leq m)$ the (i, n) submatrix of A consisting of the first i rows of A. If $\mathcal{M}_i$ denotes the linear matroid induced by A_i then $\mathcal{M}_0, \ldots, \mathcal{M}_m$ is a chain of matroid in strong map relation.

□

Strong matroid systems can be characterized via the optimality of a modified greedy algorithm. The interested reader is refered to BRYLAWSKI, DIETER [1986] and GOECKE [1986a], [1986b] for a detailed discussion of strong matroid systems.

2.37 Twisted matroids

Let T be a feasible set of a matroid $(E, \mathcal{M})$ and define

$$\mathcal{M} \triangle T := \{F \triangle T : F \in \mathcal{M}\},$$

where $F \triangle T := (F \backslash T) \cup (T \backslash F)$. We call $\mathcal{M} \triangle T$ a *twisted matroid.*

Claim 2.7: *$(E, \mathcal{M} \triangle T)$ is a greedoid.*

Proof. To show this, first observe that $\mathcal{M} \triangle T$ is hereditary. Indeed, if $F \triangle T \in \mathcal{M} \triangle T$ then $(F \triangle T) - x \in \mathcal{M} \triangle T$ for all $x \in F \cap T$ and, if $F \cap T = \emptyset, (F \triangle T) - x \in \mathcal{M} \triangle F$ for all $x \in F$.

Now suppose that $A_1, A_2 \in \mathcal{M} \triangle T$ and that $|A_1| < |A_2|$. Let $A_1 = F_1 \triangle T$ and $A_2 = F_2 \triangle T$ where $F_1, F_2 \in \mathcal{M}$. If $F_1 \cap T \not\subseteq F_2 \cap T$ then there exists $y \in (F_1 \cap T) \backslash (F_2 \cap T) \subseteq A_2 \backslash A_1$ and $A_1 \cup y = (F_1 - y) \triangle T \in \mathcal{M} \triangle T$. Now assume that $F_1 \cap T \subseteq F_2 \cap T$. If the inclusion is strict then from the identity

$$k := |F_2| - |F_1| = 2 \cdot (|F_2 \cap T| - |F_1 \cap T|)$$

we deduce that there are k elements $z \in F_2 \backslash F_1$ such that $F_1 \cup z \in \mathcal{M}$. From this follows that there exists a $z \in F_2 \backslash F_1$ such that $F_1 \cup z \in \mathcal{M}$ and $z \notin T$ implying $A_1 \cup z = (F_1 \cup z) \triangle T \in \mathcal{M} \triangle T$.

Finally we consider the case where $F_1 \cap T = F_2 \cap T$. In this case augment F_1 from F_2 by some $z \in F_2 \backslash F_1$ and $A_1 \cup z = (F_1 \cup z) \triangle T \in \mathcal{M} \triangle T$ follows directly. This shows that $(E, \mathcal{M} \triangle T)$ is a greedoid. In fact, we have already shown more than this. $(E, \mathcal{M} \triangle T)$ has the *exchange property* :

(EP) *for every $A_1, A_2 \in \mathcal{M} \triangle T$ such that $|A_1| = |A_2|$ and for every $x \in A_1 \backslash A_2$ there exists a $y \in A_2 \backslash A_1$ such that either*

$$A_1 \cup y \in \mathcal{M} \triangle T$$
$$\text{or} \quad A_1 - x \cup y \in \mathcal{M} \triangle T.$$

This follows from the arguments above since there we only used the fact that $|A_2| > |A_1|$ in the very last step where we assumed that $F_1 \cap T = F_2 \cap T$. However, this together with $|A_1| = |A_2|$ already implies that $F_1 = F_2$, and $F_1 \backslash F_2 = A_1 \backslash A_2$. It is well known from matroid theory that for every $x \in F_1 \backslash F_2$ there exists $y \in F_2 \backslash F_1 = A_2 \backslash A_1$ such that $F_1 - x \cup y \in \mathcal{M}$, and hence $A_1 - x \cup y \in \mathcal{M} \triangle T$.

□

Greedoids with the exchange property are discussed in more detail in BRYLAWSKI, DIETER [1986]. The reader may convince himself that strong matroid systems, antimatroids and undirected branching greedoids satisfy the exchange property.

2.38 Dual twisted matroids

Let $(E, \mathcal{M} \triangle T)$ be a twisted matroid and let

$$\overline{\mathcal{M} \triangle T} := \mathcal{M} \triangle T \cup \{S \subseteq E : S \text{ contains a basis of } \mathcal{M} \triangle T\}$$

be the canonical completion of $\mathcal{M} \triangle T$, then $\overline{(\mathcal{M} \triangle T)}^D := \{E - F : F \in \overline{\mathcal{M} \triangle T}\}$ is again a greedoid, which we call a *dual twisted matroid.* The proof of the greedoid axioms follows directly from the following property of twisted matroids. If $X, Y \in \mathcal{M} \triangle T$ and $|X| < |Y|$ then there exists $y \in Y \backslash X$ such that $Y \backslash y \in \mathcal{M} \triangle T$.

We get a less clumsy description of dual twisted matroids in the special case where the ground set E can be partitioned into disjoint bases, say $E = T \cup (E \backslash T)$ then

$$\mathcal{M} \triangle T = \overline{\mathcal{M} \triangle T} \text{ and } \overline{(\mathcal{M} \triangle T)}^D = \{T \triangle S : S \text{ is spanning in } \mathcal{M}\}$$

2.39 Paving greedoids

Paving greedoids are defined similarily to paving matroids. We take $\mathcal{P} \subseteq 2^E$ such that $\emptyset \notin \mathcal{P}$ and for all $A, B \in \mathcal{P}$ with $|A| = |B|$ and $|A \backslash B| = 1$ if follows that $A \cap B \in \mathcal{P}$. Then $(E, 2^E \backslash \mathcal{P})$ is a paving greedoid. We first verify that $2^E \backslash \mathcal{P}$ is hereditary. So let $F \in 2^E \backslash \mathcal{P}, F \neq \emptyset$. If $F \backslash x \in \mathcal{P}$ for all $x \in F$ we would deduce that $(F \backslash x) \cap (F \backslash y) = F \backslash \{x, y\} \in \mathcal{P}$ for all $x, y \in F$. Eventually we would deduce that $\emptyset \in \mathcal{P}$, contradicting our assumptions.

We now show the augmentation property: Let $X, Y \in 2^E \backslash \mathcal{P}, |X| < |Y|$. Assume that $X \cup y \in \mathcal{P}$ for all $y \in Y \backslash X$. If $|Y \backslash X| = 1$, then $X \cup y = Y$ for some $y \in Y$, contradiction. If $|Y \backslash X| > 1$ then there exist $y_1, y_2 \in Y \backslash X$, $y_1 \neq y_2$. Let $A = X \cup y_1$, $B = X \cup y_2$, then by our assumption $A, B \in \mathcal{P}$ and $A \cap B = X \in \mathcal{P}$, contradiction.

Note, that in a paving greedoid at most one single letter word can be infeasible. If $\mathcal{P}$ is a family of subsets of E closed under intersection, such that $\cap \mathcal{P} \neq \emptyset$, then we have a greedoid of a particular simple structure.

This general construction, in fact, allows to give a good upper bound for the number of greedoids on a given ground set E, (see BRYLAWSKI, DIETER [1986] for details).

2.40 Ear decomposition greedoids

The technique of ear decomposition is a very useful tool in graph theory. It turns out that this procedure leads to the definition of a greedoid.

Let G be a 2-connected graph, $e_0 \in E(G)$ and $E := E(G) - e_0$. $\mathcal{F}$ is the collection of all subsets X of E such that $X \cup \{e_0\}$ is a connected subgraph and every block of $X \cup \{e_0\}$, except possibly the one containing e_0, is a single edge. It can easily be seen that $(E, \mathcal{F})$ is a full greedoid.

The basic words $x_1 \ldots x_m (|E| = m)$ define an ear decomposition of G as follows: Let $H(i)$ be the one block of the subgraph induced by $\{x_1, \ldots, x_i, e_0\}$ which contains e_0, and let $1 \leq i_1 < i_2 < \ldots < i_r = m$ be those indices in $\{1, \ldots, m\}$ for which the corresponding blocks are different, i.e. $H(i_\mu) \neq H(i_{\mu-1}) (\mu = 1, \ldots, r)$. Then $H(i_\mu) := H(i_{\mu-1}) \cup P_\mu$, where P_μ is a path which has both its endpoints in common with $H(i_{\mu-1})$. Thus $G = e_0 \cup P_1 \cup \ldots P_r$ is an ear decomposition of G, and conversely, every ear decomposition arises from an appropriate basic word.

The ear decomposition greedoid (E, F) does not have the transposition property, as can be seen from Figure 2.6.

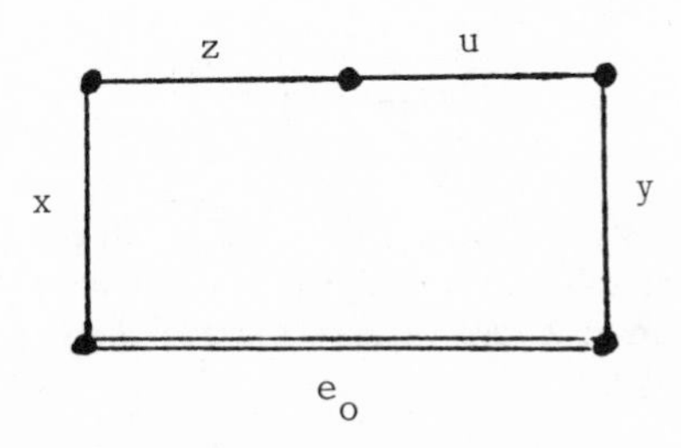

Figure 2.6

Here we have that x, y are feasible, xy is not feasible, however $xzuy$ is feasible, while $yzux$ is not.

2.41 Ear decomposition in strong digraphs

The above mentioned procedure is also applicable in strong digraphs: Let G be a strongly connected digraph and $r \in V(G)$. $\mathcal{F}$ consists of all arc sets $X \subseteq E(G)$ such that $X := X_0 \cup \{X_t : t \in V(X_0)\}$ where X_0 is a strongly connected digraph containing r and X_t is an arborescence rooted at t with $V(X_t) \cup V(X_s) = \emptyset$ for all $t, s \neq 0$, and $V(X_t) \cap V(X_0) = \{t\}$. The greedoid (E, F) does not have the transposition either.

2.42 Blossom greedoids (cf. EDMONDS [1965])

This example will be interesting for those familiar with Edmond's matching algorithm. Let G be a graph and let M be a maximum matching in G. Define a greedoid on $E = E(G) - M$ as follows: Roughly speaking, a feasible word is a sequence of edges as they enter the alternating forest to get shrunk during a matching algorithm. More precisely, let $X \subseteq E$ be feasible iff X, together with those lines of M which meet them, forms a subgraph G' with the following structure. The points of G' can be labelled "outer" and " inner" so that

(1) The outer points of G' span vertex-disjoint factor-critical (hypomatchable) subgraphs $H_1, \ldots, H_m$.

(2) The inner points of G' have degree 2 and are incident with two cut-edges of G', which connect it to different hypomatchable subgraphs H_i.

(3) M contains a near-perfect matching of each H_i and exactly one of the two edges incident with an inner point.

To prove that this is a greedoid, let us observe the following:

(i) Feasibility of a subset is an intrinsic property, i.e. if G_1 is any subgraph of G containing M and $X \subseteq E(G_1) - M$, then X is feasible in the blossom greedoid of G_1 iff it is feasible in the blossom greedoid of G.

(ii) All maximal feasible sets have the same cardinality. In fact, if the matching algorithm terminates with a subgraph G' then the set A of inner points of G' is independent of the run of the algorithm. Hence for any maximal feasible set X, we have

$$|X| = \sum_{i=1}^{m} (|E(H_i)| - \left(\frac{|V(H_i)| - 1}{2}\right)) + |A|.$$

Since using (i) we can easily see that for any $Y \subseteq E$, all maximal feasible sets contained in Y have the same cardinality, it follows that the *blossom greedoid* is indeed a greedoid. Let us denote by $\mathcal{F}(G, M)$ the blossom greedoid with respect to the graph $G = (V(G), E(G))$ and the maximum matching $M \subseteq E(G)$.

Denote by V_0 the set of vertices of G not covered by M. For $F \subseteq E$ let $\overline{F} := F \cup \{e \in M : e$ incident to an edge in $F\}$, and $V(\overline{F}) := \{v \in V : v$ is covered by an edge in $\overline{F}\}$ and $V(\emptyset) := V_0$. We now give a recursive description of $\mathcal{F}(G, M)$:

(i) $\emptyset \in \mathcal{F}(G, M)$;

(ii) suppose $F \in \mathcal{F}(G, M)$ and $x \in E \backslash F$ then

$$F \cup x \in \mathcal{F}(G, M) \text{ if and only if}$$

either x has exactly one endpoint in $V(\overline{F})$ and there exists an M–alternating path in $\overline{F} \cup \{x\}$ ending in x

or both endpoints of x are in $V(\overline{F})$ and x is a chord or an element of an M–alternating odd circuit in $\overline{F} \cup \{x\}$.

Let us mention some interesting properties of blossom greedoids which might also give some insight into Edmond's matching algorithm.

(1) For every basis $B \subseteq E$ of $(E, \mathcal{F})$ we have the following property: Let

$$V(\overline{B}) = I(\overline{B}) \cup \mathcal{O}(\overline{B})$$

be the partition into inner and outer nodes, then

$$\mathcal{O}(\overline{B}) = \mathcal{O} := \{v \in V : \text{ there exists a maximum matching not covering } v\},$$

$$\text{and } I(\overline{B}) = I := \{v \in V \backslash \mathcal{O} : v \text{ is adjacent in } G \text{ to some vertex } w \in \mathcal{O}\}.$$

This observation is a consequence of the Gallai–Edmonds structure theorem (cf. LOVÁSZ, PLUMMER [1986]).

(2) If $F, F \cup x \in \mathcal{F}$, then $\mathcal{O}(\overline{F}) \subseteq \mathcal{O}(\overline{F \cup x})$. This together with Remark (1) implies that for $\hat{E} := \{e \in E : e \in F \text{ for some } F \in \mathcal{F}\}$ we have

$$\hat{E} \subseteq E(\mathcal{O} \cup T) \backslash (E(I) \cup M)$$

(3) Let $V' := \mathcal{O} \cup I$, $E' := E(\mathcal{O} \cup I) \backslash E(I)$, and $M' := M \cap E'$, then we observe that

– M' is a maximum matching in the graph $G' = (V', E')$,

– I is covered by M',

– $\mathcal{F} = \mathcal{F}(G', M')$

(4) If M is a perfect matching for G if and only if $\mathcal{F} = \{\emptyset\}$.

(5) Let $G = (I, \mathcal{O}; E(G))$ be any bipartite graph with positive surplus from I (i.e. for all $I' \subseteq I$ the number of neighbors of I' is greater or equal $|I'| + 1$), and let $M \subseteq E(G)$ be a maximum matching. Denote by V_0 the vertex set not covered by M; observe that $V_0 \subseteq \mathcal{O}$.

From G we derive a directed graph: Direct every edge in G downward from $\mathcal{O}$ to I.

Now identify all vertices in $\mathcal{O}$ not covered by M to a single vertex V_0 and then contract every edge in M. Denote this digraph by $D = (V, E)$, where $V := (\mathcal{O} \backslash V_0) \cup \{v_0\}$ and $E := E(G) \backslash M$.

We claim that the blossom greedoid $\mathcal{F}(G,M)$ and the directed branching greedoid with respect to D and root v_0 coincide.

Since G is bipartite there do not exist odd circuits. This implies that the feasible sets in $\mathcal{F}(G,M)$ correspond to alternating forests rooted at vertices in V_0 (see (ii) in the recursive definition).

Each such forest corresponds to a directed branching in D. And, vice versa, it is easy to see that each branching can be viewed as an alternating forest.

(6) Let $G = (V(G), E(G))$ be any graph and define $\mathcal{O}$ and I as in Remark (1). Denote by $\mathcal{O}'$ the connected components of the graph induced by $\mathcal{O}$. Consider the following bipartite graph

$$G' = (\mathcal{O}', I, E')$$

where $E' = \{e \in E(G) : \; e \text{ connects a vertex in } I \text{ to a connected component of } \mathcal{O}\}$.

From the Gallai–Edmonds structure theorem we know that G' is a bipartite graph with positive surplus and that for every maximum matching M in $G, M \cap E'$ is a maximum matching for G'.

This together with Remark (5) implies that the trace of $\mathcal{F}(G,M)$ to the set $E' \backslash M$ is a directed branching greedoid and that the bases of $\mathcal{F}(G,M)$ are the bases of a suitable directed branching greedoid.

3. Algorithmic properties of greedoids

Greedoids have been introduced to provide a structural framework for the greedy algorithm. Many combinatorial algorithms, like search algorithms in graphs, poset elimination, branchings, Gaussian elimination, matching algorithms, etc. give rise to special classes of greedoids. Many of them have been explained in the previous chapter. In this chapter we would like to discuss some further algorithmic properties of greedoids. Some of those properties have been already mentioned in the literature (cf. KORTE, LOVÁSZ [1984a], [1984c]), some others are discussed here for the first time. Like in the case of matroids we characterize special combinatorial structure by algorithmic properties.

3.1 The greedy algorithm

It is impossible to give a general and precise definition of "the" *greedy algorithm,* since the greedy approach depends on the specific structure on which it is applied. Thus we shall first confine ourselves to a purely intuitive description of greediness. After this we will restrict to special optimization problems with the effect that we can give a precise description of the greedy algorithm, which will be general enough for our purposes during the rest of this chapter.

The greedy-rule is a fundamental principle in discrete as well as continuous optimization. Nearly every optimization procedure works as follows:

Starting with an initial feasible solution one successively improves this solution until either we have an optimal solution or we are at least close enough to an optimal solution.

The core of such a procedure is the rule by which one determines the next solution from the preceeding one. The *greedy rule* says that we should take that solution which is the best among all "neighboring" solutions.

The steepest unit ascent procedure in continuous optimization is a typical greedy procedure, namely, preceede in that direction where the function value decreases most. Another famous greedy procedure is known from number theory. Let $0 < \frac{a}{b} < 2$ be a rational number. We want to find different integers $n_1, \ldots, n_k$ such that $\frac{a}{b} = \frac{1}{n_1} + \ldots + \frac{1}{n_k}$. The " natural" greedy rule in this case is the following:

Suppose we already have determined $n_1, \ldots, n_i$, then choose $n_{i+1} > n_i$ such that $\frac{a}{b} - (\frac{1}{n_1} + \ldots + \frac{1}{n_i} + \frac{1}{n_{i-1}})$ is as small as possible (cf. ERDÖS and GRAHAM [1980]). It can be shown that this procedure terminates after a finite number of steps.

We now restrict ourselves to a special class of optimization problems. In fact both examples mentioned above do not fit into this framework. However many combinatorial optimization problems can be described as follows:

Let $(E, \mathcal{L})$ be a hereditary language over a finite ground set E and

$$w : \mathcal{L} \rightarrow \mathbb{R}$$

an objective function. We consider the optimization problem

(O) $max\{x(\alpha) : \alpha \text{ is a basic word in } \mathcal{L}\}$

The *greedy algorithm* for (O) goes as follows:

(1) Take $\alpha = \emptyset$ (the empty word) as an initial feasible solution;

(2) choose $x \in E$ such that $\alpha x \in \mathcal{L}$ and $w(\alpha x) \geq w(\alpha y)$ for all $y \in E$ with $\alpha y \in \mathcal{L}$, if no such x exists - STOP;

(3) replace α by αx and GOTO (2).

For completeness and for later reference we also formulate the greedy algorithm for accessible sets systems:

Let $(E, \mathcal{F})$ be an accessible set system and

$$w : \mathcal{F} \rightarrow \mathbb{R}$$

an objective function and consider the optimization problem

(O) $max\{w(F) : F \text{ basis of } (E, \mathcal{F})\}$.

The greedy algorithm goes as follows:

(1) Take $F = \emptyset$ as the initial feasible solution;

(2) choose $x \in E \backslash F$ such that $F \cup x \in \mathcal{F}$ and $w(F \cup x) \geq w(F \cup y)$ for all $y \in E \backslash F$ with $F \cup y \in \mathcal{F}$, if no such x exists - STOP;

(3) replace F by $F \cup x$ and GOTO (2).

It is clear that this algorithm always computes a basic feasible solution which we call a *greedy solution.* One should notice that the choice in step (2) is not unique. Thus we cannot talk about "the" greedy solution if we want to be precise. However, we often call a solution generated by the above mentioned precedure the greedy solution.

We call a word $\alpha \in \mathcal{L}$ *greedy with respect to* $w : \mathcal{L} \to \mathbb{R}$ if α can be expanded to a greedy solution, i.e. $\alpha\beta$ is a greedy solution for some β. Similarily we say $F \in \mathcal{F}$ is *greedy with respect to* $w : \mathcal{L} \to \mathbb{R}$ if F is the underlying set of a greedy word $\alpha \in \mathcal{L}$. In this case α is called a *greedy ordering* of F.

In general the greedy algorithm will not find an optimal solution even if $(E, \mathcal{L})$ is a greedoid. Simply imagine an objective function which puts weight 1 on exactly one basic word, and weight 0 elsewhere. Then the greedy algorithm has no chance to find an optimal solution, except by chance or if $\mathcal{L}$ contains only one basic word.

Hence to guarantee the optimality of the greedy algorithm one needs some connection between the structure of $\mathcal{L}$ and the objective function $w : \mathcal{L} \to \mathbb{R}$.

Let $(E, \mathcal{L})$ be a greedoid and $w : \mathcal{L} \to \mathbb{R}$ an objective function. We say that w has *property* (W) *with respect to* $\mathcal{L}$ if for all α, $\alpha x \in \mathcal{L}$ such that

$$w(\alpha x) \geq w(\alpha y) \text{ for all } \alpha y \in \mathcal{L}$$

the following is true:

(W1) if $\alpha\beta x\gamma \in \mathcal{L}$ and $\alpha\beta y\gamma \in \mathcal{L}$ then $w(\alpha\beta x\gamma) \geq w(\alpha\beta y\gamma)$;

(W2) if $\alpha x\beta y\gamma \in \mathcal{L}$ and $\alpha y\beta x\gamma \in \mathcal{L}$ then $w(\alpha x\beta y\gamma) \geq w(\alpha y\beta x\gamma)$.

Note that in the above definition the strings α, β, γ can also be empty.

The intuitive meaning of this definition is clear: if the letter x is the best possible choice at a certain stage then by (W 1) it should be the best choice at a later stage, and by (W 2) it also should be better to choose first x and later y than vice versa.

If $w : \mathcal{L} \to \mathbb{R}$ is *stable*, i.e. $w(\alpha) = w(\beta)$ if $\tilde{\alpha} = \tilde{\beta}$, then the above definition simplifies to:

A stable objective function $w : \mathcal{L} \to \mathbb{R}$ is stable then w satisfies condition (W) with respect to $\mathcal{L}$ if and only if for all $A, A \cup x \in \mathcal{F}(\mathcal{L})$ such that $w(A \cup x) \geq w(A \cup y)$ for all $y \in E\backslash A$, $A \cup y \in \mathcal{F}(\mathcal{L})$ the following holds:

(W3) if $B \subseteq A$ and $x \in E\backslash B, B \cup x \in \mathcal{F}(\mathcal{L})$

then $w(B \cup x) \geq w(B \cup y)$ for all $y \in E\backslash B$ with $B \cup y \in F(\mathcal{L})$.

Theorem 3.1: *Let $(E, \mathcal{L})$ be a greedoid. Then the greedy algorithm gives an optimal solution to the optimization problem (O) for all objective functions $w : \mathcal{L} \to \mathbb{R}$ satisfying condition (W).*

This theorem needs a somewhat involved proof which we prepare by the following lemmas, which by themselves give additional insight into greedoidal structures.

Lemma 3.2: (Recombination lemma) *Let $(E, \mathcal{L})$ be a greedoid, $\alpha\gamma, \beta \in \mathcal{L}$ and $\tilde{\alpha} = \tilde{\beta}$, then $\beta\gamma \in \mathcal{L}$.*

Proof. Let $\gamma = y_1 \dots y_k$, then applying the augmentation property we get from $\alpha y_1, \beta \in \mathcal{L}$ that there exists $z \in \tilde{\alpha} \cup \{y_1\}$ such that $\beta z \in \mathcal{L}$. Since $\tilde{\alpha} = \tilde{\beta}$ we have $z = y_1$, thus $\beta y_1 \in \mathcal{L}$. By repeating this procedure we get $\beta\gamma \in \mathcal{L}$.

□

Lemma 3.3: (Generalized augmentation) *Given a greedoid $(E, \mathcal{L})$ and $\alpha\beta \in \mathcal{L}$ and $\alpha z \in \mathcal{L}$ with $z \in E$. Then β can be written in the form*

$$\beta = y_1\beta_1 y_2\beta_2 \dots y_r\beta_r,$$

where β_i are strings of appropriate size (including the possibility of being the empty string), so that if we define β' to be

$$\beta' := z\beta_1 y_1\beta_2 \dots y_{r-1}\beta_r$$

then

$$\alpha\beta' \in \mathcal{L}.$$

Proof. By induction on k, the length of β. If $k = 1$, then $\beta' = z$ and nothing remains to be proved. For $k-1$ the induction hypothesis says that there are s positions $y_1 \dots y_s$ in β such that by setting

$$\beta' := z\beta_1 y_1\beta_2 y_2 \dots y_{s-1}\beta_s$$

we get $\alpha\beta' \in \mathcal{L}$. Applying the simple augmentation property on $\alpha\beta'$ and $\alpha\beta$ we get an $x \in \alpha\beta$ such that $\alpha\beta' x \in \mathcal{L}$. Since $x \notin \alpha\beta'$, we have either $x = y_s$ or $x = y_k$ (last letter of β). In the first case we set $r = t+1$ and $\beta_r = \emptyset$; in the second case we have $r = s$ and $\beta_r = \beta_s y_k$.

□

Lemma 3.4: *Given a greedoid $(E, \mathcal{L})$, and let $\alpha\beta \in \mathcal{L}$ and $\alpha\beta' \in \mathcal{L}$, where for some fixed positions $y_1 \dots y_r$ in β and $x \in E$ we have*

$$\beta = y_1\beta_1 y_2 \dots y_r\beta_r$$

and

$$\beta' = z\beta_1 y_1\beta_2 y_2 \dots y_{r-1}\beta_r.$$

Let

$$\beta' := y_1\beta_1 y_2\beta_2 \dots \beta_{t-1} z\beta_t y_t\beta_{t+1} \dots y_{t-1}\beta_t$$

for $t = 1, \dots, r$, then $\alpha\beta' \in \mathcal{L}$.

Proof. By assumption we have

$$\alpha y_1\beta_1 y_2 \dots y_{t-1}\beta_{t-1} \in \mathcal{L}$$

and

$$\alpha z\beta_1 y_1 \dots \beta_{t-1} y_{t-1} \in \mathcal{L}.$$

Augmenting these formulas with property (G), we get

$$\alpha y_1\beta_1 y_2 \dots y_{t-1}\beta_{t-1} z \in \mathcal{L}.$$

Applying the Recombination Lemma with $\gamma := \beta_t y_t\beta_{t-1} \dots y_{r-1}\beta_r$, and the two last formulas as the respective beginning sections whose underlying sets are equal, we obtain $\alpha\beta' \in \mathcal{L}$.

□

The above lemmas imply the following useful property. If $\gamma = y_1 \dots y_n$ is a basic word of a greedoid $(E, \mathcal{L})$ and $z \in \mathcal{L}$ then there exists a $1 \leq j \leq n$ such that $y_1 \dots y_{j-1} z y_{j+1} \dots y_n \in \mathcal{L}$.

We can now give the

Proof of Theorem 3.1. Let α be a greedoid solution and β an optimal solution of problem (O). Write $\alpha = \gamma\alpha'$ and $\beta = \gamma\beta'$, i.e. γ is the common beginning section, and choose β such that $|\gamma|$ is maximum. We claim that $\alpha = \beta = \gamma$. Let $x \in E$ be the first letter of α', i.e. $\alpha' = x\alpha''$. From the proceeding lemmas follows that we may write

$$\beta' = y_1\beta_1y_2\beta_2\dots y_r\beta_r$$

such that for every $t = 1, \dots, r$,

$$\gamma y_1\beta_1\dots\beta_{t-1}x\beta_ty_t\dots y_{r-1}\beta_r \in \mathcal{L}$$

The definition of the greedy algorithm implies that for every $y \in E$ with $\gamma y \in \mathcal{L}$ we have

$$w(\gamma x) \geq w(\gamma y).$$

Therefore, by property (W2) we have:

$$w(\gamma y_1\beta_1\dots x\beta_{r-1}y_{r-1}\beta_r) \geq w(\gamma y_1\beta_1\dots y_{r-1}\beta_{r-1}x\beta_r).$$

If we apply (W2) successively $(r-1)$–times, we get

$$w(\gamma x\beta_1y_1\beta_2\dots\beta_{r-1}y_{r-1}\beta_r) \geq w(\gamma y_1\beta_1\dots y_{r-1}\beta_{r-1}x\beta_r).$$

By property (W1), we can replace x in the argument of the righthand side of the above formula by y_r and get

$$\begin{aligned} &w(\gamma x\beta_1y_1\dots\beta_{r-1}y_{r-1}\beta_r) \geq \\ &w(\gamma y_1\beta_1y_2\dots\beta_{r-1}y_r\beta_r) = w(\beta). \end{aligned}$$

Since $w(\beta)$ is optimal, we must have equality. But the argument on the lefthand side is a feasible word which has a larger beginning section in common with α than β. This contradiction proves $\alpha = \beta = \gamma$ and hence the theorem.

□

Let $(E, \mathcal{L})$ be a hereditary language and $f : E \times \mathbb{N} \to \mathbb{R}$ be such that

$$f(x, k+1) \geq f(x, k) \text{ for every } (x, k) \in E \times \mathbb{N}.$$

We then define for $x_1 \dots x_k$

$$w(x_1\dots x_k) := min(f(x_1, 1), \dots, f(x_k, k)).$$

If $w : \mathcal{L} \to \mathbb{R}$ can be represented in this form we call w a *generalized bottleneck function.*

It is easily seen that a generalized bottleneck function has property (W) with respect to $\mathcal{L}$. Thus by Theorem 3.1, the greedy algorithm optimizes every generalized bottleneck function over a greedoid $(E, \mathcal{L})$. This already characterizes greedoids as the following theorem shows.

Theorem 3.5: *Let $(E, \mathcal{L})$ be a hereditary language whose basic words have the same cardinality, and assume that the greedy algorithm gives an optimal solution of the optimization problem (O) for all generalized bottleneck functions. The $(E, \mathcal{L})$ is a greedoid.*

Proof. Let $x_1 \dots x_k$, $y_1 \dots y_l \in \mathcal{L}$ with $k > l$. We have shown that $y_1 \dots y_l$ can be augmented from $x_1 \dots x_k$.

Define $Z := \{x_1, \dots, x_k\} \cup \{y_1, \dots, y_l\}$ and

$$f(z,i) = \begin{cases} 1/2 & \text{for } i \le k,\ z \notin Z \\ 1 & \text{for } i \le k,\ z \in Z \text{ and for } i > k,\ z \notin Z \\ 2 & \text{for } i > k,\ z \in Z. \end{cases}$$

Then for a basic word γ which starts with $x_1 \dots x_k$ we have $w(\gamma) = 1$. On the other hand the greedy solution $\alpha = z_1 \dots z_l z_{l+1} \dots z_k$ may start with $y_1 \dots y_l$. Since this solution is assumed to be optimal we have

$$1 = w(\gamma) \le w(\alpha) = min(f(z_1, 1)), \dots f(z_l, l)) \le f(z_{l+1}, l+1).$$

Under the assumption that the bases are of equal size the greedy solution has at least length k, i.e. the position $l+1$ will not be vacant. By the construction of $f(z,i)$ we have $z_{l+1} \in Z$ which implies $z_{l+1} \in \{x_1, \dots, x_k\}$.

□

3.2 Matroids and antimatroids

We have already mentioned that the W-condition simplifies if the objective function is stable. An important subclass of stable objective functions are linear objective functions:

We call $w : 2^E \to \mathbb{R}$ a *linear objective* function if $w(A) = \sum_{e \in A} w(\{e\})$ for $A \subseteq E$. Since matroids are also greedoids, we get the famous Edmonds–Rado theorem as a corollary from Theorem 3.1 and 3.5:

Corollary 3.6: *An independence system is a matroid if and only if the greedy algorithm gets an optimal solution to the optimization problem (O) for every linear objective function.*

The following example shows that a particular optimization problem can have more than one greedy strategies.

Example (Scheduling under precedence constraints, cf. LAWLER [1973]) Suppose E is a finite set of jobs. We want to schedule these jobs such that

(i) the schedule satisfies certain precedence constraints.

(ii) the completion cost of the schedule is minimal.

The completion cost of a schedule is the maximum of the completion cost of the individual jobs. The completion cost of job $x \in E$ is given by a non-decreasing function $f(x,i)$ depending on x and i, where $i \in \mathbb{N}$ is the position of job x in the schedule.

Let $\mathcal{L}$ be the collection of feasible (partial or complete) schedules we have to solve the problem

$$min\{max(f(x_1, 1), \dots, f(x_n, n)) : x_1 \dots x_n \text{ is a maximal word in } \mathcal{L}\}.$$

Of course, $(E, \mathcal{L})$ is a poset greedoid. A greedy strategy to find a min-cost feasible schedule could be the following:

Find a job x_1 with no precedence constraints which minimizes $\{f(x,1) : x \in \mathcal{L}\}$. Then choose $x_2 \in E$ such that $x_1\, x_2 \in \mathcal{L}$ and $f(x_2, 2)$ minimizes $\{f(x,2) : x_1 x \in \mathcal{L}\}$, and so on.

This strategy, however, does not work in general. If, for instance, $E = \{x,y\}$, $\mathcal{L} = \{\emptyset, x, y, xy, yx\}, f(x,1) = f(x,2) = 1, f(y,1) = 2, f(y,2) = 10$, then xy would be the greedy solution (costs $= 11$) while yx would be the optimal schedule with costs $= 3$.

The idea of LAWLER was the following: Instead of determining the first job then the second and so on, he chooses the best last job, then the best last but one job, etc.

First we notice that

$$\mathcal{L}^* := \{x_n \ldots x_k : \text{ there exists a basic word } x_1 x_2 \ldots x_{k-1} x_k \ldots x_n \in \mathcal{L}\}$$

is also a poset greedoid, and secondly, that for

$$f^*(x,k) := -f(x, n+1-k)$$
$$w^*(y_1, \ldots, y_k) := min(f^*(y_1, 1), \ldots, f^*(y_k, k))$$

is a generalized bottleneck function. By Theorem 3.1 we know that the greedy algorithm is optimal for the problem

$$max\{w^*(y_1 \ldots y_n) : y_1 \ldots y_n \text{ is a basic word in } \mathcal{L}^*\}.$$

This problem is equivalent to the scheduling problem stated above.

The last example was in fact the motivation to define antimatroids within the framework of greedoids as *alternative precedence structures.* However, they were introduced earlier by EDELMAN [1980] and JAMISON [1982] as combinatorial abstraction of convexity. Since antimatroids are special greedoids we know by Theorem 3.5 that for every generalized bottleneck function the greedy algorithm finds an optimal solution. Moreover, since antimatroids have more structural properties than general greedoids we would expect that the greedy algorithm is optimal for an ever larger class of objective function.

Let $(E, \mathcal{L})$ be a hereditary language and $f : E \times \mathbb{R} \to \mathbb{R}$ be such that

$$f(x,s) \leq f(x,t) \text{ for all } x \in E,\ s \leq t.$$

Assign to every $x \in E$ a processing time $t(x) \geq 0$ and define for $x_1 \ldots x_k$

$$w(x_1 \ldots x_k) := min(f(x_1, t(x_1)), \ldots, f(x_k, t(x_1) + \ldots + t(x_k))).$$

We call w a *time dependent bottleneck function.*

Example: Let $E = \{x,y,z\}$, $t(x) = t(y) = 1$, $t(z) = z$, $f(x,t) = 4$, $f(z,t) = 3$ for all $t \geq 0$;

$$f(y,t) = \begin{cases} 2 & \text{for } t \leq 2 \\ 3 & \text{for } t > 2. \end{cases}$$

Let $(E, \mathcal{L})$ be the bipartite matching greedoid with respect to the following bipartite graph

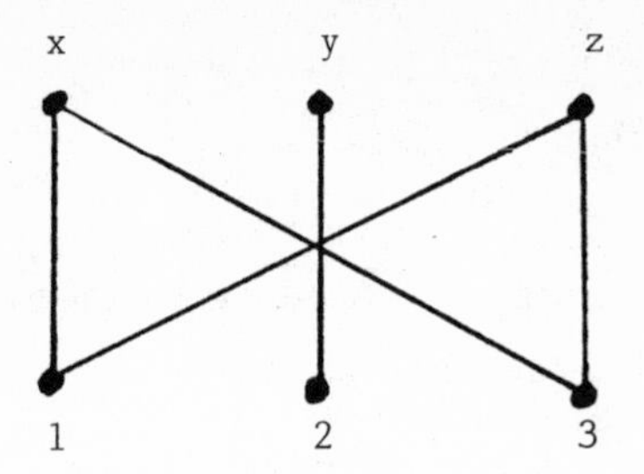

Figure 3.1

If we define $w : \mathcal{L} \to \mathbb{R}$ as above, we see that the greedy algorithm yields $xyz \in \mathcal{L}$ while $zyx \in \mathcal{L}$ is the optimal solution of (O). This shows that the greedy algorithm for time dependent bottleneck functions is not optimal for general greedoids. However, we have

Theorem 3.7: (FAIGLE [1985]) *Let $(E, \mathcal{L})$ be a hereditary language whose basic words have all length $|E|$. Then the following statements are equivalent.*

(1) $(E, \mathcal{L})$ is an antimatroid.

(2) For every time-dependent bottleneck function $w : \mathcal{L} \to \mathbb{R}$ the greedy algorithm finds an optimal solution to the optimization problem (O).

Proof.
(1) $\Rightarrow$ (2). Let $(E, \mathcal{L})$ be an antimatroid and $\alpha = x_1 \dots x_n$, $n = |E|$, be a greedy solution with respect to a time–dependent bottleneck function $w : \mathcal{L} \to \mathbb{R}$. We claim that

$$w(x_1 \dots x_n) \geq w(y_1 \dots y_k)$$

for all $y_1 \dots y_k \in \mathcal{L}$, $k \leq n$. Then, in particular, α is an optimal solution to (O).

Suppose there is a minimal k such that for a certain $\beta = y_1 \dots y_{k-1} y_k \in \mathcal{L}$ we have

$$\begin{aligned} w(x_1 \dots x_{k-1}) &\geq w(y_1 \dots y_{k-1}) \\ \text{and } w(x_1 \dots x_{k-1} x_k) &< w(y_1 \dots y_{k-1} y_k). \end{aligned}$$

Let $j \leq k$ be the smallest index such that $y_j \notin \{x_1, \dots, x_k\}$. Then $x_1 \dots x_{k-1} \in \mathcal{L}$ by the antimatroid augmentation property. Furthermore the following inequalities hold:

$$\begin{aligned} w(x_1 \dots x_{k-1} y_j) &= min(w(x_1 \dots x_{k-1}), f(y_j, t(x_1) + \dots + t(x_{k-1}) + t(y_j))) \\ &\geq min(w(y_1 \dots t_{k-1}), f(y_j, t(y_1) + \dots + t(y_j))) \\ &\geq w(y_1 \dots y_k) \\ &> w(x_1 \dots x_{k-1} x_k). \end{aligned}$$

This, however, contradicts the choice of x_k during the greedy procedure.

(2) $\Rightarrow$ (1). Let $\alpha = x_1 \dots x_k \in \mathcal{L}$, $\beta = y_1 \dots y_{l-1} y_l \in \mathcal{L}$ and assume that $\{y_1, \dots, y_{l-1}\} \subseteq \{x_1, \dots, x_k\}$ and $y_l \notin \{x_1, \dots, x_k\}$.

It suffices to prove that $\alpha y_l \in \mathcal{L}$. We define a time–dependent bottleneck function $w : \mathcal{L} \to \mathbb{R}$ such that

(i) α is greedy with respect to $w : \mathcal{L} \to \mathbb{R}$,

(ii) $w(\beta\gamma) = 1$ for all γ such that $\beta\gamma \in \mathcal{L}$,

(iii) $w(\alpha u) = 0$ for all $u \neq y_l$.

From this it follows that $\alpha y_l \in \mathcal{L}$ since otherwise the greedy algorithm would not be optimal.

Choose a function $t : E \to \mathbb{R}_+$ such that

$$t(x_1) + \ldots + t(x_n) + \epsilon = t(\beta) := t(y_1) + \ldots + t(y_l)$$

and $t(z) = \epsilon > 0$ for all $z \notin \{x_1, \ldots x_k, t_l\}$.
Now define $f(z,t) = 1$ for all $z \in \{x_1, \ldots, x_k, y_l\}$ and $t \in \mathbb{R}$
and for $z \notin \{x_1, \ldots, x_k, y_l\}$

$$f(z,t) := \begin{cases} 0 & \text{if } t \leq t(\beta) \\ 1 & \text{if } t > t(\beta). \end{cases}$$

Let $w : \mathcal{L} \to \mathbb{R}$ be the corresponding time–dependent bottleneck function. We show that (i)–(iii) hold for w.

(i): α is greedy with respect to $w : \mathcal{L} \to \mathbb{R}$ since $w(x_1) = \ldots = w(x_1 \ldots x_k) = 1$

(ii):
$$\begin{aligned} w(\beta\gamma) &= w(y_1 \ldots y_l z_1 \ldots z_m) \\ &= min(1, f(z_1, t(\beta) + t(z_1)), \ldots, f(z_{m_1}) + (\beta) + t(z_m)) \\ &= 1, \end{aligned}$$

(iii):
$$\begin{aligned} w(\alpha u) &= (1, f(u, t(x_1) + \ldots + t(x_k) + t(u)) \\ &= min(1, f(u, t(\beta))) = 0 \end{aligned}$$

for $u \neq y_l$.

□

We know that an independent system is a matroid if and only if the greedy algorithm is optimal for every linear objective function. We would like to characterize the optimality of the greedy algorithm for accessible set systems.

An accessible set system $(E, \mathcal{F})$ is called *greedy* if the greedy algorithm is optimal for every linear objective function.

Theorem 3.8: *Let $(E, \mathcal{F})$ be an accessible set system, then $(E, \mathcal{F})$ is greedy if and only if the strong exchange property holds, i.e.,*

if $A \in \mathcal{F}, B \supseteq A$ a basis of $\mathcal{F}$ and if $A \cup x \in \mathcal{F}$ for some $x \in E \backslash B$ then there exists $y \in B \backslash A$ such that $A \cup y \in \mathcal{F}$ and $B \cup x - y \in \mathcal{F}$.

Proof. Suppose first that $(E, \mathcal{F})$ is greedy and let $A \in \mathcal{F}, B \supseteq A$ a basis of $\mathcal{F}$ and $A \cup x \in \mathcal{F}$ for some $x \in E \backslash B$. Define a linear objective function $w : E \to \mathbb{R}$, as follows

$$w(e) := \begin{cases} 1 & \text{for } e \in A \cup x \\ t & \text{for } e \in B \backslash A \\ 0 & \text{else,} \end{cases}$$

with $0 < t < 1$.

Note that the set of greedy solutions is independent of t provided $0 < t < 1$. Since by assumption every greedy solution is optimal we follow that every greedy solution contains $A \cup x$ (for t sufficiently small). Thus, if B_g is any greedy solution, then for $t = 2/3$

$$w(B_g) = |A| + 1 + 2/3|B_g \cap (B \backslash A)|$$

and $w(B_g) \geq w(B) = |A| + 2/3|B \backslash A|$, hence $|B_g \cap (B \backslash A)| \geq |B \backslash A| - 1$. Since $A \cup x \subseteq B_g$ we get $B_g = B \cup x - y$ for some $y \in B \backslash A$.
So we know that

$$C := \{y \in B : B \cup x - y \in \mathcal{F}\} \neq \emptyset.$$

We claim that there is no basis $B' \in \mathcal{F}$ such that $C \cup x \subseteq B'$.

Again we define a special linear objective function

$$w(e) = \begin{cases} 1 & \text{for } e \in C \cup x \\ t & \text{for } e \in B \backslash C \\ 0 & \text{else.} \end{cases}$$

If for some basis $B' \in \mathcal{F}$ we have $C \cup x \subseteq B'$ and therefore $C \cup x \subseteq B_g$ for every greedy solution (t sufficiently small.) For t sufficiently large B would be an optimal solution, contradicting the fact that $C \cup x \not\subseteq B$. It remains to prove that $A \cup y \in \mathcal{F}$ for some $y \in C$. Let

$$w(e) = \begin{cases} 1 & \text{for } e \in A \cup C \\ 1/2 & \text{for } e = x \\ 0 & \text{else .} \end{cases}$$

If $A \cup y \notin \mathcal{F}$ for all $y \in C$ then there exists a greedy solution $B_g \supseteq A \cup x$. Since by the above $C \cup x \not\subseteq B_g$ and hence

$$w(B_g) \leq |A| + 1/2 + (|C \backslash A| - 1).$$

However, $w(B) = |A| + |C \backslash A| > w(B_g)$, contradicting the assumption that $(E, \mathcal{F})$ is greedy.

Suppose now that $(E, \mathcal{F})$ satisfies the strong exchange property and let $w : E \to \mathbb{R}$ be a linear objective function. Let B_g be a greedy solution with greedy ordering $a_1 \dots a_n$. Choose an optimal basic solution B such that $a_1, \dots, a_k \in B$ with k maximal.

Suppose $k < n$ and let $A = \{a_1, \dots, a_k\}$. By the strong exchange property there exists $y \in B \backslash A$ such that $A \cup y \in \mathcal{F}$ and $B \cup a_{k+1} - y \in \mathcal{F}$. Since $a_1 \dots a_k\ a_{k+1}$ is greedy with respect to w we know that $w(y) \leq w(a_{k+1})$ and $w(B \cup a_{k+1} - y) \geq w(B)$.

However, this implies that $B \cup a_{k+1} - y$ is again an optimal solution, contradicting the choice of B.

□

Example: Let $G = (V, E)$ be an undirected graph. Given a weight function $w : E \to \mathbb{R}$ we want to find a maximum weight branching $F \in \mathcal{F}$. The standard greedy algorithm for this problem uses the fact that collection of forests forms a matroid. Thus, adding edges of maximum weight until no edge can be added without generating a circuit yields an optimal solution. However, according to Theorem 3.8 the following strategy will also find an optimal solution:
Starting with the trivial subtree consisting only of a single vertex $r \in V$, we successively contract the maximum weight (non-loop) edge incident to the vertex until there is no non-loop left. The set of contracted edges is a maximum weight branching. To see this one has to verify that the branching greedoid (Example 2.19) has the strong exchange property.

3.3 Strong matroid systems

Let $(E, \mathcal{F})$ be an accessible set system and $w : E \to \mathbb{R}$ a linear objective function. So far we considered the problem

$$max\{w(F) : F \text{ basis in } \mathcal{F}\}$$

and characterized those structures for which the greedy algorithm solves this problem.

Now consider the problem

(O') $max\{w(F) : F \in \mathcal{F}\}$.

Since the standard greedy algorithm always determines a basis solution we have to modify the greedy algorithm for this problem. One possible modification could be to replace step (2) of the standard greedy algorithm by

(2') choose $x \in E \backslash F$, such that $F \cup x \in \mathcal{F}, w(F \cup x) \geq w(F)$ and $w(F \cup x) \geq w(F \cup y)$ for all $y \in E \backslash F$ with $F \cup y \in \mathcal{F}$ if no such x exists – STOP.

Thus this version of the greedy algorithm would stop if there is no further improvement in the objective function.

However, one can easily show that if $(E, \mathcal{F})$ is not an independence system this greedy algorithm will always fail for a particular linear objective function irrespective of what other "nice" properties $(E, \mathcal{F})$ has.

An adequate version of the greedy algorithm for accessible set systems is the following:

(1) Take $F^* = \emptyset$ as the initial feasible solution and set $F = \emptyset$;

(2) choose $x \in E \backslash F$ such that $F \cup x \in \mathcal{F}$ and $w(F \cup x) \geq w(F \cup y)$ for all $y \in E \backslash F$ with $F \cup y \in \mathcal{F}$; if no such x exists - STOP (F^* is the greedy solution);

(3) if $w(F \cup x) > w(F^*)$ then replace F^* by $F \cup x$;

(4) replace F by $F \cup x$ and GOTO (2).

Notice that in step (2) we may add x even if $w(F \cup x) < w(F)$. Furthermore, one should observe that this modified greedy algorithm is just the standard greedy algorithm if applied to an independence system.

We are going to characterize those structures for which the modified greedy algorithm computes an optimal solution for every linear objective function. For non negative linear objective functions we already have such a characterization, namely greedy set systems (cf. Theorem 3.8).

The concept of strong maps between matroids (cf. HIGGS [1968]) will provide us with adequate tools. We want to give an *algorithmic characterization* of strong map relations between matroids. The following theorem gives a very useful *algebraic characterization*.

Theorem 3.9: *Let $(E, \mathcal{M}_1)$ and $(E, \mathcal{M}_2)$ be two matroids with rank functions r_1 and r_2, then the following statements are equivalent:*

(i) *$\mathcal{M}_1$ is an elementary strong map of $\mathcal{M}_2$, i.e. $r_1(E) = r_2(E) - 1$ and every closed set in $\mathcal{M}_1$ is also closed in $\mathcal{M}_2$.*

(ii) *$r_1(E) = r_2(E) - 1$ and $r_1(Y) - r_1(X) \leq r_2(Y) - r_2(X)$ for all $X \subseteq Y \subseteq E$.*

(iii) *$r_1(E) = r_2(E) - 1, \mathcal{M}_1 \subseteq \mathcal{M}_2$, and if B is a basis of $\mathcal{M}_1$ and $B - e \cup \{x, y\}$ is a basis of $\mathcal{M}_2$, then $B \cup y$ is a basis of $\mathcal{M}_2$ or $B - e \cup y$ is a basis of $\mathcal{M}_1$.*

Proof. The equivalence of (i) and (ii) is well-known, a proof can be found in WELSH [1976].

(ii) $\Rightarrow$ (iii). By (ii) we have that

$$r_2(B \cup y) - r_2(B - e \cup y) \geq r_1(B \cup y) - r_1(B - e \cup y)$$

which implies that

$$r_2(B \cup y) + r_1(B - e \cup y) \geq r_2(B - e \cup y) + r_1(B \cup y) = 2\,|B|\,.$$

This means that $r_2(B \cup y) \geq |B| + 1$ or $r_1(B - e \cup y) \geq |B|$. Thus (iii) holds.

(iii) $\Rightarrow$ (i). We first show that every hyperplane H of $\mathcal{M}_1$ (i.e. H is a closed set of rank $r_1(E)-1$) is closed in $\mathcal{M}_2$.

Suppose not, then there exists a hyperplane H of $\mathcal{M}_1$ and $e \in E - H$ such that

$$r_2(H \cup e) = r_2(H).$$

Since H is a hyperplane of $\mathcal{M}_1$, there is a basis B of $\mathcal{M}_1$, such that $B \subseteq H \cup e$. Clearly $e \in B$ and $B - e \in \mathcal{M}_2$ since $\mathcal{M}_1 \subseteq \mathcal{M}_2$.

Since $r_2(H) = r_1(E) = r_2(E) - 1$ and $r_2(B - e) = r_1(E) - 1$ by the augmentation property there exists $y \in H \backslash (B - e) = H \backslash B$ such that $B - e \cup y \in \mathcal{M}_2$ and $x \in E - (H \cup e)$ such that $B - e \cup \{x, y\}$ is a basis of $\mathcal{M}_2$.

By assumption we know that $B \cup y$ is a basis of $\mathcal{M}_2$ or that $B - e \cup y$ is a basis of $\mathcal{M}_1$. Both possibilities, however, lead to a contradiction:

$B \cup y \subseteq H \cup e$ implies that $B \cup y$ is not a basis of $\mathcal{M}_2$. Similarily $B - e \cup y \subseteq H$ implies that $B - e \cup y$ is not a basis of $\mathcal{M}_1$. This proves that every hyperplane of $\mathcal{M}_1$ is closed in $\mathcal{M}_2$.

If now C is an arbitrary closed set of $\mathcal{M}_1$ then C can be represented as the intersection of hyperplanes of $\mathcal{M}_1$ (cf. WELSH [1976]). However these hyperplanes are closed in $\mathcal{M}_2$ and thus C is closed in $\mathcal{M}_2$ being the intersection of closed sets of $\mathcal{M}_2$.

□

Remark: Property (ii) of Theorem 3.9 can be illustrated nicely by a forbidden configuration. Let $(E, \mathcal{M}_1)$ and $(E, \mathcal{M}_2)$ be two matroids, $\mathcal{M}_1 \subseteq \mathcal{M}_2$, the rank of which are k and $k+1$ respectively. Then $\mathcal{M}_1$ is a strong map of $\mathcal{M}_2$ if and only if the configuration of Figure 3.2

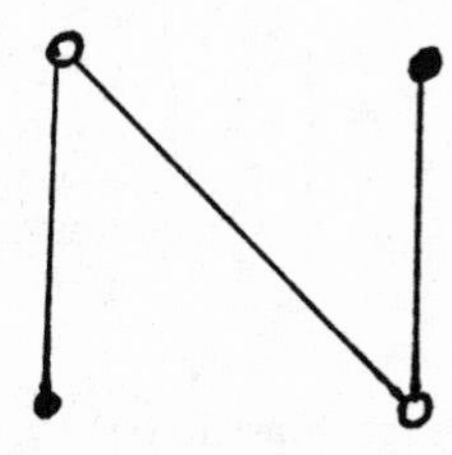

Figure 3.2

does not occur in the Hasse diagram of the Boolean algebra of the $k-$ and $(k+1)-$element subsets of E. In Figure 3.2 a full point represent bases of $\mathcal{M}_1$ and $\mathcal{M}_2$ and the light points correspond to all other k- or $(k+1)$- element subsets of E.

Theorem 3.10: *Let $(E, \mathcal{F})$ be an accessible set system, then the following statements are equivalent:*

(i) For every linear objective function $w : \mathcal{F} \to \mathbb{R}$ the modified greedy algorithm determines an optimal feasible solution for the problem

$$(O') \qquad max\{w(F) : F \in \mathcal{F}\}.$$

(ii) $\mathcal{F}$ is a strong matroid system, i.e. $\mathcal{F}$ can be represented as

$$\mathcal{F} = \bigcup_{i=0}^{m} B_i$$

where for $i = 0, \ldots, m$ B_i is the collection of bases of a matroid M_i of rank i and M_{i-1} is an elementary strong map of M_i.

(iii) $\mathcal{F}$ is a greedoid and the configuration of Figure 3.2 does not occur in the Boolean algebra over E.

Proof.
(i) $\Rightarrow$ (iii). Suppose (i) holds, and let $A, B \in \mathcal{F}$ with $|A| < |B|$. Define

$$w : E \to \mathbb{R}$$

$$w(e) := \begin{cases} 1 & \text{if } e \in A \cup B, \\ -|E| & \text{else.} \end{cases}$$

Since any feasible ordering of A is a greedy ordering it immediately follows that A can be augmented from B since otherwise the greedy solution would not be optimal. Thus we know that $(E, \mathcal{F})$ has to be a greedoid.

Suppose now that Figure 3.2 does occur in the Boolean algebra over E. This means that there exists $B \subseteq E$, $z \in B$, x, $y \notin B$ such that $B \in \mathcal{F}, B - z \cup \{x, y\} \in \mathcal{F}, B - z \cup y \notin \mathcal{F}$ and $B \cup y \notin \mathcal{F}$.

We want to show that the modified greedy algorithm does not find an optimal solution for the following weight function $w : E \to \mathbb{R}$:

$$w(e) := \begin{cases} 1 & e \in B - z \\ w_x & e = x \\ w_y & e = y \\ w_z & e = z \\ -|E| & \text{else,} \end{cases}$$

where $-|E| < w_x < w_z < w_y < 1$.

It is a trivial but important observation that the modified greedy algorithm proceedes independently of the specific values of w_x, w_y, w_z provided the inequality above is satisfied.

Let us remark that by assumption (*i*) the following is true: If A is greedy with respect to w then either A contains an optimal solution or A can be augmented to an optimal solution. We claim that B is greedy with respect to w provided $-|E| < w_x < w_z < w_y < 1$.

This is true by the remark above, since B is the unique optimal solution of the problem $max\{w(F) : F \in \mathcal{F}\}$, where $w_x = -|E| + 1$, $w_z = \frac{1}{3}$, $w_y = \frac{1}{2}$. On the other hand $B - z \cup \{x, y\}$ is the unique optimal solution of $max\{w(F) : F \in \mathcal{F}\}$, where $w_x = \frac{-1}{2}$, $w_z = \frac{-1}{3}$ and $w_y = \frac{2}{3}$.

Thus B does not contain nor can be augmented to an optimal solution. This contradicts the fact that B is greedy with respect to w, if $-|E| < w_x < w_z < w_y$.

(iii) $\Rightarrow$ (ii). Let $m := max\{|F| : F \in \mathcal{F}\}$, by induction on $t, 0 \leq t \leq m$, we will show that $\mathcal{B}_t := \{F \in \mathcal{F} : |F| = t\}$ is the set of bases of a matroid $\mathcal{M}_t$.

Using Theorem 3.9 and the subsequent remark property *(ii)* will follow directly. We can assume $t \geq 2$. Let $B_1, B_2 \in \mathcal{B}_t$ and $x \in B_1 \backslash B_2$. We have to show that there exists $y \in B_2 \backslash B_1$ such that $B_1 - x \cup y \in \mathcal{B}_t$.

Choose a feasible ordering $x_1, \ldots, x_t$ of B_1 such that the index i with $x_i = x$ is maximal.

If $i = t$, then $B_1 - x \in \mathcal{F}$ and by augmentation property (M3) we can augment $B_1 - x$ from B_2 such that $B_1 - x \cup y \in \mathcal{B}_t$ with $y \in B_2 \backslash B_1$.

Now assume that $1 \leq i < t$. Let $\hat{B}_1 := \{x_1, \ldots, x_i\}$ and choose $\hat{B}_2 \subseteq B_2$ such that $\hat{B}_2 \in \mathcal{F}$ and $|\hat{B}_2| = i$.

By induction hypothesis there exists $y \in \hat{B}_2 \backslash \hat{B}_1$ such that $\hat{B}_1 - x \cup y \in \mathcal{F}$. We also know that $y \notin B_1$ otherwise we could find a feasible ordering of B_1 with x in a higher than the i-th position by augmenting $\hat{B}_1 - x \cup y = \{x_1, \ldots, x_{i-1}, y\}$ from B_1.

We claim that $\{x_1, \ldots, x_{i-1},\ y,\ x_{i+1}, \ldots, x_t\} \in \mathcal{F}$.

Suppose not, then let $j \geq i+1$ be the smallest index such that

$$\{x_1, \ldots, x_{i-1}, y, x_{i+1}, \ldots, x_j\} \notin \mathcal{F}.$$

Define $B := \{x_1, \ldots, x_{i-1},\ y,\ x_{i+1}, \ldots, x_{j-1}\}$, then we have $B, B - y \cup \{x_i, x_j\} \in \mathcal{F}$. Furthermore $B \cup x_j \notin \mathcal{F}$, by the choice of j, and $B - y \cup x_j \notin \mathcal{F}$, by minimality of i (augment $B - y \cup x_j = \{x_1, \ldots, x_{i-1},\ x_{i+1}, \ldots, x_j\}$ from B_1).

This however contradicts *(iii)* because we have the following configuration

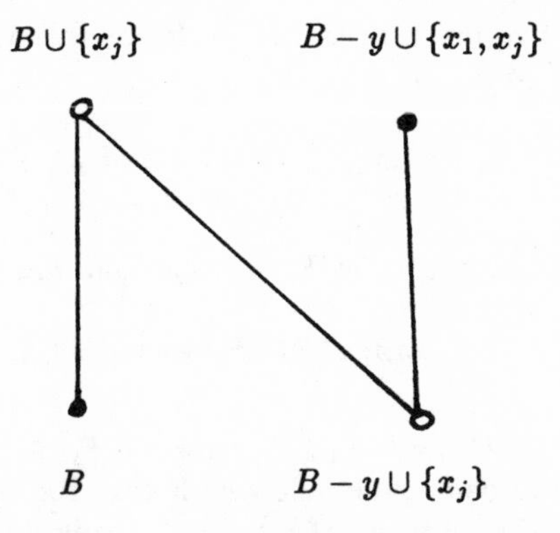

Figure 3.3

(ii) $\Rightarrow$ (i). Let $w : E \rightarrow \mathbb{R}$ be a weight function and F_w be a greedy solution determined by the modified greedy algorithm and let $x_1, \ldots, x_k$ be a greedy ordering of F_w. Among all optimal solutions of the problem $max\{w(F) : F \in \mathcal{F}\}$ choose one, say $\hat{F}$, which has the feasible ordering $y_1 \ldots y_l$ such that the index t for which holds $x_1 = y_1, x_2 = y_2, \ldots, x_t = y_t$ is maximal.

It is easy to see that $t = l$ or $t = k$ implies $F_w = \hat{F}$ proving that the greedy solution is optimal. Thus we may assume $t < min(k, l)$.

Claim: There exists $y \in \hat{F} \backslash \{x_1, \ldots, x_t\} = \{y_{t+1}, \ldots, y_l\}$ such that

$$\{x_1, \ldots, x_t, y\} \in \mathcal{F} \text{ and } \{x_i, \ldots, x_t,\ x_{t+1}, y_{t+1}, \ldots, y_l\} - y \in \mathcal{F}.$$

By (ii) we know that $\hat{F}$ is a basis of the matroid $\mathcal{M}_l$ and by the choice of $\hat{F}$ we have $x_{t+1} \notin \hat{F}$. Denote by C the unique circuit contained in $\hat{F} \cup x_{t+1}$ with respect to the matroid $\mathcal{M}_l$ and let $A := \{x_1, \ldots, x_t\}$. To prove the claim it is enough to show that

$$r_{t+1}(A \cup C - x_{t+1}) > r_{t+1}(A).$$

Here r_i denotes the rank function of $\mathcal{M}_i$ for $0 \leq i \leq m$. Suppose that $r_{t+1}(A \cup C - x_{t+1}) = r_{t+1}(A)$, then $r_{t+1}(A \cup C) = r_{t+1}(A) + 1$. Since $\mathcal{M}_{t+1}$ is a strong image of $\mathcal{M}_l$ we have

$$\begin{aligned} 1 &= r_{t+1}(A \cup C) - r_{t+1}(A \cup C - x_{t+1}) \\ &\leq r_l(A \cup C) - r_l(A \cup C - x_{t+1}) \\ &\leq r_l(C) - r_l(C - x_{t+1}) = 0, \end{aligned}$$

contradiction. This proves the claim.

By the claim there exists $y \in F\backslash\{x_1, \ldots, x_t\}$ such that $\{x_1, \ldots, x_t, y\} \in \mathcal{F}$ and $B; = \{x_1, \ldots, x_{t+1},\ y_{t+1}, \ldots, y_l\} - y \in \mathcal{F}$. By choice of x_{t+1} in the greedy algorithm it follows that $w(y) \leq w(x_{t+1})$ and $w(B) \geq w(\hat{F})$, thus B is another optimal solution, contradicting the choice of $\hat{F}$. This proves the theorem.

□

Corollary 3.11: *(BRYLAWSKI [1986]) Let $(E, \mathcal{M}_1), (E, \mathcal{M}_2)$ be two matroids and assume that the rank of $\mathcal{M}_1$ is by one smaller than the rank $\mathcal{M}_2$, then the following statements are equivalent:*

(i) $\mathcal{M}_1$ is strong map of $\mathcal{M}_2$.

(ii) For every linear objective function $w : E \to \mathbb{R}$ the following holds: If $F_1 \in \mathcal{M}_1$ is an optimal solution of the problem

$$max\{w(F) : F \text{ basis of } \mathcal{M}_1\},$$

then there exists $x \in E\backslash F_1$ such that $F_1 \cup x$ is an optimal solution of

$$max\{w(F) : F \text{ basis of } \mathcal{M}_2\}.$$

Remark: Suppose $\mathcal{F}_1, \mathcal{F}_2 \subseteq 2^E$ and $|F_1| = k$ for all $F_1 \in \mathcal{F}_1$ and $|F_2| = k+1$ for all $F_2 \in \mathcal{F}_2$. Call $(\mathcal{F}_1, \mathcal{F}_2)$ a ***greedy pair*** if the following holds: For every linear objective function $w : E \to \mathbb{R}$ every optimal solution of $max\{w(F_1) : F_1 \in \mathcal{F}_1\}$ is contained in an optimal solution of $max\{w(F_2) : F_2 \in \mathcal{F}_2\}$.

By the last corollary we know that pairs of bases of matroids which are in strong map relation are greedy pairs. BRYLAWSKI [1986] conjectures that in fact all greedy pairs arise in this way. Call a set system $\mathcal{F} \subseteq 2^E$ *equicardinal* if all sets in $\mathcal{F}$ have the same cardinality.

Conjecture: The pair $(\mathcal{F}_1, \mathcal{F}_2)$ is greedy if and only if there exists a subset $E' \subseteq E$ and matroids $\mathcal{M}_1, \mathcal{M}_2$ with bases $\mathcal{B}_1, \mathcal{B}_2$ on E' which are in elementary strong map relation such that

$$\begin{aligned} \mathcal{F}_1 &= \{B_1 \cup F : B_1 \in \mathcal{B}_1, F \in \mathcal{F}\} \\ \mathcal{F}_2 &= \{B_2 \cup F : B_2 \in \mathcal{B}_2, F \in \mathcal{F}\} \end{aligned}$$

for some equicardinal set system $\mathcal{F} \subseteq 2^{E \setminus E'}$.

Let $\mathcal{F}_1, \mathcal{F}_2 \subseteq 2^E$ be equicardinal such that $|F_1| = |F_2| - 1$ for all $F_i \in \mathcal{F}_i$. Define a bipartite graph $G = (\mathcal{F}_1, \mathcal{F}_2, A)$ where each $F_i \in \mathcal{F}_1 \cup \mathcal{F}_2$ represents a node and $F_1 \in \mathcal{F}_1$ and $F_2 \in \mathcal{F}_2$ are linked by an edge iff $F_1 \subseteq F_2$. With this we reformulate the above conjecture: If $G = (\mathcal{F}_1, \mathcal{F}_2, A)$ is connected then $\mathcal{F}_1$ and $\mathcal{F}_2$ are the bases of two matroids $\mathcal{M}_1, \mathcal{M}_2$ which are in strong map relation.

References

BJÖRNER, A. [1985]: *On matroids, groups and exchange languages.* in: L. Lovász and A. Recski (eds): Matroid Theory and its Applications. Conference Proceedings, Szeged, September 1982. Colloquia Mathematica Societatis Janós Bolyai, Vol. 40, North Holland, Amsterdam, Oxford, New York, 1985, pp. 25–40.

BJÖRNER, A. , KORTE, B. and LOVÁSZ, L. [1985]: *Homotopy properties of greedoids.* Advances in Applied Mathematics 6 (1985), pp. 447-494.

BLAND, R.G. [1977]: *A combinatorial abstraction of linear programming.* J. Comb. Th. Series B 23 (1977), pp. 33-57.

BRYLAWSKI, Th.H. [1986]: *Greedy families of linear objective functions.* Preprint, Dept. of Mathematics, University of North Carolina 1986.

BRYLAWSKI, Th.H. and DIETER, E. [1986]: *Exchange systems.* Preprint, Dept. of Mathematics, University of North Carolina, Chapel Hill, USA.

CRAPO, H. [1982]: *Selectors. A theory of formal languages, semimodular lattices and shelling processes.* Advances in Mathematics 54 (1984), pp. 233-277.

DIETRICH, B. [1985] Lecture at the III. International Conference on Combinatorial Mathematics, New York Academy of Sciences.

DUFFUS, D. and RIVAL, I. [1976]: *Crown on dismantable partially ordered sets.* in: Colloquia Mathematica Societatis Janós Bolyai, 18. Combinatorics. Keszthely, 1976, pp. 271-292.

EDELMAN, P.H. [1980]: *Meet–distributive lattices and anti–exchange closure.* Algebra Universalis 10, 1980, pp. 290-299.

EDELMAN, P.H. and JAMISON, R.E. [1985]: *The theory of convex geometries.* Geometriae dedicata 19, 1985, pp. 247-270.

EDMONDS, J. [1965]: *Paths, trees and flowers.* Canadian Journal of Mathematics 17 (1965), pp. 449-467.

EDMONDS, J. [1970]: *Submodular functions, matroids and certain polyhedra.* in: R. Guy, H. Hanani, N. Sauer, J. Schonheim (eds.): Combinatorial Structures and their Applications. Gardon and Breach (1970), pp. 69-87.

EDMONDS, J. [1971]: *Matroids and the greedy algorithm.* Math. Programming 1 (1971), pp. 127-136.

EDMONDS, J. and FULKERSON, D.R. [1970]: *Bottleneck extrema.* J. Comb. Th. 8 (1970), pp. 299-306.

ERDŐS, P. and GRAHAM, R.L. [1980]: *Old and new problems in combinatorial number theory.* Monographie No 28 de L'Enseignement Mathématic, Université de Genéve, Genéve 1980.

FAIGLE, U. [1979]: *The greedy algorithm for partially ordered sets.* Discrete Mathematics 28 (1979), pp. 153-159.

FAIGLE, U. [1980]: *Geometries on partially ordered sets.* J. Comb. Theory, Series B, 28 (1980), pp. 26-51.

FAIGLE, U. [1985]: *Submodular combinatorial structures.* Habilitationsschrift Universität Bonn 1985.

FAIGLE, U., GOECKE, O. and SCHRADER, R. [1986]: *Church-Rosser decomposition in combinatorial structures.* Preprint, Institut für Operations Research, Universität Bonn.

GOECKE, O. [1986a]: *A greedy algorithm for hereditary set systems and a generalization of the Rado-Edmonds characterization of matroids.* Preprint, Institut für Operations Research, Universität Bonn 1986; to appear in: Discrete Applied Mathematics.

GOECKE, O. [1986b]: *Eliminationsprozesse in der kombinatorischen Optimierung – ein Beitrag zur Greedoid–Theorie.* Dissertation, Universität Bonn.

GOECKE, O. and SCHRADER, R. [1986]: *Minor characterization of undirected branching greedoids – a short proof.* Preprint, Institut für Operations Research, Universität Bonn.

GOLUMBIC, M.C. [1980]: *Algorithmic Graph Theory and Perfect Graphs.* New York, London, San Francisco: Academic Press (1980).

HIGGS, D.D. [1968]: *Strong maps of geometries.* J. Comb. Th. 5 (1968), pp. 185-191.

JAMISON, R.E. [1982]: *A perspective on abstract convexity: Classifying alignments by varieties.* in: D.C. Kay and M. Breen (eds.): Convexity and related combinatorial geometry. New York: Marcel Dekker (1982), pp. 113-150.

KORTE, B. and LOVÁSZ, L. [1981]: *Mathematical structures underlying greedy algorithms.* in: F. Gécseg (ed.): Fundamentals of Computation Theory. Lecture Notes in Computer Science 117. Berlin/Heidelberg/New York: Springer (1981), pp. 205-209.

KORTE, B. and LOVÁSZ, L. [1983]: *Structural properties of greedoids.* : Combinatorica 3 (1983), pp. 359-374.

KORTE, B. and LOVÁSZ, L. [1984a]: *Greedoids, a structural framework for the greedy algorithm.* in: W.R. Pulleyblank (ed.): Progress in Combinatorial Optimization. Proceedings of the Silver Jubilee Conference on Combinatorics, Waterloo, June 1982, pp. 221-243. Academic Press, London/New York/San Francisco 1984.

KORTE, B. and LOVÁSZ, L. [1984b]: *Shelling structures, convexity and a happy end.* in: B. Bolobás (ed.): Graph Theory and Combinatorics, Proceedings of the Cambridge Combinatiorial Conference in Honour of Paul Erdös, Academic Press, London, 1984, pp. 219–232.

KORTE, B. and LOVÁSZ, L. [1984c]: *Greedoids and linear objective functions.* SIAM Journ. Algebr. Discr. Methods 5 (1984), pp. 229–238.

KORTE, B. and LOVÁSZ. L. [1985a]: *A note on selectors and greedoids.* : European Journal of Combinatorics 6 (1985), 59–67.

KORTE, B. and LOVÁSZ, L. [1985b]: *Posets, matroids and greedoids.* in: L. Lovász and A. Recski (eds): Matroid Theory and its Applications. Conference Proceedings, Szeged, September 1982. Colloquia Mathematica Societatis Janós Bolyai, Vol. 40, North–Holland, Amsterdam, Oxford, New York, 1985, pp. 239–265.

KORTE, B. and LOVÁSZ, L. [1985c]: *Polymatroid greedoids.* : Journal of Combinatorial Theory B, 38 (1985), pp. 41–72.

KORTE, B. and LOVÁSZ, L. [1985d]: *Relations between subclasses of greedoids.* Zeitschrift für Operations Research, Series A 29 (1985), pp. 249–267).

KORTE, B. and LOVÁSZ, L. [1985e]: *Basis graphs of greedoids and two-connectivity.* in: R.W. Cottle (ed): Essays in Honor of George B. Dantzig, Part I. Mathematical Programming Study No. 24, North–Holland, Amsterdam 1985, pp. 158–165.

KORTE, B. and LOVÁSZ, L. [1985f]: *Polyhedral results for antimatroids.* Preprint, Institut für Operations Research, Universität Bonn, to appear in: Annals of the New York Academy of Sciences. Proceedings of the Third International Conference on Combinatorial Mathematics.

KORTE, B. and LOVÁSZ, L. [1985g]: *Intersections of matroids and antimatroids.* Preprint, Institut für Operations Research, Universität Bonn. to appear in: Discrete Mathematics.

KORTE, B. and LOVÁSZ, L. [1986a]: *Homomorphisms and Ramsey properties of antimatroids.* Discrete Applied Mathematics 15 (1986), pp. 283-290.

KORTE, B. and LOVÁSZ, L. [1986b]: *Non–interval greedoids and the transposition property.* Discr. Math. 59 (1986), pp. 297–314.

LAWLER, E.L. [1973]: *Optimal sequencing of a single machine subject to precedence contraints.* Management Science 19 (1973), pp. 544-546.

LOVÁSZ, L. and PLUMMER, M.D. [1986]: *Matching Theory.* Akademiai Kiadó, Budapest 1986.

SCHMIDT, W. [1985a]: *A characterization of undirected branching greedoids.* Institut für Operations Research, Universität Bonn, to appear in: J. Comb. Th..

SCHMIDT, W. [1985b]: *Greedoids and searches in directed graphs.* Preprint, Institut für Operations Research, Universität Bonn.

SCHMIDT, W. [1985c]: *A min-max theorem for greedoids.* Preprint, Institut für Operations Research, Universität Bonn.

SCHRADER, R. [1986]: *Structural Theory of Discrete Greedy Procedures.* Habilitationsschrift Universität Bonn, 1986.

SEYMOUR, P. [1985]:*private communication* .

WELSH, D.J.A. [1976]: *Matroid theory.* London, New York, San Francisco: Academic press (1976).

Combinatorial Structures and Combinatorial Optimization

Eugene L. Lawler
Computer Science Division
University of California
Berkeley, CA 94720
USA

"Discrete optimization", "integer programming", and "combinatorial optimization", all refer to more or less the same subject matter; in some contexts these terms are almost interchangeable. However, there are crucial distinctions:

Discrete optimization encompasses all types of optimization problems for which discrete solutions are sought. Integer programming and combinatorial optimization are special types of discrete optimization, with contrasting goals and methodologies.

Integer programming deals with mathematical programming problems in which at least some of the variables are required to be integers. In particular, integer *linear* programming problems are of the form

minimize cx subject to $Ax \geq b$, x nonnegative integer.

Integer programming algorithms generally exploit only very weak assumptions about the characteristics of problem instances. For example, the constraint matrix A in may be assumed to be sparse, or to contain only 0's and 1's. As a consequence, the hallmark of integer programming is generality.

By contrast, *combinatorial optimization* deals with problems with significant combinatorial structure, often of a type that is awkward or inappropriate to formulate in a standard form such as $Ax \geq b$. Combinatorial optimization algorithms are designed to to exploit the specific structure of problem instances. It follows that combinatorial optimization is characterized by specialization.

In these notes we consider three types of combinatorial structures and the specialized algorithms that have been developed to deal with them. Part I deals with matroids and submodular functions. We show that a very natural elaboration and generalization of the classical network flow model of Ford and Fulkerson allows one to formulate and solve a wide variety of problems in polymatroid optimization. This material is adapted and condensed from [1].

In Part II, we consider various problems in subgraph optimization. Many of these problems are NP-complete for general graphs. However, as we show, many of these same problems admit of linear time algorithms when the problem instances are restricted to certain interesting subclasses of graphs, such as trees and series parallel graphs. This material is drawn from [2].

Finally, in Part III, we deal with certain special cases of the Traveling Salesman Problem. Emphasis is placed on problems that can be solved by the method of "subtour patching". This material is extracted from [3], to which the reader is referred for a discussion of still other special cases.

References

[1] E. L. Lawler and C. U. Martel, "Flow Network Formulations of Polymatroid Optimization Problems," *Annals Discrete Math.*, 16 (1982) 189-200.

[2] M. W. Bern, E. L. Lawler and A. L. Wong, "Why Certain Subgraph Computations Require Only Linear Time," *Proc. 26th Annual IEEE Symposium on Foundations of Computer Science,* October 1985.

[3] P. C. Gilmore, E. L. Lawler and D. B. Shmoys, "Well-Solved Special Cases," chapter in *The Traveling Salesman Problem: A Guided Tour of Combinatorial Optimization,* E. L. Lawler, J. K. Lenstra, A. H. G. Rinnooy Kan and D. B. Shmoys (eds.), J. Wiley, 1985.

Part I

Polymatroid Network Flows

1. Introduction

In the 'classical' network flow model, flows are constrained by the capacities of individual arcs. In the 'polymatroidal' network flow model, flows are constrained by the capacities of *sets* of arcs, where these capacities are imposed by polymatroid rank functions on the sets of arcs directed into and out of each node. Yet, as the authors have shown in another paper [5] the essential features of the classical model are retained; the augmenting path theorem, the integral flow theorem, and the max-flow min-cut theorem all yield to straight-forward generalization. Moreover, a maximal integral flow can be computed efficiently, provided that there is a 'feasibility' oracle available for each capacity function.

We believe that the polymatroidal network flow model provides a satisfying generalization and unification of both network flow theory and much of the theory of (poly)matroid optimization. In this paper we demonstrate the usefulness of the model by providing network flow formulations of (poly)matroid intersection, matroid partitioning, problems of gammoids and linking systems, and problems formulated by Iri, Krogdahl and Fujishige. In fact, virtually every known problem in (poly)matroid optimization known to the authors can be easily formulated in terms of their network flow model, with the conspicuous exception of the polymatroid matching problem, solved by Lovász for the case of linearly represented polymatroids [6]. (Polymatroid matching cannot be formulated as a polymatroidal network flow problem, just as (nonbipartite) graphic matching cannot be formulated as an ordinary network flow problem.)

We assert that, when the maximal flow algorithm described in [5] is applied to each of the networks described in this paper, it either specializes to a known algorithm or is competitive with special algorithms which have been developed for the problems in question. We shall not elaborate on this point, but instead emphasize the ease with which min-max theorems can be proved, as corollaries of the generaly max-flow min-cut theorem proved in [5].

2. The Polymatroidal Network Flow Model

A *polymatroid* (E,ρ) is defined by a finite set of E *and a rank function* $\rho : 2^E \to \mathbf{R}+\cup\{\infty\}$ satisfying the following properties:

$$\rho(\varnothing)=0, \tag{2.1}$$

$$\rho(X)\le\rho(Y) \quad (X\subseteq Y\subseteq E), \tag{2.2}$$

$$\rho(X\cup Y)+\rho(X\cap Y)\le\rho(X)+\rho(Y) \quad (X\subseteq E, Y\subseteq E). \tag{2.3}$$

Inequality (2.2) states that the rank function is monotone and (2.3) asserts that it is submodular. If also ρ is integer-valued and $\rho(\{e\})=0$ or 1 for all $e\in E$ then the polymatroid is a *matroid.* If (E,ρ) is matroid and $I\subseteq E$ is such that $|I|=\rho(I)$, then I is an *independent* set.

A *polymatroidal flow network* is a directed multigraph with a *source s*, a *sink t* and two *capacity functions* α_j and β_j for each node j. Each function α_j (β_j) satisfies properties (2.1) - (2.3) with respect to the set of arcs A_j (B_j) directed out from (into) node j. Thus (A_j,α_j) and (B_j,β_j) are polymatroids. (Comment: We permit multiple arcs from one node to another. Hence A_j and B_j may be arbitrarily large finite sets.)

A *flow* in the network is an assignment of real numbers to the arcs of the network. Thus a flow is given by a function $f : E\to\mathbf{R}$, where E is the set of arcs. Such a function can be extended to subsets of arcs in a natural way, i.e.,

$$f(\varnothing)=0, \tag{2.4}$$

$$f(X) \ \sum_{e\in X} f(e) \quad (\varnothing\ne X\subseteq E).$$

(We continue to write $f(e)$ instead of the more cumbersome $f(\{e\})$.) A flow is said to be *feasible* if

$$f(A_j) = f(B_j), \quad j \neq s, t, \tag{2.5}$$

$$f(X) \leq \alpha_j(X) \quad \text{for all } j \text{ and } X \subseteq A_j, \tag{2.6a}$$

$$f(X) \leq \beta_j(X) \quad \text{for all } j \text{ and } X \subseteq B_j, \tag{2.6b}$$

$$f(e) \geq 0 \quad \text{for all arcs } e. \tag{2.7}$$

Equation (2.5) imposes the customary flow conservation law at each node other than the source and sink. Inequalities (2.6a) and (2.6b) assert that capacity constraints are satisfied on sets of arcs, and (2.7) simply demands that the flow through each arc be nonnegative. Our objective is to find a feasible flow which has maximum value, i.e., one which maximizes

$$v = f(A_s) - f(B_s) = f(B_t) - f(A_t).$$

Comment. It is possible to generalize the polymatroidal network flow model to provide for costs on arc flows and lower bounds on arc flows as imposed by supermodular set functions. However these generalizations are unnecessary for our present purposes.

3. Integrality and Max-Flow Min-Cut Theorems

We now state two theorems proved in [5]. Strictly speaking, the statements of these theorems should allow for the possibility of the nonexistence of a maximal flow (which occurs if the maximum flow value is unbounded). However, we ignore this possibility, since it does not occur in any of the problems considered here.

Integrality Theorem. *If all capacity functions are integer-valued, then there exists a maximal flow which is integral.*

An *arc-partitioned cut* (S, T, L, U) is defined by a partition of the nodes into two sets S and T, with $s \in S$ and $t \in T$, and by a partition of the forward arcs across the cut into two sets L and U. The capacity of an arc partitioned cut is defined as

$$c(S, T, L, U) = \sum_{i \in S} \alpha_i(U \cap A_i) + \sum_{j \in T} \beta_j(L \cap B_j).$$

As in the case of ordinary flow networks, the value v of any feasible flow f is equal to the net flow across any cut, i.e.,

$$v = f(L) + f(U) - f(B),$$

where B is the set of backward arcs, and clearly

$$v \leq c(S, T, L, U).$$

Max-Flow Min-Cut Theorem. *The maximum value of a flow is equal to the minimum capacity of an arc-partitioned cut.*

4. (Poly)matroid Intersection

The (unweighted) *matroid intersection* problem is as follows: Given two matroids (E, ρ_1) and (E, ρ_2), find a largest set $I \subseteq E$ such that I is independent in each of the matroids.

This problem can be formulated and solved as a flow problem as shown in Figure 1. There are two nodes, s and t, and each arc from s to t corresponds to an element $e_i \in E$. The two capacity functions are determined by the two matroid rank functions: $\alpha_s = \rho_1$, $\beta_t = \rho_2$. Since these capacity functions are integer-valued, there exists a maximal flow which is integral. Any such integral maximal flow corresponds to a solution to the matroid intersection problem.

Any partitioned cut $(S, T, L\ U)$ must have $S = \{s\}, T = \{t\}$ and is thus determined simply by a partition of E into two subsets L and $U = E - L$. The duality theorem for this problem thus follows as an immediate corollary of the max-flow min-cut theorem.

Matroid Intersection Duality Theorem. $\max |I| = \min_{L \subseteq E} \{\rho_1(E - L) + \rho_2(L)\}$.

In the *polymatroid intersection* problem, the capacity functions are simply the rank functions of the two polymatroids, and the duality theorem generalizes in an obvious way.

5. Matroid Partitioning

A general version of the *matroid partitioning problem* is as follows: Given k matroids $M_j = (E, \rho_j)$, $j = 1,...,k$, find a largest set $I \subseteq E$ such that I can be partitioned into k sets $I_1,...,I_k$, where I_j is independent in (E, ρ_j).

A flow network for this problem is shown in Figure 2. There is a node for each element $e_i \in E, i = 1,...,n$, and a node for each matroid M_j. The flow into each node M_j is constrained by the capacity function $\beta_j = \rho_j$. Each arc (s, e_i) has unit capacity; for convenience we assume that these unit capacities are imposed by the capacity function α_s, where $\alpha_s(X) = |X|$, for all $X \subseteq A_s$. There are no other capacity constraints.

It follows from the integrality theorem that there exists a maximal flow which corresponds to an optimal solution to the partitioning problem. In order to determine an optimal dual solution we reason as follows. An arc-partitioned cut with finite capacity must have all nodes M_j in T. Let us denote by A the subset of nodes e_i in S; the nodes in $E - A$ are in T. For given S and T, L and U must be chosen as follows, in order for the arc-partitioned cut to have finite capacity:

$$L = \{(e_i, M_j) \mid e_i \in A\}, \quad U = \{(s, e_i) \mid e_i \in E - A\},$$

as shown in Figure 3. Thus the minimum capacity of an arc-partitioned cut is

$$\min_{A \subseteq E} \left\{ | E - A | + \sum_j \rho_j(A) \right\}.$$

This capacity is strictly less than $|E|$ if and only if

$$|A| > \sum_j \rho_j(A),$$

for some $A \subseteq E$. We have thus proved the following well-known result.

Matroid Partitioning Duality Theorem. (Edmonds and Fulkerson [1]). *Let I be a feasible solution to the matroid partitioning problem.* Then

$$\max |I| = \min_{A \subseteq E} \left\{ | E - A | + \sum_j \rho_j(A) \right\}.$$

Moreover, E is a feasible solution if and only if

$$|A| \leq \sum_j \rho_j(A),$$

for all $A \subseteq E$.

6. Independent Assignments

The unweighted version of the 'independent assignment' problem considered by Iri and Tomizawa [4] is as follows: Given a bipartite graph $G = (V_1, V_2; E)$ and two matroids $M_1 = (V_1, \rho_1), = (V_2, \rho_2)$, find a matching in G such that the vertices $I_1 \subseteq V_1, I_2 \subseteq V_2$ covered by the matching are independent in M_1 and M_2.

This problem, which can also be formulated as a matroid intersection problem, is clearly a special case of the problem of Fujishige, where $V = V_1 \cup V_2$, and $c_{ij} = \infty$ for each arc $(i,j) \in E$, $i \in V_1$, $j \in V_2$. In order for an arc-partitioned cut (S, T, L, U) to have finite capacity, there can be no arcs (i,j) such that $i \in S$, $j \in T$. Thus the vertices $V'_1 \cup V'_2$, where $V'_1 = T \cap V_1$, $V'_2 = S \cap V_2$, must provide a covering of the arcs of G. Define the *rank* of such a covering $V'_1 \cup V'_2$ as $\rho_1(V'_1) + \rho_2(V'_2)$. We then have the following result.

Theorem 6.1. *The maximum size of an 'independent assignment' is equal to the minimum rank of a covering of edges by vertices.*

The well-known König-Egervary theorem is clearly a corollary, for the case that (V_j, ρ_j), $j = 1,2$, are free matroids, i.e., $\rho_j(X) = |X|$, for all $X \subseteq V_j$.

7. The Rado-Hall Problem

Let $M = (E, \rho)$ be a matroid and let $\boldsymbol{E} = (E_1,...,E_k)$ be a family of subsets of E (with repetitions allowed). The problem is to find a largest partial transversal which is independent in M.

This is a matroid intersection problem, in which the two matroids are M and the transversal matroid induced by $\boldsymbol{E}$. the problem can also be viewed as a special case of the independent assignment problem in which the graph $G = (V_1, V_2; A)$ and the two matroids $M_1 = (V_1, \rho_1)$, $M_2 = (V_2, \rho_2)$ are as follows: $V_1 = E, V_2 = E, (i,j) \in A$ if and only if $e_i \in E_j, \rho_1 = \rho$, $\rho_2(V'_2) = |V'_2|$ for all $V'_2 \subseteq V_2$. It is now easy to obtain the following result.

Rado-Hall Theorem. (Rado [7]). *The size of a largest independent partial transversal is*

$$\min_{\boldsymbol{E} \subseteq \boldsymbol{E}} \{|\boldsymbol{E} - \boldsymbol{E}'| + \rho(E')\}$$

where

$$E' = \bigcup_{E_j \in \boldsymbol{E}'} E_j .$$

Moreover, there exists an independent transversal if and only if the union of any k sets in $\boldsymbol{E}$ *has matroid rank at least k.*

Since this theorem specializes to the well-known Hall theorem in the case that M is a free matroid, it is known as the Rado-Hall theorem.

8. Concluding Remarks

We believe that the formulations given in this paper demonstrate the usefulness of the polymatroidal network flow model. It should be emphasized that other models, e.g., those of Fujishige [3] and Edmonds and Giles [2], are equivalent in the sense that any problem which can be formulated and solved in terms of one model can be formulated and solved in terms of one of the others. (We assert, but do not prove here, that the Edmonds-Giles model can be reformulated in terms of the polymatroidal network flow model, with costs on the arcs.) We believe that the advantage of our model is that it permits problem formulations to be particularly simple and transparent. For example, although one can, in principle, reformulate the matroid partition problem as a matroid intersection problem, it seems simpler and more direct to formulate it directly in terms of the network given in Section 5.

References

[1] J. Edmonds and D. R. Fulkerson, "Transversals and Matroid Partition", *J. Res. Nat. Bur. Standards,* B69 (1965) 147-153.

[2] J. Edmonds and R. Giles, "A Min-Max Relation for Submodular Functions on Graphs", *Ann. Discrete Math.,* 1 (1977) 185-204.

[3] S. Fujishige, "Algorithms for Solving the Independent Flow Problems", *J. Oper. Ress. Soc. Japan,* 21 (1978) 189-204.

[4] M. Iri and N. Tomizawa, *"An Algorithm for Finding an Optimal Independent Assignment", J. Oper. Res. Soc. Japan,* 19 (1976) 32-57.

[5] E. L. Lawler and C. U. Martel, "Computing Maximal 'Polymatroidal' Network Flows", *Math. Oper. Res.,* 7 (1982) 334-347.

[6] L. Lovász, "The Matroid Matching Problem", in: Algebraic Methods in Graph Theory, *Colloq. Math. Soc. János Bolyai,* 25 (1978) 495-517.

[7] R. Rado, "A Theorem on Independence Relations", *Quart. J. Math.,* (Oxford) 13 (1942) 83-89.

Part II

Optimal Subgraphs*

1. Introduction

One of the most common types of combinatorial optimization problems is that of finding an optimal subgraph of a given graph. Such problems include the maximum matching problem, the Traveling Salesman Problem, and the Steiner network problem. Sometimes a feasible subgraph may consist only of a subset of vertices, as in the case of the maximum independent set and minimum dominating set problems.

Quite typically, optimal subgraph problems are NP-complete if there is no restriction on the set of problem instances, but are solvable in polynomial-time when the graphs are restricted to an appropriate special class. In fact, the polynomial time bound may actually be linear, as in the case of:

- the minimum dominating set problem for trees [CGH],
- the minimum total dominating set problem for trees [LPHH],
- the maximum disjoint triangle problem for series parallel graphs [TNS],
- the Traveling Salesman Problem for Halin networks [CNP],
- the Steiner problem for outerplanar networks [WC82,WC83],

*Research supported in part by NSF grant MCS-8311422

- the Steiner problem for Halin networks [W],
- the Traveling Salesman Problem for bandwidth-limited networks [MS],
- the "Steiner Tour" problem for "rectilinear" networks [RR],

and a great variety of similar problems.

Each of the cited problems is of the general form: Given a graph G drawn from a class Γ find, from among all subgraphs H of G that satisfy a certain property P, a subgraph that is optimum (where optimum may be in the sense of minimum number of vertices, maximum total weight of vertices and edges with respect to a given weighting, etc.). We assert that each of the problems has two key features that allow it to be solved in linear time: (1) the class Γ of problem instances is definable in terms of certain rules of composition and (2) the property P is "regular" with respect to these rules of composition. In this paper we show how these two key features provide the basis for a systematic methodology for the construction of linear-time algorithms (though not necessarily the same linear-time algorithms presented in the referenced papers). [TNS] treats a wide class of combinatorial problems on series parallel graphs (including recognizing any subclass of series parallel graphs defined by a finite number of forbidden subgraphs) with a similar though less developed methodology. We shall illustrate the application of our methods by constructing a linear-time algorithm for computing the irredundance number of a tree. This is a problem for which no polynomial-time algorithm was previously known.

2. Graphs Definable by Composition

Let us consider how each of the following classes of graphs can be defined recursively in terms of rules of composition: rooted trees, series parallel graphs, outerplanar graphs, Halin graphs, and bandwidth-limited graphs.

A rooted tree is a tree $T=(V,E)$ with one vertex r of V designated as its root. If T_1 and T_2 are two rooted trees, then placing an edge between their roots creates a new tree. This suggests the following definition.

Definition 1. (Rooted Trees)

(1) The graph K_1 consisting of a single vertex r is a rooted tree and r is its root.

(2) If $T_1=(V_1,E_1)$ and $T_2=(V_2,E_2)$ are rooted trees with roots r_1 and r_2, respectively, then $T=T_1 \circ T_2=(V_1 \cup V_2, E_1 \cup E_2 \cup \{(r_1,r_2)\})$ is a rooted tree with root r_1. We say that T_1 is the *parent* tree and T_2 is the *child* tree.

The type of series parallel graph we are interested in is "two terminal" series parallel or "edge" series parallel, as opposed to "vertex" series parallel [TNS].

Definition 2. (Series Parallel Graphs)

(1) The graph $G = (\{u,v\}, \{(u,v)\})$ is series parallel, with terminals u and v.

(2) If $G_1 = (V_1, E_1)$ and $G_2 = (V_2, E_2)$ are series parallel graphs, with terminals u_1, v_1 and u_2, v_2, respectively, then:

(a) The graph obtained by identifying u_1 and u_2 is a series parallel graph, with v_1 and v_2 as its terminals. This graph is the *series* composition of G_1 and G_2.

(b) The graph obtained by identifying u_1 and u_2 and also v_1 and v_2 is a series parallel graph, the *parallel* composition of G_1 and G_2. This graph has u_1 $(=u_2)$ and v_1 $(=v_2)$ as its terminals.

Outerplanar graphs can be viewed as a special kind of series parallel graph, namely one in which at least one of the graphs entering into a parallel composition operation must be a primitive graph, i.e., a single edge extending between two (terminal) vertices. We leave details to the reader.

Halin graphs were devised to provide examples of edge minimal planar 3-connected graphs. The usual method of construction of a Halin graph is described as follows: Start with a tree T, in which each nonleaf vertex has degree at least three. Embed the tree in the plane and then connect the leaves with a cycle in such a way that the planar embedding remains valid. We shall relax the condition that the tree vertices have degree at least three and give a recursive definition of "protoHalin" graphs with three terminals (t, l, r), a root terminal and left and right "leaf" terminals. A Halin graph is a protoHalin graph with one additional edge between its left and right terminals.

Definition 3. (protoHalin Graphs)

(1) The graph $K_1 = (\{v\}, \emptyset)$ is a protoHalin graph; v is its root, left, and right terminals.

(2) If $G_1 = (V_1, E_1)$, $G_2 = (V_2, E_2)$ are protoHalin graphs with terminals (t_1, l_1, r_1) and (t_2, l_2, r_2), respectively, then:

(a) If $G_1 = K_1$, then $G = (\{v\} \cup V_2, E_2 \cup \{(v, t_2)\})$ is a protoHalin graph, with terminals (v, l_2, r_2).

(b) If $G_1 \neq K_1$, then $G = (V_1 \cup V_2, E_1 \cup E_2 \cup \{(t_1, t_2), (r_1, l_2)\})$ is a protoHalin graph, with terminals (t_1, l_1, r_2).

A graph of bandwidth k is normally defined as one whose adjacency matrix $A = (a_{ij})$ has all its nonzero entries within a band of width k about the diagonal (possibly after a renumbering of the vertices). That is, if $|i-j| > k$ then $a_{ij} = 0$. We choose to define "complete" undirected graphs of bandwidth k, i.e., graphs that are maximal with respect to the bandwidth bound. A recursive definition is as follows:

Definition 4. (Bandwidth Limited Graphs)

(1) The complete graph K_k with vertices $v_1, v_2, \ldots, v_k$ is a graph of bandwidth k, and $(v_1, v_2, \ldots, v_k)$ is an ordered set of terminals.

(2) If G is a graph of bandwidth k with terminals $(v_1, v_2, \ldots, v_k)$, then the graph obtained by creating a vertex v and edges (v_1, v), (v_2, v), . . . , (v_k, v) is also a graph of bandwidth k and $(v_2, v_3, \ldots, v_k, v)$ is its ordered set of terminals.

By rectilinear graphs we mean graphs whose edges are considered to be either horizontal or vertical edges of a grid-like pattern. Many different classes of such rectilinear graphs can be constructed. In [RR] such a graph is used to model the aisles of a particular warehouse. Rectilinear graphs can be given recursive definitions, but we shall omit giving an example here.

Note that in each case, the definition is recursive, involving the definition of certain *primitive graphs* and instructions about how to manufacture further graphs by certain rules of composition. Each graph has a special set of vertices called *terminals* and these sets of terminals are of bounded size. The terminals are used as attachment points to join graphs together by composition. The terminals of the composition are a subset of the terminals of the composing graphs. It is important to note that our type of recursive definition does not encompass other types of recursive definitions, e.g., a well known recursive characterization of chordal graphs.

3. Homomorphisms

Suppose we have a class Γ of graphs defined by rules of composition. For simplicity, let us assume that Γ is defined in terms of a single rule of composition $\circ$. Let P be a computable predicate on graph$\times$subgraph pairs (G,H). To further simplify matters, let us suppose that we are concerned only with subgraphs that consist of subsets of vertices (in which case we will denote the subgraph by S). For example, the predicate P might be true for a pair (G,S) if and only if S is a dominating set of vertices of G. In this restricted case of vertex subsets, we can unambiguously extend the definition of the operator $\circ$ to graph$\times$subgraph pairs by writing $(G_1, S_1) \circ (G_2, S_2)$ for $(G_1 \circ G_2, S_1 \cup S_2)$. In general, extending composition operators to graph$\times$subgraph pairs may require an increase in the number of operators in order to allow the option of adding attachment edges to the composed subgraph. We give an example illustrating this case later in this paper. Let C be a set that has a multiplication operation (not necessarily associative nor commutative) defined on it and denote the operator by the same symbol $\circ$. We say that a mapping h from graph$\times$subgraph pairs onto C defines a *homomorphism* with respect to the class Γ and the property P if

$$(1)\ h(G_1, S_1) = h(G_2, S_2) \Rightarrow P(G_1, S_1) = P(G_2, S_2),$$
$$(2)\ h((G_1, S_1) \circ (G_2, S_2)) = h(G_1, S_1) \circ h(G_2, S_2).$$

More intuitively, a homomorphism is a mapping that preserves the property and respects the graph$\times$subgraph composition operation. In general, if the set of graph$\times$subgraph pairs has k composition operations, C should also have k composition operations, and (2) above must hold for all operations.

Two pairs (G_1,S_1), (G_2,S_2) are said to be *equivalent* if $h(G_1,S_1)=h(G_2,S_2)$. Equivalence is clearly an equivalence relation. If C, which is the range of h since h is an onto mapping, is finite, then the number of equivalence classes is finite and and one can write an explicit multiplication table for elements of C. (Equivalently, the multiplication table for C can be viewed as defining operations on equivalence classes. In later sections of this paper, we use this treatment and speak of multiplying equivalence classes.) Thus, for the very simple case of rooted trees and independent sets, we have the table in Figure 1, in which equivalence classes are expressed by diagrams of representative elements. In these diagrams, the symbol $\otimes$ represents a "chosen" vertex, that is, a member of the subgraph (which is simply a vertex subset), and the symbol $\bigcirc$ represents an "unchosen" vertex. The three classes of tree$\times$subset pairs (G,S) consist of, respectively, (1) those pairs in which S is an independent set and the root r of G is chosen, (2) those in which S is an independent set and r is unchosen, and (3) those in which S is not independent. Thus, the entry in the first row and first column shows that the composition of a parent subtree of class 1 with a child subtree of the same class is an element of class 3, that is, the induced subset in the composed tree is not independent. It is straightforward to verify the other multiplication table entries. To determine the class of an arbitrary tree$\times$subset pair, one traverses the tree in postorder and repeatedly applies homomorphism equation (2) above.

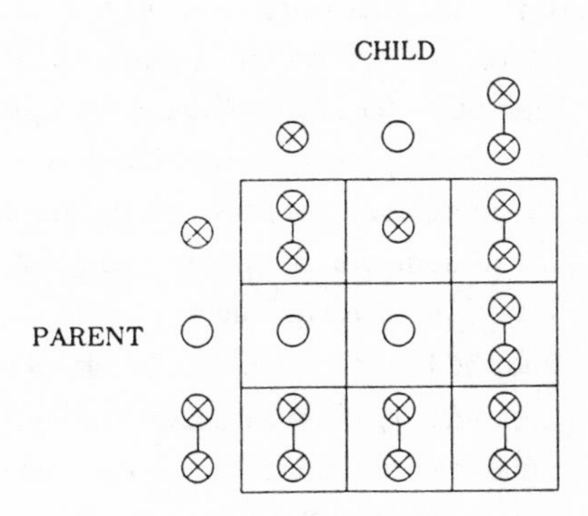

Figure 1. Multiplication table for independence in trees.

The multiplication table of a homomorphism is similar to the state transition table of a finite state automaton, the principal difference being that instead of having entries for state$\times$input pairs, there are entries for $(G,S)\times(G,S)$ pairs. (In finite automaton terms, the input alphabet and the set of states are identical.) In the example of independence for rooted trees, classes 1 and 2 are the *accepting* classes because they contain (G,S) pairs with the property of independence. A procedure, similar to that of state reduction, can be applied to simplify a given homomorphism to obtain an equivalent homomorphism with a minimum number of equivalence classes.

4. A Linear-Time Dynamic Programming Algorithm

We say that a property or predicate P is *regular* with respect to a class Γ of graphs if it admits of a homomorphism (with respect to Γ) with a finite range and therefore a finite number of equivalence classes. The term "regular" is proposed by analogy with the notion of regular sets. One of our principal results is the following.

Theorem 1. If P is a regular property with respect to a class Γ, then (assuming there is a linear-time algorithm to decompose graphs in Γ) there is a linear-time algorithm for finding optimal (either minimum or maximum cardinality) subgraphs satisfying property P.

Proof: We prove the theorem by indicating how to construct a linear-time algorithm for solving the subgraph problem, given the multiplication table for a finite homomorphism. The algorithm is basically a dynamic programming algorithm. The structure of the graph G can be represented by means of a decomposition tree in which each leaf is identified with a primitive graph and each internal node with a composition operation. That is, each internal node is identified with the graph resulting from the composition operation applied to the graphs identified with its children. One then carries out a computation in an order corresponding to a postorder traversal of the nodes of the decomposition tree. At the leaves, an optimal representative of an equivalence class will be the optimal primitive graph$\times$subgraph pair in the class. (A primitive graph$\times$subgraph pair is any pair (G,H) in which G is a primitive graph for the class Γ.) At each interior node v of the tree, we may assume that one has computed an optimal representative of each homomorphism equivalence class for the graphs identified with the children of v. From these, one is able to compute an optimal representative for each equivalence class for the graph identified with v. For each equivalence class C_k, we consider all pairs C_i, C_j such that $C_k = C_i \circ C_j$, and find the best such pair. If we are interested only in the number (for example, cardinality in the case of a maximum cardinality problem) of an optimal subgraph, we need only remember the optimal number achieved for each equivalence class. If we are interested in recovering the optimal subgraph itself, we use the usual dynamic programming technique of maintaining "back pointers"; that is, for each node and each class we would remember which pair C_i, C_j achieved the optimum. The optimal solution is the best of the optimal representatives of the accepting classes at the root of the decomposition tree.

To show that this algorithm is linear in the size $//G//$ of G, we note the following. First, the decomposition tree for G is linear in $//G//$ (by assumption) and can be traversed in linear time. Second, the time required to process a node is constant since this depends only on the size of the multiplication table which in turn depends only on the number of equivalence classes and the number of operators. These latter two numbers are both constant with respect to $//G//$ by the definitions of regular property and decomposable graphs. $\square$

As a simple example, consider the problem of finding the maximum cardinality of an independent set in a rooted tree G. Without loss of generality, we may assume G is given in the form of a decomposition tree for its construction. (If it is not, we can always construct such a tree.) We recall that the three equivalence classes for tree × subset pairs (G, S) are (1) the set of those pairs in which S is an independent set and the root of G is chosen, (2) the set of those pairs in which S is an independent set and the root of G is not chosen, and (3) the set of those pairs in which S is not independent. Leaves of the decomposition tree are identified with single vertex trees. For a single vertex tree K_1, the maximum cardinality of a vertex subset S such that (K_1, S) is in class 1 is 1, such that (K_1, S) is in class 2 is 0, and such that (K_1, S) is in class 3 is $-\infty$. Let T be the tree identified with some interior node of the decomposition tree and let T_1 and T_2 be the trees identified with the children of that node. Referring to the multiplication table in Figure 1, we see that the only way to produce class 1 is by multiplying class 1 with class 2. Thus the maximum cardinality of S such that (T, S) is in class 1 is the sum of the maximum cardinalities achieved by a T_1 pair from class 1 and a T_2 pair from class 2. (By a T_1 pair, we mean any pair (T, S) in which T is T_1.) Similarly, there are two possible ways to produce an element of class 2: (class 2) $\circ$ (class 1), or (class 2) $\circ$ (class 2). The cardinality of the best representative of class 2 for T is the maximum of the two possible sums in this case. More generally, cardinalities of best representatives for the graph identified with a node can be computed by considering the sums of best cardinalities of appropriate classes for its parent and child subgraphs. We choose the way that produces a tree × subset pair with the largest subset. Ties are broken arbitrarily. In order to compute the best representative of class 3, we must take the maximum over 6 possible compositions. (Actually this particular computation is unnecessary, as a subtree of class 3 can never lead to acceptance.) Finally, at the root of the decomposition tree, the solution to the maximum independent set problem is given by the maximum of the accepting classes, that is, classes 1 and 2.

So far, we have primarily considered problems that deal only with subgraphs that do not contain edges. We shall indicate by example how to deal with subgraphs that do contain edges. Consider the problem of finding a minimum cardinality maximal matching in a rooted tree. (This is NP-complete for general graphs [GJ].) To show that the maximal matching property is regular with respect to the class of rooted trees, we must construct a finite set of equivalence classes and a homomorphism. First, note that since the composition operation $\circ$ for rooted trees creates a new edge, the extension of $\circ$ from trees to tree × subgraph pairs must specify whether the new edge is chosen to be in the subgraph. We do this by creating two operators for tree × subgraph pairs: $\circ_1$ in which the new edge is not chosen and $\circ_2$ in which the new edge is chosen. (In our subgraph version of matching, if an edge is chosen to be in the matching, then so must both its endpoints.) Examples of each of these operations are given in Figure 2. Vertices in the matching are denoted by ⊗ and edges in the matching are denoted by wiggly lines. Note that for a class of decomposable graphs, it may be necessary to extend a single composition operator $\circ$ to as many as 2^k operators on graph × subgraph pairs, where k is the number of attachment edges created by $\circ$.

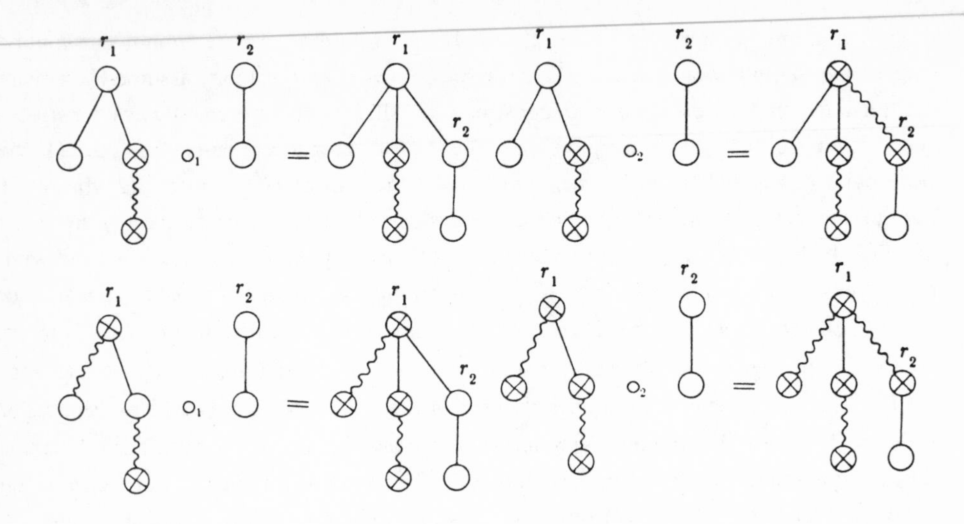

Figure 2. Examples of $\circ_1$ and $\circ_2$.

We now define the following set of four equivalence classes for the maximal matching property: class C_1 is the set of those pairs (G,H) in which the root r is unmatched and H is a maximal matching for G; class C_2 is the set of those pairs in which r is matched and H is a maximal matching for G; class C_3 is the set of those pairs in which r is unmatched, H is a matching for G but not a maximal one, and H is a maximal matching for the subgraph of G induced by the vertices in $V-\{r\}$, where V is the vertex set of G; and finally, class C_4 is the set of remaining pairs. Intuitively, C_1, C_2 and C_3 contain pairs that can appear as subgraphs of maximally matched pairs and C_4 contains pairs that can never appear as such. Classes C_1 and C_2 contain maximally matched pairs and are therefore the accepting classes. The multiplication tables with respect to each composition operator are given in Figure 3. The algorithm given in the proof of Theorem 1 can now be modified slightly to solve a subgraph problem in which the subgraph contains edges. The only change necessary is in the computation of best representatives of each equivalence class for the graph associated with an internal vertex of the decomposition tree. This computation must now be modified to include both composition operations. For example, consider determining the best representative of C_2. By inspection of the two multiplication tables, we note that there are altogether six ways of producing C_2, namely: $C_2 \circ C_1$, $C_2 \circ C_2$, $C_1 \oslash C_1$, $C_1 \oslash C_3$, $C_3 \oslash C_1$, and $C_3 \oslash C_3$. We choose from these the way that gives the best (minimum number of edges in this case) representative.

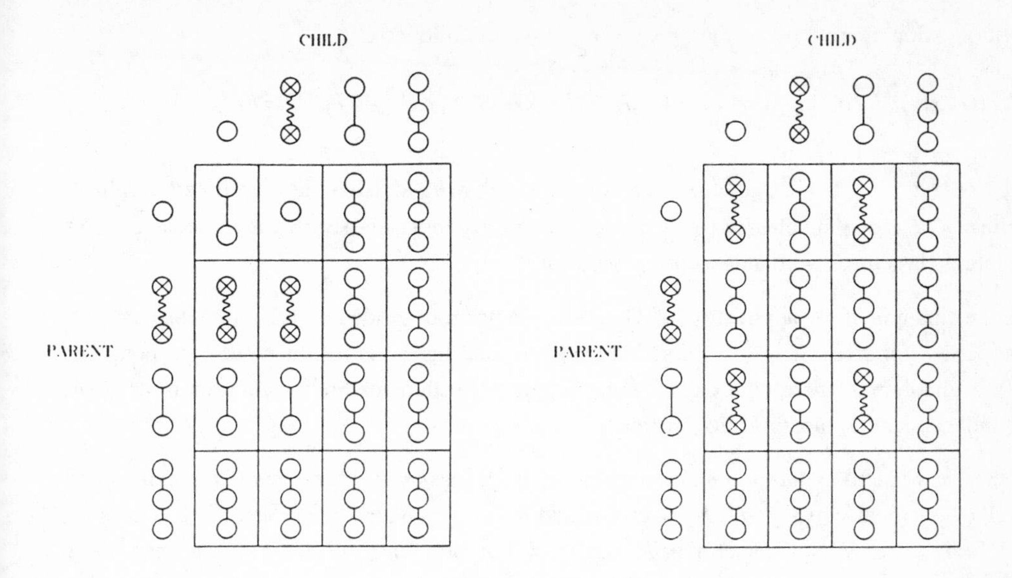

Figure 3. Maximal Matching

5. Operations on Regular Properties

It is well known that regular sets are closed under various operations, such as union, intersection, relative complement, reversal, Kleene closure, etc. The same is true for regular properties.

Theorem 2. The properties that are regular with respect to a given class Γ are closed under Boolean operations (and, or, not).

Proof: The operation "not" follows immediately from definitions. Its set of accepting classes is the complement of the set of accepting classes for the original property.

"And" and "or" are shown by direct product constructions. We sketch the proof for "and"; "or" is similar. Suppose we have two properties P and Q (defined on the same class Γ) with finite homomorphisms h_1 and h_2, respectively. We think of h_1 and h_2 as mapping graph$\times$subgraph pairs to equivalence classes of graph$\times$subgraph pairs. We assume that we have multiplication tables for each of h_1 and h_2 for the same set of graph$\times$subgraph pair composition operations. (This may require creating superfluous multiplication tables for one or both properties.) To create a homomorphism h_3 with respect to both P and Q consider all pairwise intersections of equivalence classes in the range of h_1 with equivalence classes in the range of h_2. Under h_3, a graph$\times$subgraph pair (G, H) maps to the set $h_1(G, H) \cap h_2(G, H)$. We define the multiplication operations for the range of h_3 to be the following:

$$h_3(G_1,H_1) \circ h_3(G_2,H_2) = [h_1(G_1,H_1) \circ h_1(G_2,H_2)] \cap [h_2(G_1,H_1) \circ h_2(G_2,H_2)],$$

for each operator $\circ$. It is straightforward to confirm that h_3 satisfies the two homomorphism properties. The accepting classes for P and Q are exactly those classes that are intersections of an accepting class for P with an accepting class for Q. □

As an example of the application of Theorem 2, since independence and dominance are regular properties for trees, it follows that independent domination is also a regular property for trees. By Theorem 1, there then exists a linear-time algorithm for finding minimum (or maximum) independent dominating sets for trees.

Other very useful and important operations on properties are *max* and *min* (for maximal and minimal). Specifically, *maxP* (G,H) = true if and only if $P(G,H)$ = true and there is no subgraph $H' \neq H, H \subset H' \subseteq G$, such that $P(G,H')$ = true, and *minP* (G,H) = true if and only if $P(G,H)$ = true and there is no subgraph $H' \neq H, H' \subset H \subseteq G$ such that $P(G,H')$ = true.

Theorem 3. If P is a regular property, then so are *maxP* and *minP*.

Proof: Consider only *maxP*; the proof for *minP* is analogous. In order to simplify matters, we consider only the case in which H consists of a subset of vertices (which we shall denote S); the proof for the general case is similar though notationally more complicated.

For any (G,S) consider the effect of adding any single vertex v of $V-S$ to S. (V is the vertex set of G.) Notice that *maxP* (G,S) = true if and only if $P(G,S)$ = true and $P(G,S \cup \{v\})$ = false for each v *in* $V-S$. This characterization motivates the following construction of a finite homomorphism for *maxP*.

Suppose we have a finite homomorphism h_1 for P. We can create a homomorphism h_2 for *maxP* by considering a particular refinement of the equivalence classes in the range of h_1. The image $h_2(G,S)$ of a pair (G,S) will be characterized by the h_1 equivalence class of (G,S) and by the set of h_1 equivalence classes that are obtainable from (G,S) by the addition of a single vertex to S. If we denote the h_1 equivalence classes by C_i, for $i = 1, 2, \cdots, k$, then we can denote the h_2 equivalence classes by pairs $[C_i, A]$ where A is a subset of $\{C_1, C_2, \cdots, C_k\}$. Notice that there are no more than $k \cdot 2^k$ h_2 equivalence classes. (In practice, many of these may be empty or redundant.)

We must now verify that the two defining properties of homomorphisms hold for h_2. In order to confirm the first property, that is, $h_2(G_1,S_1) = h_2(G_2,S_2) \Rightarrow$ *maxP* (G_1,S_1) = *maxP* (G_2,S_2) for any two pairs (G_1,S_1) and (G_2,S_2), observe that the equivalence classes of h_2 for which *maxP* is true are exactly those classes $[C_i, A]$ for which P is true for every pair in C_i and P is false for every pair in every class in A. Thus, any member (G,S) of such a class has property P, and if any vertex is added to S, it no longer has P.

The second property, that is, $h_2((G_1,S_1)\circ(G_2,S_2))=h_2(G_1,S_1)\circ h_2(G_2,S_2)$ for any two pairs (G_1,S_1) and (G_2,S_2), is the more interesting. Suppose $h_2(G_1,S_1)$ is the class $[C_1,A_1]=[h_1(G_1,S_1),A_1]$ and $h_2(G_2,S_2)$ is the class $[C_2,A_2]=[h_1(G_2,S_2),A_2]$. Then we assert that both $h_2((G_1,S_1)\circ(G_2,S_2))$ and $h_2(G_1,S_1)\circ h_2(G_2,S_2)$ must be the class denoted by

$$[C_1\circ C_2,\ \{C_i\circ C_2 \mid C_i\in A_1\}\cup\{C_2\circ C_j \mid C_j\in A_2\}]. \qquad (*)$$

Proving this assertion will confirm property (2) and, moreover, show how to construct a multiplication table for the equivalence classes of h_2. To confirm the first part of (*), we must show that both $h_2((G_1,S_1)\circ(G_2,S_2))$ and $h_2(G_1,S_1)\circ h_2(G_2,S_2)$ are subsets of the h_1 equivalence class $C_1\circ C_2$. Both inclusions are immediate from property (2) for the homomorphism h_1. In order to confirm the second part of (*), notice that any vertex v added to $S_1\cup S_2$ must come from either G_1 or G_2. Assume v comes from G_1. Then

$$\begin{aligned} h_1(G_1\circ G_2,S_1\cup S_2\cup\{v\}) &= h_1((G_1,S_1\cup\{v\})\circ(G_2,S_2)) \\ &= h_1(G_1,S_1\cup\{v\})\circ h_1(G_2,S_2) \\ &= C_i\circ C_2 \text{ for some } C_i\in A_1. \end{aligned}$$

Similarly, adding any vertex v from G_2 yields a class that is a product of C_1 with an element of A_2. Conversely, all products of classes in A_1 with C_2 and all products of C_1 with classes in A_2 are obtainable from the graph $G_1\circ G_2$ by the addition of some vertex. ▯

It may be an interesting exercise for the reader to carry out a similar construction for regular sets. That is, given a language L over the alphabet $\{0,1\}$, let us say that a string x is *maximal* in L if and only if there is no distinct string y *in* L, $|y|=|x|$, such that y has a 1 in each position that x has a 1 ($|x|$ is the length of x). Let us denote by *maxL* the set of all maximal strings in L. The assertion to be proved is that if L is regular, then *maxL* is regular too.

Note that in order to solve the maximal matching problem for trees, we could have first solved the matching problem by constructing an appropriate set of equivalence classes and then applied Theorem 3. However, this problem was simple enough to solve directly; in the next section, we shall consider a more difficult problem involving maximality.

6. The Irredundance Problem for Trees

We now apply our theoretical machinery to the irredundance number problem. A vertex v is said to *dominate* (or *cover*) itself and each of its neighbors. A subset S of the vertices of a graph G is said to be *irredundant* if there is no proper subset S' of S that dominates the same set of vertices as S; otherwise S is *redundant*. An irredundant set, even a maximal irredundant set, need not be a dominating set, in that certain vertices may be left uncovered. The *irredundance number* of a graph is defined as the minimum cardinality of a maximal irredundant set [CFPT]. This number has received a certain amount of attention because of several inequalities relating it to domination and independence [BC79, BC84, CFPT].

Computing the irredundance number of a general graph or even a bipartite graph is an NP-complete problem [P]. The computation of the irredundance number for the case of trees was for several years an open question [HHL]. We now indicate how to construct a linear-time algorithm for this problem, starting with a finite homomorphism for simple irredundance. Figure 4 gives a multiplication table for the property of irredundance, which is regular with respect to the class of rooted trees. The accepting classes, that is, the classes with the property of irredundance, are the first five in the table. The sixth class, though not accepting, contains elements that can appear as subgraphs of irredundant tree × subtree pairs. The seventh class contains elements that can never appear as subgraphs of irredundant tree × subtree pairs.

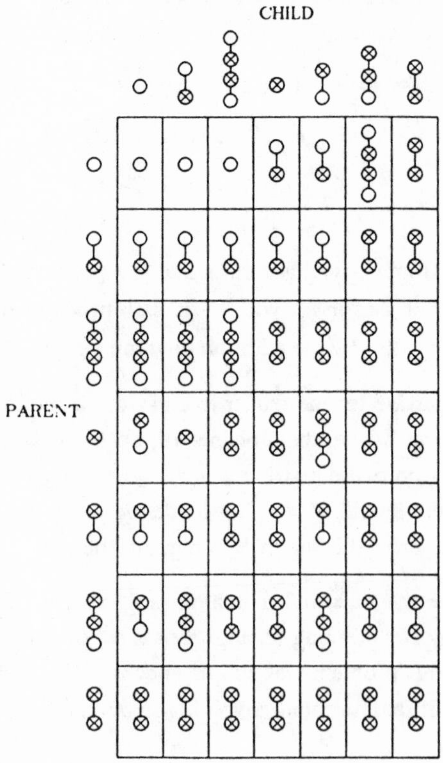

Figure 4. Irredundance

Applying the construction indicated in the proof of Theorem 3 to the table in Figure 4, we obtain a multiplication table for maximal irredundance. The table we have given in Figure 5 is one obtained by simplifying a table with $7 \cdot 2^7$ equivalence classes. Accepting classes are indicated with an asterisk. It would be very difficult to construct such a table without the bootstrapping afforded by Theorem 3. Perhaps worth mentioning is that actually constructing representatives of each equivalence class as we have done is not necessary. In order to apply the dynamic programming algorithm of section 4, it is sufficient to know the multiplication table or tables, the classes that contain the primitive graph × subgraph pairs and the classes that possess the property of maximal irredundance.

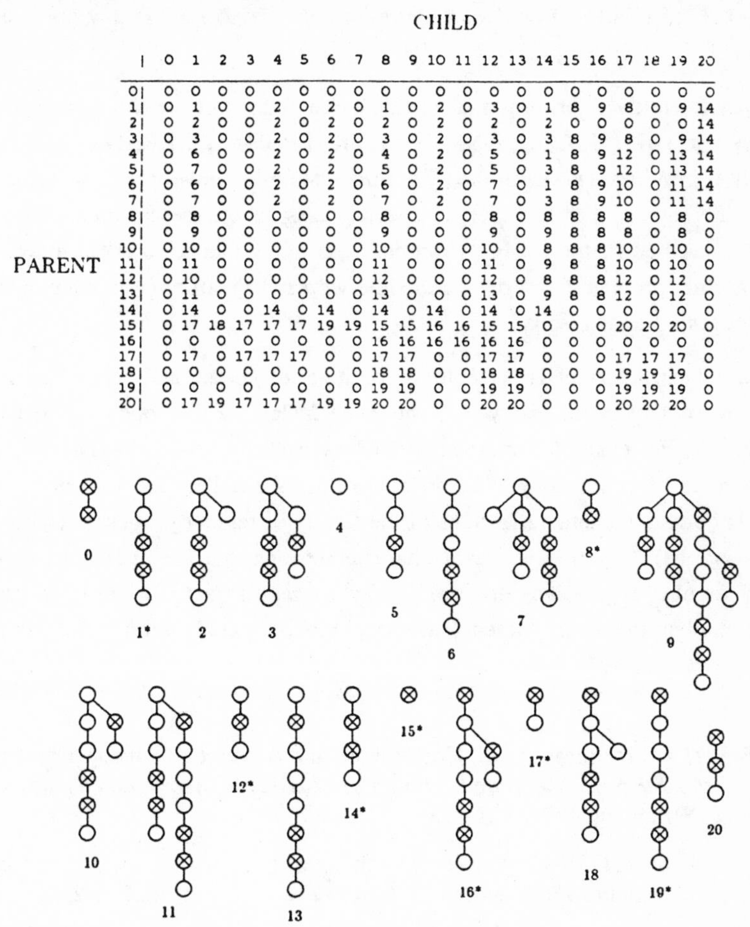

CHILD

PARENT	0	1	2	3	4	5	6	7	8	9	10	11	12	13	14	15	16	17	18	19	20
0	0	0	0	0	0	0	0	0	0	0	0	0	0	0	0	0	0	0	0	0	0
1	0	1	0	0	2	0	2	0	1	0	2	0	3	0	1	8	9	8	0	9	14
2	0	2	0	0	2	0	2	0	2	0	2	0	2	0	2	0	0	0	0	0	14
3	0	3	0	0	2	0	2	0	3	0	2	0	3	0	3	8	9	8	0	9	14
4	0	6	0	0	2	0	2	0	4	0	2	0	5	0	1	8	9	12	0	13	14
5	0	7	0	0	2	0	2	0	5	0	2	0	5	0	3	8	9	12	0	13	14
6	0	6	0	0	2	0	2	0	6	0	2	0	7	0	1	8	9	10	0	11	14
7	0	7	0	0	2	0	2	0	7	0	2	0	7	0	3	8	9	10	0	11	14
8	0	8	0	0	0	0	0	0	8	0	0	0	8	0	8	8	8	8	0	8	0
9	0	9	0	0	0	0	0	0	9	0	0	0	9	0	9	8	8	8	0	8	0
10	0	10	0	0	0	0	0	0	10	0	0	0	10	0	8	8	8	10	0	10	0
11	0	11	0	0	0	0	0	0	11	0	0	0	11	0	9	8	8	10	0	10	0
12	0	10	0	0	0	0	0	0	12	0	0	0	12	0	8	8	8	12	0	12	0
13	0	11	0	0	0	0	0	0	13	0	0	0	13	0	9	8	8	12	0	12	0
14	0	14	0	0	14	0	14	0	14	0	14	0	14	0	14	0	0	0	0	0	0
15	0	17	18	17	17	17	19	19	15	15	16	16	15	15	0	0	0	20	20	20	0
16	0	0	0	0	0	0	0	0	16	16	16	16	16	16	0	0	0	0	0	0	0
17	0	17	0	17	17	17	0	0	17	17	0	0	17	17	0	0	0	17	17	17	0
18	0	0	0	0	0	0	0	0	18	18	0	0	18	18	0	0	0	19	19	19	0
19	0	0	0	0	0	0	0	0	19	19	0	0	19	19	0	0	0	19	19	19	0
20	0	17	19	17	17	17	19	19	20	20	0	0	20	20	0	0	0	20	20	20	0

Figure 5. Maximal Irredundance

7. Comments and Conclusions

In this paper we have given a systematic technique for constructing linear-time algorithms for various optimal subgraph problems. This amounts to the following:

(1) Obtain a suitable recursive definition for the class of graphs to which the algorithm is to be applied. This definition should employ rules of composition that make it easy to decompose any graph in the class. Attachment points in a composed graph must be chosen from among the

attachment points in the composing subgraphs. (Note: Occasionally, as in [RR] the algorithm is to be applied to only one graph, with different edge weights or other data.)

(2) Obtain a finite homomorphism for the property that a subgraph must satisfy in order to be a feasible solution. If this homomorphism cannot be obtained directly, try to obtain it from a combination of simpler homomorphisms, by appropriate application of Theorems 2 and 3.

(3) Apply the basic linear-time dynamic programming algorithm to any given decomposed problem instance.

This approach yields algorithms that are different from, but are thought to be at least as efficient as those in [BPHM, CGH, CNP, FP, KYK, LPHH, MS, RR, W]. The merit of this approach is that it provides a uniform theory for understanding how optimal subgraph problems can be solved efficiently in a wide variety of special cases, where these problems had previously been attacked by ad hoc methods. For example, it is no challenge to create linear-time algorithms for the Steiner problem for series parallel networks or the minimum independent domination problem for outerplanar graphs.

As with any such theory, it is possible to continue to abstract and generalize the approach beyond the point of any value. However, there are at least a few open problems that seem worthy of attention. For example, can the approach be usefully applied to classes of graphs that are defined by rules of composition that do not admit of efficient decomposition algorithms? Are there good alternative characterizations of the family of regular properties (with respect to a given class of graphs)? By this is meant characterizations that are analogous to right linear grammars and regular expressions for the family of regular sets. Finally, of course, what significant and challenging open problems might be solved by application of the theory?

References

[BPHM] T. Beyer, A. Proskurowski, S. Hedetniemi, and S. Mitchell, "Independent Domination in Trees," *Proc. 8th S. E. Conf. on Combinatorics, Graph Theory, and Computing*, Utilitas Mathematica, Winnipeg, 1977, 321-328.

[BC79] B. Bollobás and E. J. Cockayne, "Graph Theoretic Parameters Concerning Domination, Independence, and Irredundance," *J. Graph Theory*, 3 (1979), 241-250.

[BC84] B. Bollobás and E. J. Cockayne, "The Irredundance Number and Maximum Degree of a Graph," *Discrete Math.*, 49 (1984), 197-199.

[CFPT] E. J. Cockayne, O. Favaron, C. Payan, and A. Thomason, "Contributions to the Theory of Domination, Independence, and Irredundance in Graphs," *Discrete Math.*, 33 (1981), 249-258.

[CGH] E. J. Cockayne, S. E. Goodman, and S. T. Hedetniemi, "A Linear Algorithm for the Domination Number of a Tree," *Infor. Processing Lett.*, 4 (1975), 41-44.

[CNP] G. Cornuéjols, D. Naddef, and W. R. Pulleyblank, "Halin Graphs and the Traveling Salesman Problem," *Math. Programming*, 26 (1983), 287-294.

[FP] A. M. Farley and A. Proskurowski, "Computing the Maximum Order of an Open Irredundant Set in a Tree," manuscript, 1983.

[GJ] M. R. Garey and D. S. Johnson, *Computers and Intractability: A Guide to the Theory of NP-Completeness*, W. H. Freeman and Co., 1979.

[HHL] S. M. Hedetniemi, S. T. Hedetniemi, and R. Laskar, "Domination in Trees: Models and Algorithms," *Proc. 5th International Conference on Graph Theory*, Kalamazoo, 1984.

[KYK] T. Kikuno, N. Yoshida, and Y. Kakuda, "A Linear Algorithm for the Domination Number of a Series-Parallel Graphs," *Discrete Appl. Math.*, 5 (1983) 299-311.

[LPHH] R. Laskar, J. Pfaff, S. M. Hedetniemi, and S. T. Hedetniemi, "On the Algorithmic Complexity of Total Domination," *SIAM J. Alg. Disc. Methods*, 5 (1984), 420-425.

[MS] D. Monien and I. H. Sudborough, "Bandwidth Constrained NP-Complete Problems," *Proc. 13th Ann. Symposium on the Theory of Computing*, 1981, 207-217.

[P] J. Pfaff, Ph.D. thesis, Clemson University.

[RR] H. D. Ratliff and A. S. Rosenthal, "Order Picking in a Rectangular Warehouse: A Solvable Case of the Traveling Salesman Problem," *Operations Res.*, 31 (1983), 507-521.

[TNS] K. Takamizawa, T. Nishizeki, and N. Saito, "Linear-Time Computability of Combinatorial Problems on Series-Parallel Graphs," *JACM*, 29 (1982), 623-641.

[WC82] J.A. Wald and C.J. Colburn, "Steiner Trees in Outerplanar Graphs," *Proc. 13th Southeastern Conf. Combinatorics, Graph Theory and Computing*, 1982, 15-22.

[WC83] J.A. Wald and C.J. Colburn, "Steiner Trees, Partial 2-Trees, and Minimum IFI Networks," *Networks*, 13 (1983), 159-167.

[W] P. Winter, unpublished manuscript, 1985.

Part III

Special Cases of the Traveling Salesman Problem

1. Introduction

The Traveling Salesman Problem (TSP) has a well deserved reputation for computational intractability. Nevertheless, many interesting special cases can be solved efficiently, i.e., in polynomial time. We shall give some examples, each of which is solved by "subtour patching". In this method, an assignment problem is solved for the distance matrix of the TSP. Then the subtours of the assignment ϕ are then "patched" together to yield an optimal tour.

We will make use of concepts related to the symmetric group of permutations on n elements. It is expected that the reader has some familiarity with the concepts, so we briefly note only some of the more important points. Consider an arbitrary permutation ϕ; $\phi(i)=j$ denotes that i is mapped to j by ϕ. Since the set of all permutations on n elements forms a group, for any two permutations τ and ϕ there exists a unique permutation ψ such that $\tau=\phi\psi$. As is customary, $\rho\phi(i)$ denotes $\rho(\phi(i))$ and also $\tau^{-1}(j)=i$ is equivalent to $\tau(i)=j$. It is well known that every permutation ϕ has a unique set of disjoint factors. We write permutations in their factored form, e.g. if $\tau(1)=2$, $\tau(2)=1$, $\tau(3)=4$, and $\tau(4)=3$ we write that $\tau=(1,2)(3,4)$.

There is a one-to-one correspondence between permutations and feasible solutions to the assignment problem; $\phi(i)=j$ has the interpretation that element c_{ij} is used in the assignment. Therefore, tours correspond to permutations where all of the elements (cities) are contained within one factor, that is, cyclic permutations or cycles. Also note that cycles on disjoint sets of elements commute, in contrast to the case of two arbitrary permutations. For an assignment that is not a tour, factors correspond to subtours. The cost of a permutation ϕ is

$$c(\phi)=\sum_{i=1}^{n} c_{i\phi(i)}.$$

Often we will modify one assignment ϕ by multiplying it by another permutation ψ; we will be interested in the additional cost of this permutation

$$c\phi(\psi) = c(\phi\psi) - c(\phi)$$

above the original cost of ϕ. When the bottleneck criterion is used we will denote the bottleneck cost

$$\bar{c}(\phi)=\max_i c_{i\phi(i)}$$

and, parallel to the case above, we let

$$\bar{c}\phi(\psi)=\bar{c}(\phi\psi)-\bar{c}(\phi).$$

2. Upper Triangular Matrices

We say that C is *upper triangular* if $i \geq j$ implies $c_{ij}=0$. In this section we shall show that the TSP for upper triangular matrices is essentially as easy as the assignment problem [Lawler, 1972].

Lemma 1 Let C be upper triangular and ϕ be an assignment that is optimal subject to the constraint that $\phi(n)=1$. Then $c(\phi)$ is a lower bound on the length of an optimal tour.

Proof: Call an arc (i,j) *backward* if $i \geq j$; for such an arc $c_{ij}=0$. Let τ be an optimal tour. Remove each backward arc from the part of τ that extends from city n to city 1. The result is a set of paths, each extending from a city j to a city i, with $j \le i$. Now turn each of these paths into a cycle by adding a backward arc from i to j. The result is an assignment ϕ with $\phi(n)=1$ and $c(\phi)=c(\tau)$. □

Theorem 2 Let C be upper triangular and ϕ be an assignment that is optimal subject to the constraint that $\phi(n)=1$. Then $c(\phi)$ is equal to the length of an optimal tour τ that can be easily constructed from ϕ.

Proof: Let ϕ be an assignment that is optimal subject to the constraint that $\phi(n)=1$. If ϕ is not a tour then it consists of s subtours, where $s \ge 2$, each of each contains at least one backward arc (as defined in the proof of Lemma 4). Remove arc $(n,1)$ from the subtour containing cities 1 and n and any one backward arc from each of the other subtours. The result is a set of paths, one extending from city 1 to city n, and the others from j_1 to i_1, j_2 to i_2, ..., j_{s-1} to i_{s-1}, where $j_1 > j_2 > \cdots > j_{s-1}$ and $i_1 \ge j_1 > j_2$, $i_2 \ge j_2 > j_3$, ..., $i_{s-1} \ge j_{s-1} > 1$. Now add backward arcs (n, j_1), (i_1, j_2),..., (i_{s-2}, j_{s-1}), $(i_{s-1}, 1)$ to obtain a tour τ with $c(\tau)=c(\phi)$. Since, by Lemma 4, $c(\phi)$ is a lower bound on the length of an optimal tour, τ is optimal. □

Note that an assignment ϕ that is optimal subject to the constraint that $\phi(n)=1$ is easily obtained by applying any algorithm for the assignment problem to the $(n-1)\times(n-1)$ matrix C' that results from the deletion of column 1 and row n from C. Standard assignment algorithms require no more than $O(n^3)$ time. The construction of an optimal tour, as indicated in the proof of Theorem 2 requires considerably less time. The reader may be interested in verifying that the construction requires no more than $O(n)$ time.

As a simple example, let

$$C = \begin{bmatrix} 0 & -1 & 7 & -20 & 3 & -2 & 5 \\ 0 & 0 & 12 & 8 & 16 & 9 & 8 \\ 0 & 0 & 0 & 3 & 7 & 6 & 2 \\ 0 & 0 & 0 & 0 & 4 & 4 & 9 \\ 0 & 0 & 0 & 0 & 0 & -18 & -1 \\ 0 & 0 & 0 & 0 & 0 & 0 & 3 \\ 0 & 0 & 0 & 0 & 0 & 0 & 0 \end{bmatrix}$$

Then

$$C' = \begin{bmatrix} -1 & 7 & -20 & 3 & -2 & 5 \\ 0 & 12 & 8 & 16 & 9 & 8 \\ 0 & 0 & 3 & 7 & 6 & 2 \\ 0 & 0 & 0 & 4 & 4 & 9 \\ 0 & 0 & 0 & 0 & -18 & -1 \\ 0 & 0 & 0 & 0 & 0 & 3 \end{bmatrix}$$

and an optimal solution to the assignment problem is indicated by the encircled entries. This is converted to an optimal solution to the traveling salesman problem.

3. Graded Matrices

We say that a matrix C is *graded across its rows* if $c_{ij} \leq c_{i,j+1}$ for all i,j and *graded up its columns* if $c_{ij} \geq c_{i+1,j}$, for all i,j. A matrix is *doubly graded* if it is graded both across its rows and up its columns.

The TSP is NP-hard for graded, even doubly graded, matrices since any matrix can be made doubly graded by a linear admissible transformation; that is, by adding constants to its rows and columns. However, it is possible to obtain a useful approximation result for graded matrices, as we show below. In Section 10 we show that there is a polynomial bounded algorithm for obtaining an exact solution to the bottleneck TSP for graded matrices.

Theorem 3. Let C be nonnegative and graded up its columns. Given an optimal assignment ϕ it is easy to construct a tour τ such that

$$c(\tau) \leq c(\phi) + \max_j \{c_{1j}\}.$$

Proof: If ϕ is a tour, let $\tau = \phi$. Else choose one city from each of the $m \geq 2$ subtours of ϕ, and let these cities be $i_1, i_2, ..., i_m$, with $i_1 < i_2 < ... < i_m$. Remove arcs $(i_1, \phi(i_1)),... (i_m, \phi(i_m))$ from ϕ and substitute arcs $(i_2, \phi(i_1))$, $(i_3, \phi(i_2))$, ..., $(i_{m-1}, (i_{m-2}))$, $(i_m, \phi(i_{m-1}))$ and $(i_1, \phi(i_m))$. From

$$\begin{aligned} c_{i_1, \phi(i_1)} &\geq c_{i_2, \phi(i_1)}, \\ c_{i_2, \phi(i_2)} &\geq c_{i_3, \phi(i_2)}, \\ &\vdots \\ c_{i_{m-1}, \phi(i_{m-1})} &\geq c_{i_m, \phi(i_{m-1})}, \end{aligned}$$

it follows that

$$\begin{aligned} c(\tau) &\leq c(\phi) + c_{i_1, \phi(i_m)} - c_{i_m, \phi(i_m)} \\ &\leq c(\phi) + \max_j \{c_{ij}\}. \end{aligned}$$

□

As a very simple example, let

$$C = \begin{bmatrix} 5 & 4 & 3 & 2 & 1 & 0 \\ 4 & 3 & 2 & 1 & 0 & 0 \\ 3 & 2 & 1 & 0 & 0 & 0 \\ 2 & 1 & 0 & 0 & 0 & 0 \\ 1 & 0 & 0 & 0 & 0 & 0 \\ 0 & 0 & 0 & 0 & 0 & 0 \end{bmatrix} \tag{3}$$

with an optimal assignment ϕ indicated by the encircled entries. Note that ϕ has three subtours: (1, 6), (2, 5), and (3, 4). Letting $i_1 = 1, i_2 = 2, i_3 = 3$, we convert ϕ to a tour τ. This gives us $c(\tau) = c(\phi) + c_{31} - c_{61} = c(\phi) + 3$.

4. The Theory of Subtour Patching

In this section we shall consider the following strategy for finding an optimal tour: First find an optimal assignment ϕ. If ϕ is a tour, it is clearly an optimal tour, and we are done. Otherwise, it consists of several cycles or "subtours." Modify ϕ so as to "patch" these subtours together to yield a single tour τ that is optimal. Thus our strategy is: given an *optimal* assignment ϕ, find a ψ such that $\phi\psi$ is an *optimal* tour.

Our first task is to investigate conditions on ψ under which $\phi\psi$ is a tour. We shall assume that the reader is somewhat familiar with the notion of a hypergraph. (A hypergraph is like an ordinary graph except that its "hyper" edges may be incident to arbitrary subsets of vertices, instead of to only subsets of size two.)

Let $P = \{\rho_1 \rho_2 ..., \rho_m\}$ be a set of (not necessarily disjoint) cycles on subsets of $N = \{1,2,...,n\}$. Let $H(N,P)$ be a hypergraph with vertex set N and with hyperedges corresponding to cycles in P. The hyperedge for ρ_i is incident to exactly those elements of N on which ρ_i acts. For example, if we have $\rho_1 = (2,3,4)$, $\rho_2 = (1,5)$, $\rho_3 = (1,3,2,4)$, then $H(N,P)$. Note that $\tau = \rho_1 \rho_2 \rho_3 = (1,2,3,4,5)$ is a tour.

A hypergraph $H(N,P)$ is *disconnected* if it is possible to partition its node set N into two nonempty parts S and T such that no hyperedge is incident to a node in S and also to a node in T. A hypergraph is *connected* if it is not disconnected.

Theorem 4 If $\tau = \rho_1 \rho_2 ... \rho_m$ is a tour, then $H(N,P)$ is connected.

Proof: Consider the contrapositive. Suppose $H = (N,P)$ is disconnected. Then there exist nonempty S and T such that $\rho_i(j) \in S$ if and only if $j \in S$, for all cycles ρ_i and cities j. Thus $\tau(j) \in S$ if and only if $j \in S$, where $\tau = \rho_1 \rho_2 ..., \rho_m$, and τ cannot be a tour. □

Corollary 5 Let ϕ, ψ have factors ϕ_i, $i = 1,2,...,r$, and ψ_j, $j = 1,2,...,s$, respectively. If $\phi\psi$ is a tour then $H(N, \{\phi_i\} \cup \{\psi_j\})$ is connected.

Corollary 5 gives a necessary condition on ψ for $\phi\psi$ to be a tour. We now seek a sufficient condition.

A hypergraph $H(N,P)$ with n vertices and m edges $\rho_1,...,\rho_m$ is a *tree* if it is connected and if

$$\sum_{i=1}^{m} (|\rho_i| - 1) = n - 1,$$

where $|\rho_i|$ is the number of nodes incident to ρ_i. Note that this definition generalizes the well known condition for a graph G to be a tree, i. e. G is connected and $m = n - 1$. Also note that this definition allows a tree to have edges that are incident to a single node and that such edges (self loops) can be added to or deleted from the hypergraph without affecting its status as a tree. It is a straightforward exercise to show that if $H(N,P)$ is connected, then

$$\sum_{i=1}^{m} (|\rho_i| - 1) \geq n - 1.$$

It follows that if $H(N,P)$ is a tree, then the deletion of any edge ρ_i, with $|\rho_i| \geq 2$, disconnects H. Moreover, any edge ρ_i must be incident to exactly one vertex in each of the subtrees formed by its deletion. With these observations, it is not difficult to provide an inductive proof of the following.

Theorem 6 If $H(N,P)$ is a tree, then $\tau = \rho_1 \rho_2 ... \rho_m$ is a tour (where the order in which the cycles ρ_i are multiplied to obtain τ is immaterial).

Corollary 7 Let ϕ, ψ have factors ϕ_i, $i = 1,2,...,r$, and ψ_j, $j = 1,2,...,s$, respectively. If $H(N, \{\phi_i\} \cup \{\psi_j\})$ is a tree then $\phi\psi$ is a tour.

We can now reinterpret the procedure for upper triangular matrices presented in Section 5. The permutation ψ that is found is in the form of a single cycle acting on exactly one city in each subtour of ϕ. It follows that $H(N, \{\phi_i\} \cup \{\psi_j\})$ is a tree and, by Corollary 6, $\phi\psi$ is a tour. The tour $\phi\psi$ is optimal because $c(\phi)$ is a lower bound on the length of a tour and $c(\phi\psi) = c(\phi)$. (Recall that in this case ϕ is not necessarily an optimal assignment, but is only optimal subject to the condition that $\phi(n) = 1$; the fact that $c(\phi)$ is a lower bound is a nontrivial result.)

The case of upper triangular matrices suggests that we investigate conditions under which there exists a ψ such that $H(N, \{\phi_i\} \cup \{\psi_j\})$ is a tree and $\phi\psi$ is an optimal tour either of the ordinary or the bottleneck variety. For simplicity, when considering the bottleneck TSP, we shall always assume that $\bar{c}(\phi) = 0$, in order to have $\bar{c}\phi(\psi) = \bar{c}(\phi\psi)$. (If this is not so, apply the transformation $c_{ij} := max\ \{0,\ c_{ij} - k\}$, where k is the value of a bottleneck optimal assignment.)

Theorem 8 Let ψ have factors ψ_j, $j = 1,2,...,s$. Then

$$c\phi(\psi) = \sum_j c\phi(\psi_j)$$

$$\bar{c}\phi(\psi) = max_j\ \{\bar{c}\phi(\psi_j)\}\ .$$

Proof: Note that

$$c\,\phi(\psi_j) = c(\phi\,\psi_j) - c(\phi)$$

$$= \sum_i c_{i,\phi\psi_j(i)} - \sum_i c_{i,\phi(i)}$$

$$= \sum_{i:\psi_j(i)\neq i} c_{i,\phi\psi_j(i)} - \sum_{i:\psi_j(i)\neq i} c_{i,\phi(i)} .$$

It follows that

$$\sum_j c\ \phi(\psi_j) = \sum_{i:\psi(i)\neq i} c_{i,\phi\psi(i)} - \sum_{i:\psi(i)\neq i} c_{i,\phi(i)}$$

$$= \sum_i c_{i,\phi\psi(i)} - \sum_i c_{i,\phi(i)}$$

$$= c\ (\phi\psi) - c\ (\phi)$$

$$= c\ \phi(\psi) .$$

The proof for $\bar{c}\ \phi\ (\psi)$ is similar. □

Theorem 8 tells us that we can deal with the factors of ψ independently of each other. This fact is useful, but by itself does not help us much in finding a ψ such that $\phi\psi$ is an optimal tour. We shall adopt the approach of building up ψ as the product of transpositions (cycles of length two). Of necessity, these transpositions will generally not be factors of ψ (i. e. they will not be disjoint), so Theorem 8 will not apply.

As an example, suppose we have an 8-city problem, with $\phi = (1,2,3), (4,5), (6,7), (8)$. The hypergraph $H = (N, \{\phi_i\})$. If we add edges for the transpositions (2,4), (5,7) and (7,8), we obtain the hypergraph tree. By Theorem 6, post-multiplication of ϕ by the transpositions (2,4), (5,7), (7,8), in any order, results is a tour. The order in which the transpositions are multiplied determines ψ. For example, we can have either

$$\psi = (2,4)\ (5,7)\ (7,8) = (2,4)\ (5,7,8)$$

or

$$\psi = (2,4)\ (7,8)\ (5,7) = (2,4)\ (5,8,7) .$$

But no matter what order the transpositions are multiplied, the cyclic factors ψ correspond to the connected components of the graph of transpositions.

With respect to a given optimal assignment ϕ, let us assign a *length* to each transposition (i,j):

$$c\phi((i,j)) = c_{i\phi(j)} + c_{j\phi(i)} - c_{i\phi(i)} - c_{j\phi(j)},$$

or

$$\bar{c}\phi((i,j)) = max\ \{c_{i\phi(j)}, c_{j\phi(i)}\}.$$

We can find a minimum length set of transpositions $\rho_1 \rho_2 ..., \rho_t$ such that the hypergraph $H(N, \{\phi_i\} \cup \{\rho_j\})$ is a tree, by solving a minimum spanning tree problem for the (multi)graph that is obtained by contracting each of the hyperedges of $H(N, \{\phi_i\})$. (Note that after contraction of the subtours (1,2,3), (4,5), (6,7) the transpositions (2,4), (5,7), (7,8) form a tree. We shall say that a set of transpositions is a *minimum spanning tree* (with respect to ϕ) if it is an optimal solution to such a spanning tree problem.

What relationship is there between the length of a minimum spanning tree and the length of an optimal tour? In the following two theorems we state conditions under which a minimum spanning tree yields a lower bound or an upper bound on the length of an optimal tour.

Theorem 3 Given an optimal assignment ϕ with respect to matrix C, suppose for any cyclic permutation ρ there exists a set T of transpositions connecting the same subtours in which there are cities on which ρ acts, such that $c\phi(T) \le c\phi(\rho)$ (or $\bar{c}\phi(T) \le \bar{c}\phi(\rho)$). Then if T is a minimum spanning tree, $c\phi(T) \le c\phi(\tau)$ (or $\bar{c}\phi(T) \le \bar{c}\phi(\tau)$), where τ is an optimal tour.

Proof: Let τ be an optimal tour and let $\psi = \phi^{-1}\tau$ have factors $\psi_1, ..., \psi_s$. Then $c\phi(\psi) = \sum c\phi(\psi_j)$, by Theorem 8. By hypothesis, there exists a connecting set of transpositions T_j for each ψ_j, with $c\phi(T_j) \le c\phi(\psi_j)$. Moreover, there exists a tree $T \subseteq T_1 \cup T_2 \cup ... \cup T_s$ that spans all the subtours of ϕ with $c\phi(T) \le c\phi(\psi)$. The proof of the bottleneck case is similar. □

Theorem 16 Given an optimal assignment ϕ with respect to matrix C, suppose for any set T of transpositions there exists a cyclic permutation ρ, acting on the same cities acted on by the transpositions in T such that $c\phi(\rho) \le c\phi(T)$ (or $\bar{c}\phi(\rho) \le c\phi(T)$). Then if T is a minimum spanning tree, $c\phi(T) \ge c\phi(\tau)$ (or $\bar{c}\phi(T) \ge \bar{c}\phi(\tau)$), where τ is an optimal tour.

Proof: Let T be a minimum spanning tree. Consider the graph with node set N and edge set T. Each connected component of this graph is a tree T_j and by hypothesis there exists a cyclic permutation ψ_j, acting on the same cities spanned by T_j, for which $c\phi(\psi_j) \le c\phi(T_j)$. By Corollary 7, $\phi\psi$ is a tour, where the permutation ψ has the ψ_j as its factors. By Theorem 8, $c\phi(\psi) = \sum_j c\phi(\psi_j)$. Hence there exists a tour $\phi\psi$ whose length is bounded from above by $c\phi(T)$. The proof of the bottleneck case is similar. □

In the sections that follow we shall consider some special classes of matrices for which the hypotheses of Theorems 9 or 10, or both, are satisfied.

5. The Bottleneck TSP for Graded Matrices

We shall now apply the theory developed in the previous section to obtain an efficient algorithm for solving the bottleneck TSP for graded matrices. (Recall that in Section 3 we gave an algorithm for obtaining an approximate solution for the ordinary TSP for the same class of matrices.)

Let C be graded up its columns and let ϕ be a bottleneck optimal assignment. (Such an assignment can be found in $O(n^2)$ time; see Exercises.) Without loss of generality, assume for all i, j

$$c_{ij} \geq \bar{c}(\phi), \qquad \text{for all } i, j. \tag{14}$$

If (14) does not hold, then apply the transformation

$$c_{ij} := \max\{0, c_{ij} - \bar{c}(\phi)\},$$

(This transformation preserves grading.)

Suppose we now permute the columns of C into the order $\phi(1), \phi(2), \ldots \phi(n)$. Assuming that (14) holds, we now have a *permuted* upper triangular matrix C^{ϕ} that is graded up its columns. (Note the c_{ij} still refers to the i, j^{th} entry of the original matrix; $c_{i\phi(j)}$ designates the i, j^{th} entry in the permuted matrix.) As an example, consider the permuted version of the matrix (3):

$$C^{\phi} = \begin{bmatrix} 0 & 1 & 2 & 3 & 4 & 5 \\ 0 & 0 & 1 & 2 & 3 & 4 \\ 0 & 0 & 0 & 2 & 2 & 3 \\ 0 & 0 & 0 & 0 & 1 & 2 \\ 0 & 0 & 0 & 0 & 0 & 1 \\ 0 & 0 & 0 & 0 & 0 & 0 \end{bmatrix} \tag{15}$$
$$\begin{matrix} 6 & 5 & 4 & 3 & 2 & 1 \end{matrix}$$

Here, as suggested by the listing of indices at the bottom of the matrix, $\phi(1) = 6, \phi(2) = 5, \ldots, \phi(6) = 1$.

We first want to show that the lower bound property of Theorem 9 holds. We must show that, for any cyclic permutation ρ, there exists a tree T of transpositions, spanning the same subtours of ϕ in which there are cities on which ρ acts, such that $\bar{c}\phi(T) \leq \bar{c}\phi(\rho)$. So let ρ be an arbitrary cycle. Because C^{ϕ} is a permuted upper triangular matrix,

$$\bar{c}\,\phi(\rho) = \max\,\{\, c_{i,\phi\rho(i)} \mid i < \rho(i) \,\}\,.$$

Because C is graded up its columns, we have for all $i \le i' < \rho(i)$,

$$\bar{c}\,\phi((i',\rho(i))) = c_{i',\phi\rho(i)} \le c_{i,\phi\rho(i)} = \bar{c}\,\phi((i,\rho(i))).$$

It follows that the set of transpositions

$$T = \{\,(i',\rho(i)) \mid i \le i' < \rho(i)\,\} \qquad (16)$$

connects the cities acted on by ρ, with $\bar{c}\,\phi(T) = \bar{c}\,\phi(\rho)$. If this set of transpositions is not a tree then transpositions can be removed from the set to obtain a tree T satisfying the hypotheses of the theorem.

In our example (15), ϕ has three subtours: (1,6), (2,5), (3,4). Let $\rho = (1,6,2,4)$, with $\bar{c}\,\phi(\rho) = \bar{c}\,\phi((1,6)) = c_{11} = 5$, the set defined by (16) is

$$T = \{\,(1,6), (2,6), (3,6), (4,6), (5,6), (2,4), (3,4)\,\}\,.$$

From this set we can select, for example, (4,5) and (5,6) to obtain a tree. Note that $\bar{c}\,\phi((4,5)) = c_{42} = 1$ and $\bar{c}\,\phi((5,6)) = c_{52} = 0$, so the bottleneck length of this tree is unity.

Now let us show that the upper bound property of Theorem 10 holds. We must show that for any tree T of transpositions there exists a cyclic permutation ρ, acting on the same cities spanned by T, such that $\bar{c}\,\phi(\rho) \le \bar{c}\,\phi(T)$. Let us define a partial order "$\le$" on transpositions: $(i,j) \le (i',j')$ if $i' \le i$ and $j \le j'$, where $i<j$ and $i'<j'$. Now remove from T all transpositions that are not maximal with respect to the partial order. The remaining transpositions are of the form $(i_1,j_1),(i_2,j_2),\ldots,(i_r,j_r)$, where $i_1<i_2<\cdots<i_k$, and i_1, j_r are the cities of smallest and largest index spanned by T. Because the transpositions in T connect the cities on which they act, we must have $j_1 \ge i_2, j_2 \ge i_3, \ldots, j_{r-1} \ge i_r$. Let $k_1>k_2>\cdots>k_s$ be the cities different from $i_1, i_2, \ldots, i_r$, $j_1, j_2, \ldots, j_r$ that are acted on by transpositions in T.

For example, suppose $T = \{\,(1,3), (2,4), (3,7), (4,5), (5,6)\,\}$. The transpositions in T that are maximal with respect to "$\le$" are $(i_1, j_1) = (1,3)$, $(i_2, j_2) = (2,4)$, $(i_3, j_3) = (3,7)$, and $k_1 = 6, k_2 = 5$. Note that $j_1 = 3 > i_2 = 2$, $j_2 = 4 > i_3 = 3$.

We assert that if we take the sequence $(i_1, j_1, i_2, j_2, \ldots\ i_r, j_r, k_1, \ldots, k_s)$, strike out the second occurrence of any index from within it, and treat the result as a cycle ρ, we have $\bar{c}\,\phi(\rho) \le \bar{c}\,\phi(T)$, as required. (The reader is asked to verify that this fact as an exercise.) In the case of our running example (15), suppose $T = \{(4,5),(5,6)\}$. Then we obtain $\rho = (4,5,6)$, with $\bar{c}\,\phi(\rho) = 1 \le \bar{c}\,\phi(T) = 1$.

The algorithm for solving the bottleneck TSP for graded matrices is as follows:

(1) Find a bottleneck optimal assignment ϕ. This requires $O(n^2)$ time.

(2) Determine the subtours of ϕ. This can be done in $O(n)$ time.

(3) Find a minimum spanning tree T of transpositions spanning the subtours of ϕ. This can be done in $O(n^2 \text{loglog} n)$ time.

(4) For each connected component T_j of the graph $G(N,T)$, find a cyclic permutation ψ_j, with $\bar{c}\,\phi(\psi_j) \le \bar{c}\,\phi(T_j)$, as described above. This requires at most $O(n \log n)$ time.

(5) Multiply ϕ by ψ in $O(n)$ time.

It is seen that the overall running time is $O(n^2 \text{loglog} n)$, the time required for the minimum spanning tree computation. As we shall see, this time bound can be reduced for the more specialized case of permuted doubly graded matrices.

Recall that a matrix is *doubly graded* if it is both graded across its rows and up its columns, i.e. both $c_{ij} \le c_{i,j+1}$ and $c_{ij} \ge c_{i+1,j}$.

Theorem 11 If C is doubly graded then a bottleneck optimal tour is given by the permutation $(1,2,\ldots,n-1,n)$.

Proof: Suppose we apply the nearest neighbor rule, starting at city 1. We shall show by induction that the path $1,2,\ldots,k$ contains no arc longer than the longest arc in a bottleneck optimal tour. Combining this result for $k=n$ with the fact that $(n,1)$ must be as short as any arc from n, we get the desired result. The basis is easy, since $(1,2)$ must be as short as any arc leaving 1. Suppose, by the inductive assumption, the path $1,2,\ldots,j$ contains no arc longer than the longest arc in a bottleneck optimal tour. Since arc $(j,j+1)$ is as short as any arc from the subset of cities $\{1,2,\ldots,j\}$ to the subset $\{j+1,\ldots,n\}$ the path may be extended to $j+1$. □

A matrix C is *permuted doubly graded* if there exists a permutation ϕ such that both $c_{i\phi(j)} \le c_{i\phi(j+1)}$ and $c_{i\phi(j)} \ge c_{i+1,\phi(j)}$.

Theorem 12 If C is permuted doubly graded with respect to ϕ, then ϕ is a bottleneck optimal assignment.

Proof: Left as an exercise. □

The principal difference between the ordinary graded case, discussed above, and the permuted doubly graded case is that in the latter case there exists a minimum spanning tree composed of only transpositions of the form $(i,i+1)$. Since there are at most $n-1$ transpositions to consider the time bound for the spanning tree computation can be reduced to $O(n \text{loglog} n)$.

In order to prove the assertion of the previous paragraph, it is sufficient to show that

$$\bar{c}\phi(i',i'+1) \leq \bar{c}\phi(i,j),$$

for all $i \leq i' < j$. We leave the details to the reader as an exercise.

6. An Application: Cutting Wallpaper

The problem dealt with in this section was formulated and solved in the context of reading records from a rotating storage device. [Fuller, 1972]. A much more specialized version of this problem was dealt with in [Garfinkel, 1977].

Suppose we are to cut $n-1$ sheets of wallpaper from a very long roll of stock with a pattern that repeats at intervals of one unit. Sheet i, $i = 1,2,\ldots,n-1$, starts at s_i (mod 1) and finishes at f_i (mod 1), with reference to the zero point of the pattern. If we cut sheet i from the roll immediately before sheet j, the intersheet waste is the distance from f_i to s_j, i. e.

$$c_{ij} = \begin{cases} s_j - f_i & \text{if } f_i \leq s_j \\ 1 + s_j - f_i & \text{otherwise} \end{cases} \tag{17}$$

$$= s_j - f_i \quad (\text{mod } 1)\,.$$

Suppose the roll begins at the zero point on the pattern and after cutting our $n-1$ sheets from the roll we must make one more cut to restore the roll to the zero point. In other words, we must minimize the total intersheet waste, rounded up to the nearest pattern unit. To formulate this problem as a TSP, we introduce an nth dummy sheet with $s_n = f_n = 0$.

For example, suppose we wish to cut four sheets of wallpaper, with $s_1 = .1, s_2 = .8, s_3 = .6, s_4 = .4$ and $f_1 = .8, f_2 = .7, f_3 = .7, f_4 = .2$. After creating a dummy fifth sheet with $s_5 = f_5 = 0$, we obtain a five-city TSP with

$$C = \begin{bmatrix} .3 & 0 & .8 & .6 & .2 \\ .4 & .1 & .9 & .7 & .3 \\ .4 & .1 & .9 & .7 & .3 \\ .9 & .6 & .4 & .2 & .8 \\ .1 & .8 & .6 & .4 & 0 \end{bmatrix}$$

Let us apply a linear admissible transformation to the matrix $C = (c_{ij})$ by adding f_i to row i and subtracting s_j from column j. The result is a $(0,1)$ matrix C', where

$$c'_{ij} = \begin{cases} 0 & \text{if } f_i \leq s_j \\ 1 & \text{otherwise} \end{cases}. \tag{18}$$

If the sheets are indexed so that $f_1 \geq f_2 \geq \ldots \geq f_n$, the matrix C' is graded up its columns. In the case of our example, we now have

$$C' = \begin{bmatrix} 1 & 0 & 1 & 1 & 1 \\ 1 & 0 & 1 & 1 & 1 \\ 1 & 0 & 1 & 1 & 1 \\ 1 & 0 & 0 & 0 & 1 \\ 0 & 0 & 0 & 0 & 0 \end{bmatrix}$$

For any tour τ, $c(\tau) = c(\tau') + \sum s_j - \sum f_i$. In other words, the cost of an optimal tour for the TSP for C' differs from the cost of an optimal tour for the TSP for C by a constant, $\sum s_j - \sum f_i$.

Since the matrix C' is graded up its columns, we can apply Theorem 9 and solve an assignment problem over C' to obtain an approximate solution to the TSP whose length differs from that of an optimal tour by no more than the value of the largest element in C', namely one unit. However, we can do better than this. In the following we shall show how to obtain a strictly optimal solution and, moreover, to obtain it in $O(n \log n)$ time.

There is a trick that simplifies matters and this involves changing the zero point of the pattern. Suppose we add a constant δ to each of the f_i and s_j values and redefine the problem in terms of $f_j{}'$ and $s_j{}'$ values, where

$$f_j{}' = f_i + \delta \quad (\text{mod } 1)$$

$$s_j{}' = s_j + \delta \quad (\text{mod } 1) .$$

Such a translation does not affect the matrix C, as defined by (17), since

$$s_j{}' - f_j{}' = s_j - f_i \quad (\text{mod } 1)$$

However, this translation does change the matrix C' as defined in (18).

In the case of our example, if we take $\delta = .3$, we obtain $s_1{}' = .4, s_2{}' = .1, s_3{}' = .9, s_4{}' = .7$, $s_5{}' = .3$, and $f_1{}' = .1, f_2{}' = 0, f_3{}' = 0, f_4{}' = .5, f_5{}' = .3$. The matrix C' then becomes

$$C' = \begin{bmatrix} 0 & 0 & 0 & 0 & 0 \\ 0 & 0 & 0 & 0 & 0 \\ 0 & 0 & 0 & 0 & 0 \\ 1 & 1 & 0 & 0 & 1 \\ 0 & 1 & 0 & 0 & 0 \end{bmatrix}$$

Suppose it is possible to find a δ such that C' can be made upper triangular and doubly graded after (independent) permutations of rows and columns. If such permutations exist, they can be effected by renumbering so that $f_1' \geq f_2' \geq \ldots \geq f_n'$ and then applying a permutation of ϕ to the columns so that $s'_{\phi(1)} \geq s'_{\phi(2)} \geq \ldots \geq s'_{\phi(n)}$. Since the permuted matrix is upper triangular, we have $c'(\phi) = 0$. Since C' is a (0,1)-matrix, it follows from Theorem 9 that an optimal tour τ is such that either $c'(\tau) = 0$ or $c'(\tau) = 1$. But the case $c'(\tau) = 0$ holds if and only if there is a tour with bottleneck length zero. And we know how to find a bottleneck optimal tour, in $O(n \log \log n)$ time, by the methods of the previous section.

Now all that remains is to show that it is always possible to find a δ such that the matrix C' becomes upper triangular and doubly graded after permutations of its rows and columns. This will be achieved if we can find a δ such that the largest s_j' is no smaller than any of the n f_i' values, the second largest s_i' is no smaller than $n - 1$ of the f_i' values, and so forth.

To see that there is such a δ, we adapt a problem and its solution from Lovász, *Combinatorial Problems and Exercises*, (Problem 21, p. 27). Suppose we are to walk around a circle on which there are n points f_i at which we are paid one dollar and n points s_j at which we must pay one dollar. Is there any point on the circle at which we can start with an empty wallet and never be financially embarrassed? And if so, how can we find this point?

The solution: Take a wallet full of money and start walking around the circle, starting at any point. Since we take in $\$n$ and give out $\$n$, we have the same amount of money in our wallet when we return to the starting point. Now remember where on the circle we had the least money. This was surely in an interval between an s_j and an f_i. Make that f_i our new starting point, i.e. set $\delta = -f_i$ and we shall have accomplished our objective.

To summarize the procedure for the wallpapering problem:

(1) Sort the f_i and s_i values, in $O(n \log n)$ time.

(2) Find the value δ and an optimal assignment ϕ in $O(n)$ time.

(3) Apply the algorithm for solving the bottleneck TSP for permuted doubly graded matrices. This requires $O(n \log \log n)$ time.

(4) If the bottleneck optimal tour found in (3) has zero length, it is an optimal solution to the problem. Else apply the approximation method of Theorem 9, to obtain an optimal tour. In this case, since an optimal assignment ϕ is already known, only $O(n)$ additional time is required.

The running time for solving the problem is dominated by the $O(n \log n)$ time required to sort the f_i and s_i values.

In the case of our example, an optimal solution is given by the tour $(1, 4, 3, 2, 5)$, with a length of 1.5.

References

[Fuller, 1972] Samuel H. Fuller, "An Optimal Drum Scheduling Algorithm," *IEEE Trans. Computers*, C-21 (1972) 1153-1165.

[Garfinkel, 1977] R.S. Garfinkel, Minimizing Wallpaper Waste, part I: a class of traveling salesman problems. *Oper. Res.* 25 (1977) 741-751.

[Lawler, 1972] Eugene L. Lawler, "A Solvable Case of the Travelling Salesman Problem," *Math. Programming* 1 (1971) 267-269.

A polynomial algorithm for partitioning line-graphs

Claudio Arbib
Dipartimento di Ingegneria Elettronica, Università di Roma "Tor Vergata"
v. O. Raimondo - 00173 Roma, Italy

Abstract

Two NP-complete unweighted graph partitioning problems are considered: Simple Max Partitioning (SMP) and Uniform Graph Partitioning (UGP). For both problems, polynomial-time algorithms are available for special classes of graphs. In the present paper, the class of line-graphs is considered and a polynomial algorithm is proposed to solve both UGP and SMP in this class.

1. Notations and preliminary definitions

In the following, a graph will be denoted by a pair (V,E), where V represents the set of vertices and E the set of edges. We also denote by V(G) and E(G) the set of vertices and the set of edges of G, respectively. If necessary, the cardinality of V(G) appears as a subscript, for example G_n. A complete graph with n vertices (clique) is denoted by K_n; a cycle is denoted by C_n. If (V,E) is a graph and $x \in V$, we set $S(x) \equiv (\{x\},E')$, with $E' \equiv \{e \in E \mid e=(x,y)\}$.

The term *graph partition* indicates a partition of the vertex-set of a graph. Any graph partition $<V_1,V_2>$ defines a *cut* C, which is the subset of edges in E having one extreme in V_1 and the other one in V_2. The cardinality of the cut associated with p will be denoted by $|p|$. A graph partition p is *maximum* if $|p| \geq |q|$ for any graph partition q.

If G=(V,E) is a graph, $p=<V_1,V_2>$ a partition of V and G'=(V',E') a subgraph of G, we say that p *induces* on G the partition $p'=<V' \cap V_1, V' \cap V_2>$ and we write $p' = p_{|G'}$. If $\wp(V)$ is the set of all the partitions $<V_1,V_2>$ of V into two subsets, let us define a function $\beta: \wp(V) \rightarrow \mathbb{Z}^+$ by $\beta(<V_1,V_2>) = ||V_1| - |V_2||$. A partition p of V will be called *uniform* if $\beta(p) \leq 1$.

2. Two Unweighted graph partitioning problems

On an unweighted graph, the UNIFORM GRAPH PARTITIONING problem (UGP) can be formulated as follows:

Given a symmetric graph G=(V,E), find a uniform partition of V having maximum cardinality.

Such a graph partition is called *UGP-optimal*. Any uniform partition of V is called *UGP-feasible*. If we relax the uniformity constraint, we get the problem of SIMPLE MAX PARTITIONING (SMP)[1] .

The Uniform Graph Partitioning problem is NP-complete: Garey, Johnson and Stockmeyer proved in fact [6] that UGP can be polynomially reduced from Simple Max Cut (i.e. given a graph (V,E), s,t $\in$ V and a positive integer k, find a partition p of V separating s from t and such that $|p| \geq k$), which is in turn as difficult as 3-Satisfiability. As for the complexity of SMP, it is easy to see that

2.1 - THEOREM: Simple Max Partitioning is NP-complete.

The proof is trivially obtained by reduction from Simple Max Cut, and hence is omitted.

3. Preliminary results

In this Section we shall give some preliminary results.

3.1 - LEMMA: For any partition p of K_n we have

$$(3.1.1) \qquad |p| = \frac{1}{4}(n^2 - \beta^2(p))$$

As a consequence, we have $|p| > |q|$ on a clique if and only if $\beta(p) > \beta(q)$. In particular, we immediately see that any uniform partition of a clique is both SMP- and UGP-optimal. Furthermore

3.2 - THEOREM: A graph G is a clique if and only if all the uniform partitions of G are SMP-optimal.

Observe that the "only if" part follows from Lemma 3.1. The "if" part can be easily proved by induction (see [1]).

3.3 - LEMMA: Let G be a graph and $\Gamma \equiv \{G^1,...,G^k\}$ a set of k subgraphs of G such that

$$(i) \qquad V(G) = \bigcup_{i=1}^{k} V(G^i)$$

$$(ii) \qquad E(G) = \bigcup_{i=1}^{k} E(G^i)$$

[1] Other Authors (see e.g. [5,6,8]) prefer to state UGP as a minimization problem. In this case, however, the notion of disconnecting set instead of cut is more indicated.

(iii) $$\bigcap_{i=1}^{k} E(G^i) = \emptyset$$

Assume that a partition p of V(G) exists such that $p_{/G^i}$ is SMP*-optimal for each G^i. Then p is also* SMP*-optimal for G.*

PROOF: Since the set $\{G^i\}$ induces a partition on E(G) (hypotheses (ii) and (iii)), it follows immediately that

(3.3.1) $$| p_{/G} | = \Sigma_{i=1..k} | p_{/G^i} |$$

As p removes the maximum number of edges from each G^i, we can conclude that p is also SMP-optimal for the whole graph G

◇

It can also be proved [1] that, if such partition exists, then all SMP-optimal partitions of G are of this form.

3.4 - LEMMA: If G^i is a clique and $p_{/G^i}$ is uniform for each G^i, then $p_{/G}$ is SMP*-optimal.*

PROOF: The assertion follows at once from Lemmas 3.2 and 3.3

◇

Recall that a graph H is called the *line-graph* of another graph R if a bijection μ: E(R) → V(H) exists such that for any e_1, e_2 in E(R), $e_1 \cap e_2 \equiv \emptyset$ if and only if $(\mu(e_1),\mu(e_2)) \in E(H)$.

The graph H clearly represents the incidence relation defined on E(R): this relation is non-reflexive, symmetric and non-transitive. In [9], the graph R is referred to as the *root graph* of H. The application **L** associating with any graph the corresponding line-graphs transforms the class Σ of all undirected graphs into a proper subset of itself.

Several characterization of **L**(Σ) have been given; Beineke [2] proved that a graph is a line-graph if and only if it does not contain any of nine forbidden graphs as induced subgraphs. As none of these subgraphs has more than six vertices, Beineke's Theorem leads in fact to an $O(n^6)$ algorithm for line-graph detection.

The most efficient algorithm known to date is due to Lehot [9], and runs in $O(n^2)$ time. When the input graph is in **L**(Σ), this algorithm also returns one of the corresponding root graphs.

A characterization of line-graphs in terms of their clique structure is given by the following theorem:

3.5 - THEOREM: A graph is a line-graph if and only if there exists a set of cliques Γ satisfying properties (i) through (iii) of Lemma 3.3 and such that every vertex belongs to at most two cliques in Γ.

PROOF: See [7].

Basically, there is a one-to-one correspondence between cliques in Γ and stars of the root graph. Such a set of cliques Γ associated with a line-graph H will be referred to as an *L-decomposition* of H. For example, $C_n \in \mathbf{L}(\Sigma)$ and, for n>3, the associated L-decomposition consists of n copies of K_2. Obviously, $K_n \in \mathbf{L}(\Sigma)$ for any n.

4. Line-graph partitioning

The purpose of this Section is to show that $\mathbf{L}(\Sigma)$ is a polynomially solvable set of instances for both SMP and UGP.

Let H=(V,E) be a line-graph and $\Gamma=\{H^i\}$ an L-decomposition corresponding to one of its root graphs, say R=(U,V). Consider then the following algorithm:

```
algorithm PARTITION

        input           R: graph;
        output          p: edge_partition;

        begin
           if R is eulerian and x in V(R)
           then p := colour (R) starting from x
           else begin
                    construct the graph R' by adding a vertex u_o to V(R)
                    and linking u_o to all vertices having odd degree in V(R);
                                                        /* R' turns out to be eulerian */
                    p' := colour (R') starting from u_o;
                    p := p'|E(R)
                end
        end.
```

where *colour (G) starting from u* is a procedure yielding an alternate bicolouring of the edges of the eulerian graph G along a eulerian circuit starting from vertex u. If $\pi = \{u_1,e_1,u_2,...,e_n,u_1\}$ is a eulerian circuit of G, an *alternate bicolouring* p=<Blue,Red> of π is a partition of $E(\pi) \equiv E(G)$ such that $e_i \in$ Blue if and only if $e_{i+1} \in$ Red. By definition of line-graph, the above algorithm returns a partition of V(H). The following theorems hold:

*4.1 - THEOREM: Algorithm PARTITION returns an SMP-optimal solution to the problem instance H=**L**(R).*

PROOF: Let us distinguish between two cases:

a. R has an even number of edges or is not eulerian;

b. R is eulerian and has an odd number of edges.

Case a:

We shall prove the assertion by showing that a set $\Gamma=\{H^i\}$ of subgraphs of H satisfying properties (i) through (iii) of Lemma 3.3 and a partition p of V(H) exist, such that $p_{|H^i}$ is uniform on each H^i. Then, by Lemma 3.4, the thesis will follow.

Let $\Gamma=\{H^i\}$ be the L-decomposition associated with R. Assume R to be eulerian and $|E(R)|=n$ even. Let $\pi = \{u_1,e_1,u_2,...,e_n,u_1\}$ be a eulerian circuit of R and p=<Blue,Red> an alternate bicolouring of π. Let us show that, for any $x \in U$, the bicolouring p is uniform on E(S(x)), where S(x) is defined as in Section 1. In fact, for any blue edge entering x there is a red one leaving it and vice versa, and this holds for every x, since $|E(R)|$ is even. Since each clique H^i is associated with a star in R, and the bijection μ preserves uniform partitions, we see that p is uniform in each clique $H^i \in \Gamma$.
Assume now that R is not eulerian. In other words, there exists a non-empty set $K \subseteq U$ of vertices with odd degree ($|K|$ even). Introduce a new vertex u_o and a set V_o of $|K|$ edges from u_o to K; then set $U' := U \cup \{u_o\}$, $V' := V \cup V_o$. The graph R'=(U',V') is eulerian.
Let now $\pi' = \{u_1,e_1,u_2,...,e_{n+|K|},u_1\}$ be a eulerian circuit of R and p' an alternate bicolouring of E(R'). Notice that, since n is not necessarily even, e_1 and $e_{n+|K|}$ might have the same colour: in other words, p may turn out to be non-uniform only on $S(u_o)$ (see fig. 4.1).
Set $p = p'_{|E(R)}$; p is clearly uniform on each star with center in E(R)\K. On the other hand, any star of R with center in K differs from the corresponding star of R' by a single edge only: hence p turns out to be uniform on all stars of R. As above, we can conclude that p is also uniform on each clique of the L-decomposition of H associated with R.

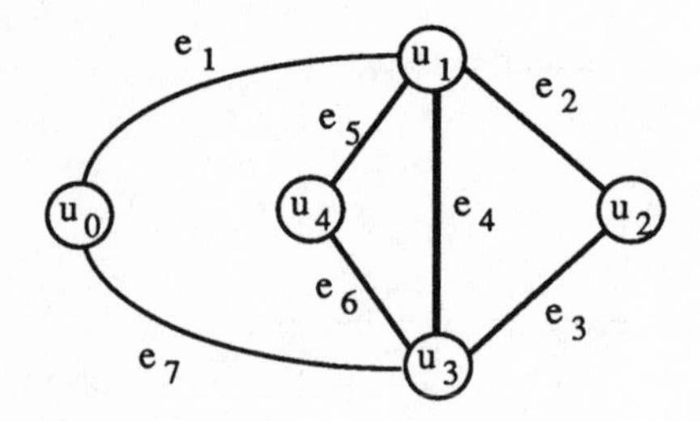

Fig. 4.1 - An alternate bicolouring gives the same colour to edges e_1 and e_7.

Case b:

If R is eulerian and $|V|$ is odd, the above argument does not apply. So we shall consider this case separately.
As previously pointed out, for any choice of a eulerian circuit $\pi = \{u_1,e_1,u_2,...,e_n,u_1\}$ and for any alternate bicolouring <Blue,Red> of π, the fact that n is odd implies that e_1 and e_n will be given the same colour. In other words, there exists no eulerian circuit π' which enters every star with a blue edge and leaves it with a red one. In view of Lemma 3.1, it is also easy to see that $|p|$ does not depend on the choice of the circuit π.

Let q=<Blue,Red> be an optimal partition of H. By Lemma 3.4, we have that $|q| \leq |p|+1$. Assume $|q| = |p|+1$. Again, by Lemma 3.4, $q_{|K^i}$ must be uniform for every $K^i \in \Gamma$, where Γ is the L-decomposition of H associated with R. A eulerian circuit π' of R can therefore be constructed by finding a collection of walks on H, each one obtained by "jumping" from a clique of Γ to another (observe that, since R is eulerian, every vertex of H is shared by two cliques at least). Each of these walks is constructed by alternately choosing blue and red vertices, which is always possible since $q_{|K^i}$ is uniform for every $K^i \in \Gamma$: π' is then obtained by considering the associated edges of R and applying Euler's Theorem.

By construction, q=<Blue,Red> is an alternate bicolouring of π' and is such that π' enters every star with a blue edge and leaves it with a red one or vice versa, which is a contradiction

◇

4.2 - THEOREM: The Simple Max Partitioning problem for a line-graph can be solved in $O(n^2)$ time.

PROOF: Recall that Lehot's algorithm recognizes line-graphs in $O(n^2)$ time. Furthermore, given a line-graph H, the algorithm yields a root graph R at no extra cost. On the other hand, the time required by Algorithm PARTITION is an $O(|E(R)|) = O(n)$, and hence the overall time required is dominated by $O(n^2)$

◇

4.3 - THEOREM: The Uniform Graph Partitioning problem for a line-graph can be solved in $O(n^2)$ time.

PROOF: It sufficient to show that p is a uniform partition of E(R). In fact, if R is eulerian, any alternate bicolouring of a eulerian circuit π of R is a uniform partition of $E(\pi) \equiv E(R)$. Since μ preserves uniform partitions, the thesis follows. If R is not eulerian, R' is eulerian by construction: consequently, any alternate bicolouring $p' = \langle V_1, V_2 \rangle$ of E(R') is uniform. Using the same notations as in Theorem 4.1 and taking into account the fact that $|V_o| = |K|$ is even, one sees that $|V_1 \cap V_o| = |V_2 \cap V_o|$. Hence, $p = p'_{|V} = \langle V_1 \backslash V_o, V_2 \backslash V_o \rangle$ is uniform too

◇

As an example, consider the line-graph of fig. 4.2.a. A corresponding root graph is shown in fig. 4.2.b, where the new vertex u_o and six added edges from e_{11} to e_{16} (dotted lines) are also indicated. The solution shown in fig. 4.2.a corresponds to the eulerian circuit $p = \{e_{11}, e_1, e_2, e_6, e_8, e_9, e_{15}, e_{16}, e_{10}, e_{14}, e_{13}, e_7, e_5, e_4, e_3, e_{12}\}$. Notice that not all the maximum partitions of a line-graph are also uniform. A counterexample is given in fig. 4.3, which shows a non-uniform maximum partition of the above graph. Hence, the proposed algorithm cannot produce, in general, all optimal solutions to SMP.

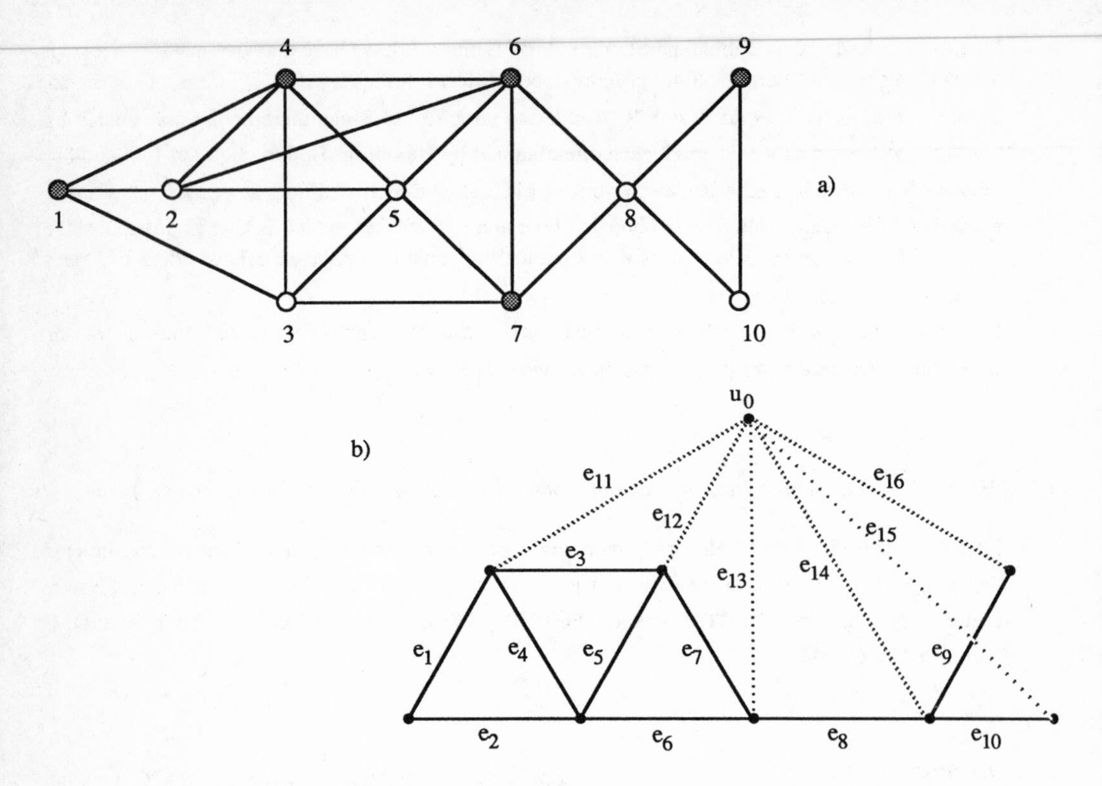

Fig. 4.2 - a) A simple max partition of a line-graph b) The associated root graph.

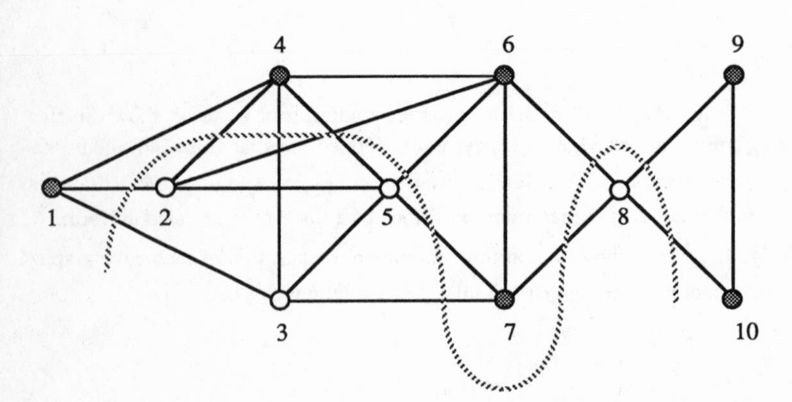

Fig. 4.3 - A non-uniform simple max partition of a line-graph.

5. A dynamic programming algorithm for solving the Simple Max Partitioning problem on line-graphs

Let be given a hypergraph $H=(V,\mathfrak{I})$, with $\mathfrak{I}$ family of subsets of V, and an edge $\xi\in\mathfrak{I}$ as the *root* of H. A hypergraph with root ξ is called *ξ-rooted*, or simply *rooted*. Given two rooted hypergraphs H', H" with roots ξ' and ξ'' respectively, define the hypergraph H = H'*H" as follows:

$H = H'*H''$ *if*

(i) $V(H) \equiv V(H') \cup V(H'') \cup \{u_o\}$

(ii) $\{u_o\} \cap (V(H') \cup V(H'')) \equiv \phi$

(iii) $E(H) \equiv (E(H') \setminus \{\xi'\}) \cup (E(H'') \setminus \{\xi''\}) \cup \{\psi', \psi''\}$

where $\psi' = \xi' \cup \{u_o\}$, $\psi'' = \xi'' \cup \{u_o\}$.

Actually, the hypergraph H is obtained by adding a new vertex u_o and enlarging the two roots ξ' and ξ'' so as to include $\{u_o\}$. Notice that H is not rooted.

Let us now consider the set Ω of all pairs (H,p), where H is a ξ-rooted hypergraph and p=<Blue,Red> is a bicolouring of V(H) satisfying the following property:

P_o: *For every* $\zeta\in E(H)$, $\beta(p_{|\zeta}) \le 2$ *if* $\zeta=\xi$ *and* $\beta(p_{|\zeta}) \le 1$ *otherwise.*

where the function β is defined as usual and $p_{|\zeta}$ is the bicolouring induced by p on ζ.

The elements of Ω can be divided into exactly five classes, according to whether the root is uniformly coloured or not:

P_1: *The root contains an odd number of vertices, the majority of which are blue.*

P_2: *Same as* P_1, *but the majority of the vertices are red.*

P_3: *The root is an even set and is uniformly coloured.*

P_4: *The root is an even set, and* $|Blue| - |Red| = 2$.

P_5: *The root is an even set, and* $|Red| - |Blue| = 2$.

Let us define the "*" operation over $\Omega\times\Omega$ so that Ω is closed with respect to "*": when composing two pairs (H',p') and (H",p"), we get a pair (H'*H",p) with p obtained by separating the blue vertices in H'*H" by the red ones and by giving vertex u_o a colour such that only one of the above five cases arises. Observe that, when ξ' is of type 4 and ξ'' is of type 5 (or vice versa), we must preliminarily exchange the colours of the vertices in one of the two hypergraphs (as a consequence, one can actually regard types 4 and 5 as a single class). The root of (H'*H") is chosen as the less uniformly coloured edge between ξ' and ξ''. If we use the symbols "♦","◇","Θ" to denote pairs satisfying properties P_1 to P_3 respectively, and the symbol "Ø" only for pairs satisfying property P_4 or P_5, we can write down the following commutative table:

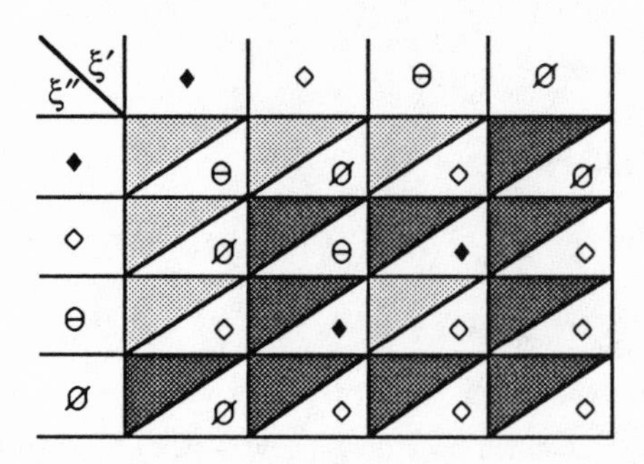

ξ'' \ ξ'	♦	◇	Θ	Ø
♦	Θ	Ø	◇	Ø
◇	Ø	Θ	♦	◇
Θ	◇	♦	◇	◇
Ø	Ø	◇	◇	◇

Tab. 5.1 - '*' operation.

where the shadowed areas indicate the colour to be given to u_o (light for red and dark for blue) and the symbol gives the equivalence class of the resulting pair.

Given $\zeta \in E(H')$, let us now define on the class ℵ of hypergraphs a unary operation $*_\zeta$ as follows:

$H = *_\zeta H'$ *if H' is ξ-rooted and*

(i) $\quad V(H) \equiv V(H') \cup \{u_o\}$

(ii) $\quad \{u_o\} \cap V(H') \equiv \phi$

(iii) $\quad E(H) \equiv (E(H') \setminus \{\xi,\zeta\}) \cup \{\psi',\psi''\}$

where $\psi' = \xi \cup \{u_o\}$, $\psi'' = \zeta \cup \{u_o\}$.

Similarly to the "*" operation, one can extend the definition of the "$*_\zeta$" operation to pairs in Ω so that Ω is closed with respect to "$*_\zeta$". If H is coloured so as to belong to one of the above equivalence classes and the hypotheses of application of "$*_\zeta$" are the same as "*", the following table can be applied:

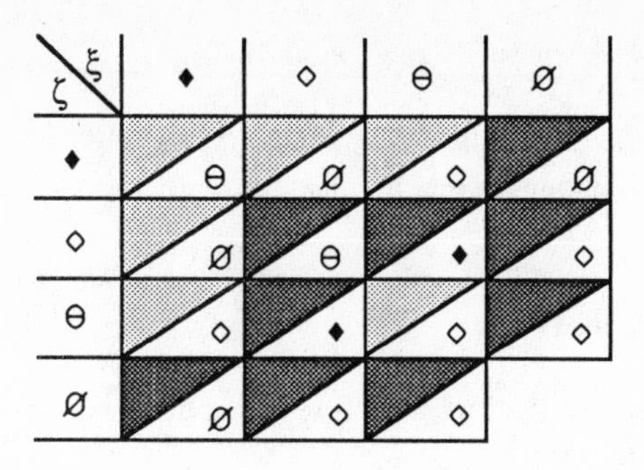

Tab. 5.2 - "$*_\zeta$" operation.

Notice that the "$*_\zeta$" operation table is identical to Tab. 5.1, except for the fact that, by assumption, ξ and ζ cannot be both non-uniformly coloured even sets.

It is immediately seen that the set Ω of all pairs satisfying P_0 can be characterized as follows:

(i) The empty hypergraph, with colouring $\{\phi,\phi\}$, belongs to Ω.

(ii) If (H',p') and (H",p") belong to Ω, then (H',p')(H",p") belongs to Ω.*

(iii) If (H,p) belongs to Ω and $\zeta \in E(H)$, then $*_\zeta(H,p)$ belongs to Ω.

An application of Theorem 3.5 shows that the projection of Ω on the class $\aleph$ of hypergraphs is nothing else than the class $\mathbf{L}(\Sigma)$ of line-graphs, provided that one substitutes a clique K_n for each n-edge. In this way, every pair (H,p) is mapped onto a pair (G,p) where G is a line-graph and p a partition of G. The four classes "♦","◇","ϴ","Ø" do not cover the set of all pairs (G,p) such that p is a partition of G. Lemma 3.4, however, guarantees that (H,p^*), with p^* SMP-optimal for the line-graph G associated with H, satisfies P_0 and can be obtained by successive compositions of elementary pairs satisfying properties P_1 to P_5. Given that, a homomorphism h mapping Ω onto the four classes "♦","◇","ϴ","Ø" exists, which preserves the properties satisfied by the graph $\times$ solution pairs and respects the defined composition operations. According to the general theory presented in [4], the above considerations lead to a linear-time dynamic programming algorithm for solving the Simple Max Partitioning problem on line-graphs.

References

[1] C. Arbib: "A polynomial characterization of some graph partitioning problems" - Inf. Proc. Letters (1987/88) pp. 223-230.

[2] L.W. Beineke: "Derived graphs of digraphs" in *Beiträge zur graphentheorie*, Sachs et al. eds. (Teubner, Leipzig 1978) pp. 17-33.

[3] C. Berge: *Graphes et Hypergraphes* - (Dunod, Paris 1970).

[4] M.W. Bern, E.L. Lawler, A.L. Wong: "Why certain subgraphs computations require only linear time" - Proc. of the 26th Ann. IEEE Symp. on Found. Comp. Sci.

[5] M.R. Garey, D.S. Johnson: *Computers and Intractability - A Guide through NP-completeness* (W.H. Freeman & Co., S. Francisco, 1979).

[6] M.R. Garey, D.S. Johnson, L. Stockmeyer: "Some simplified NP-complete graph problems" - Th. Comp. Sci. 1 (1976) pp. 240-243.

[7] F. Harary: *Graph Theory* - (Addison Wesley, Reading, 1969) p. 81.

[8] B.W. Kernighan, S. Lin: "An efficient heuristic procedure for partitioning graphs" - Bell Syst. Tech. J. 49, 2 (1970) pp. 292-307.

[9] P.G.H. Lehot: An optimal algorithm to detect a line graph and output its root graph - J. ACM 21, 4 (1974) pp. 569-575.

Acknowledgement

The author is grateful to Mario Lucertini, Antonio Sassano and Bruno Simeone for their encouragement and useful suggestions.

Structural Dependence and Systems of Equations

by

J. Bisschop (*), B. Dorhout (*), and A. Drud (**)

(*) Department of Applied Mathematics
Twente University of Technology
P.O. Box 217
7500 AE Enschede
The Netherlands

(**) Development Research Department
The World Bank
Washington D.C. 20433
U.S.A.

Abstract

Structural dependence in the incidence matrix associated with a system of equations may be used to detect misspecification of the underlying system. The incidence matrix can be represented as a bipartite graph G = (X,Y,E), with the node sets partitioned into the sets X1, X2, Y1 en Y2, such that (a) there are no edges between X1 en Y2, and (b) the cardinality of X1 + Y2 is maximum. Such a partition corresponds with the detection of complete and maximal complementary subsets of structurally dependent rows and columns. This problem is equivalent to finding a maximum independent vertex set or a minimum edge cover in thebipartite graph under the assumption of no isolated vertices. The problem of finding minimal and complete subsets of structurally dependent rows or columns is shown to be NP-complete. A heuristic approach for the detection of these minimal subsets based on Hall's transversal algorithm has worked satisfactorily in practice.

Key words

Structural dependence, structural rank, structural nullity , matching, bipartite graph, transversal, system of distinct representatives, incidence matrix, NP-completeness.

1. Structural Properties of Sparse Systems of Equations.

In this section we will examine structural properties of systems of equations induced by the underlying incidence matrix. We will begin by providing some problem background to motivate the subject of this paper. This is followed by definitions and observations.

1.1 Problem background.

In the areas of operations research and econometrics it is not unusual that the solution of large sparse systems of algebraic equations is needed during the investigation of some real-world problem. Typical systems involve several thousands of equations, and are usually very sparse in that the average number of variables occuring in an equation is at most twenty. With the current modeling technology such systems of equations are described in symbolic form using a compact algebraic notation. The input to a solution algorithm is then generated automatically from this symbolic representation by a modeling system (see e.g. [1], [3] and [5]).
This technology greatly speeds up the modeling process, and it can also help to avoid errors related to algebra. Take the case of models where the number of equations must be the same as the number of variables, and the solution must be (locally) unique. The model builder, however, can make conceptual mistakes such as specifying equations that can be derived from other equations in the model, or specifying too many or too few variables as exogenous. Several of these errors can be detected without any numerical work. This is done by looking for structural dependencies in the incidence matrix associated with the system. The motivation behind this paper is not only to find the number of mistakes of under- and overspecification (in short: structural deficiencies) in a model, but also to find a compact error diagnostic for the model builder. In large sparse systems structural deficiencies usually relate to subblocks of the incidence matrix. Unfortunately, these subblocks are not uniquely determined, and can be very large in size. It is therefore desirable that the modeling system provides the model builder with the smallest subblock that still contains structural deficiencies. This leads us to the topic examined in this paper.

1.2 Definitions and properties.

The definitions in this section we will be accompanied with the corresponding interpretation in terms of bipartite graphs. This is done because several of the proofs are dependent on existing results for bipartite graphs.

<u>Definition 1.1.</u> The *incidence matrix* associated with a system of m equations in n variables is an (mxn)-matrix N in which element N(i,j) is one whenever variable j occurs in equation i, and zero otherwise.

This corresponds with the bipartite graph G = (X,Y,E) where the vertices in X represent the row labels of N, the vertices in Y the column labels of N, and the edges (i,j) in E the nonzero elements of N.
Throughout the remainder of this paper any reference to a matrix will always mean a reference to an incidence matrix.

<u>Definition 1.2.</u> An incidence matrix N is said to be *structurally row dependent* if there is a subset of k rows wich has nonzero elements in less than k columns.

This corresponds with the existence of a subset of k vertices in X, say X(k), such that the cardinality $| N(X(k)) | < k$, where $N(X(k)) \subset Y$ denotes the set of neighbors of X(k).

Similarly, N is said to be *structurally column dependent* if there is a subset of k columns wich has nonzero elements in less than k rows.

This corresponds with the existence of a subset of k vertices in Y, say Y(k), such that the cardinality $| N(Y(k)) | < k$, where $N(Y(k)) \subset X$ denotes the neighbors of Y(k).

From definition 1.2 follows that a nonsquare matrix N always contains row- or column dependencies.

<u>Definition 1.3.</u> A *transversal* of a matrix N is the set of nonzero diagonal elements N(i,i).

This corresponds with the edge set (i,i) in E , which is a matching in G.

<u>Definition 1.4.</u> The *structural rank* of the (mxn)-matrix N is equal to the cardinality of its largest transversal under row and/or column permutations.

This corresponds with the maximum cardinality matching in (X,Y,E).

<u>Definition 1.5.</u> The *structural row nullity* of the (mxn) matrix N is equal to m minus the structural rank of N, and the *structural column nullity* of N is

equal to n minus the structural rank of N.

This corresponds with (| X | - | M |) and (| Y | - | M |) respectively, where M denotes a maximum cardinality matching.

<u>Definition 1.6.</u> A set of k structurally dependent rows of an (mxn)-matrix N with nonzero elements in exactly p columns is said to be *complete* if the structural rank of this (kxp)-submatrix is p and the structural row nullity (k-p).

This corresponds with the existence of a subset of k vertices in X, say X(k), such that the cardinality | N(X(k)) | = p, and the cardinality of the maximum cardinality matching in the subgraph of G, induced by X(k) and N(X(k)), is also p, with k > p.

Similarly, a set of k structurally dependent columns of an (mxn)-mátrix N with nonzero elements in exactly p rows is said to be complete if the structural rank of this (pxk)-submatrix is p and the structural column nullity is (k-p).

This corresponds with the existence of a subset of k vertices in Y, say Y(k), such that the cardinality | N(Y(k)) | = p, and the cardinality of the maximum cardinality matching the subgraph of G, induced by N(Y(k) and Y(k)), is also p, with k > p.

<u>Definition 1.7.</u> A *maximal complete subset* of structurally dependent rows in N is any complete subset of rows for which the structural row nullity is equal to the structural row nullity of the entire matrix N.

Similarly, a maximal complete subset of structurally dependent columns in N is any complete subset of columns for which the structural column nullity is equal to the structural column nullity of the entire matrix N.

Some of the above definitions can be illustrated with a simple example.
Consider the following incidence matrix N with ones at the positions indicated in the following scheme.

	1	2	3	4	5
1	x				
2	x				
3	x				
4	x	x	x		
5	x	x	x	x	x

Each of the sets of rows {1,2}, {1,3}, {2,3}, {1,2,3} and {1,2,3,4} are subsets of structurally dependent rows, while each of the sets of columns {4,5}, {2,3,4}, {2,3,5}, {3,4,5} and {2,3,4,5} are subsets of structurally dependent columns.

The structural rank of N is 3, and the structural nullity of N is 2.

Each of the sets of rows {1,2}, {1,3}, {2,3}, {1,2,3} is complete, while the set {1,2,3,4} is not complete. The subsets of structurally dependent columns are all complete.

Only the subset of rows {1,2,3} and the subset of columns {2,3,4,5} are maximal and complete. The identification of these complementary maximal and complete subsets of structurally dependent rows and columns is addressed in Section 2.

The subsets of rows {1,2}, {1,3}, {2,3} and the subset of columns {4,5} are all minimal and complete. The identification of these minimal and complete subsets of structurally dependent rows and columns is addressed in Section 3.

Before addressing these problems there are several observations that can be made at this point, some of which are captured in the form of a theorem. The first theorem links the definitions of structural dependency and structural rank for square systems of equations.

<u>Theorem 1.1.</u> An (nxn)-matrix is structurally dependent, if and only if its structural rank is less than n.

<u>Proof.</u> This theorem is a direct consequence of Hall's theorem concerning systems of distinct representatives (see [7]). Consider the bipartite graph (X,Y,E), which corresponds with the square matrix. Let N(x(i)) be the set of neighbors of the row vertex x(i). Then a sequence of vertices v(1), v(2),..., v(n) in Y is said to form a system of distinct representatives for the family of finite sets N(x(i)), if v(i) is in N(x(i)) for each i, and v(i) $\neq$ v(j) whenever i $\neq$ j. This corresponds to a transversal of cardinality n, which in turn corresponds with a structural rank of n. Hall's theorem states that there is NO system of distinct representatives for the family of finite sets N(x(i)) if and only if there is a subset I of {1,2,3,...,n} such that | UNION of { N(x(i)) } for i in I | < | I | . This latter condition states that there is a subset of k row vertices X(k) such that | N(X(k)) | < k , which is the definition of structural row dependence. This shows the equivalence between Hall's theorem and the above Theorem 1.1.

Corollary. An (nxn)-matrix is structurally dependent, if and only if both the structural row nullity and structural column nullity are positive.

Theorem 1.2. Structural dependence implies numerical dependence, but numerical dependence does not imply structural dependence.

Proof. By definition, a structurally dependent set of k rows contains nonzeros in fewer than k columns. Numerical dependency follows from this observation, because the numeric row rank is less than k. The same argument holds for a structurally dependent set of columns. The fact that numerical dependence does not imply structural dependence follows from the following example. Think about the (2x2)-matrix with all ones. This matrix is structurally nonsingular, but numerically singular with all the values equal to one.

In a structurally dependent system of equations there are clearly too many equations and/or variables. In a numerically dependent system of which the associated incidence matrix is not structurally dependent, there are either redundant equations or, in the case of nonlinear equations, isolated singular points in the domain of definition of the variables.

The following theorem states that certain portions of the incidence matrix are irrelevant for the detection of structurally row- or column dependent subsets. This may be of use for the development of heuristics to solve the problem of detecting minimal and complete structurally dependent subsets.

Theorem 1.3. Consider a maximal complete subset of r structurally dependent rows of the (mxn)-matrix N with nonzeros in p columns. Then the submatrix SN of N which corresponds to the intersection of the remaining (m-r) rows and (n-p) columns does not contain a subset of structurally dependent rows. The same statement holds true for a maximal complete subset of structurally dependent columns.

Proof. Assume that the submatrix SN contains structurally dependent rows. Select the smallest subset of rows that is still structurally dependent. By construction, this set of k rows with nonzero elements in k-1 columns is complete with a structural row nullity of 1. If we consider the corresponding rows of N and append these to the maximal complete subset of r structurally dependent rows of N, then we get a new complete subset of structurally dependent rows with row nullity equal to (r+k-p-(k-1)), which is one greater than(r-p).
This contradicts the assumption of maximality for the r structurally dependent

rows of N.

<u>Corollary.</u> The remaining (m−r) rows of matrix N in Theorem 1.3 are structurally independent.

<u>Theorem 1.4.</u> For any (mxn)-incidence matrix N the structural nullity of a maximal and complete subset of dependent rows differs from the structural nullity of a maximal and complete subset of dependent columns by (m−n).

<u>Proof.</u> Consider any maximal complete subset of r structurally dependent rows with nonzeros in (r−nr) columns. By assumption of completeness the structural rank is (r−nr), and the structural nullity is nr. By assumption of maximality the remaining (n−(r−nr)) columns have then nonzero elements in exactly (m−r) rows. By the previous theorem, the structural rank of these (n−r+nr) columns is (m−r), and they also form a maximal and complete subset of structurally dependent columns in N. Appending any subset of the first (r−nr) columns to these remaining (n−r+nr) columns will only increase the structural rank of these columns, but not their structural nullity. The structural nullity nc of the (n−r+nr) columns is therefore (n−r+nr)−(m−r) = (n−m)+nr. Since the above reasoning holds for any maximal complete subset, we have proved that for all maximal and complete subsets of structurally dependent rows and columns in N, the identity (nr−nc) = (m−n) holds.

<u>Corollary.</u> For any (nxn)-incidence matrix N the structural nullity of a maximal complete subset of dependent rows is equal to the structural nullity of a maximal complete subset of dependent columns.

The above corollary corresponds with the intuitive notion that in a square structurally dependent system of equations the same number of variables and equations has to be removed for a structurally independent system to remain.

In the next section we will examine the problem of finding a pair of maximal and complete subsets of structurally dependent rows and columns in an (mxn) incidence matrix N.

2. Maximal and Complete Structural Dependency.

Theorem 1.3 stated that certain portions of the incidence matrix are irrelevant for the detection of structurally dependent subsets of rows and columns, and that these portions are related to the partitioning of the incidence matrix into maximal and complete subsets of rows and columns. That is why we devote a separate section to the topic of maximal and complete structural dependency, while in fact we are ultimately only interested in the topic of minimal and complete structural dependency.

The problem of finding maximal and complete subsets of structurally dependent rows and columns in an (mxn)-incidence matrix N can be formulated with the following observation in mind. As was noted in the proof of Theorem 1.4, for every maximal and complete subset of structurally dependent rows there is a corresponding maximal and complete subset of structurally dependent columns. For any such pair we obtain, after the appropriate permutation of rows and columns, a block with all zero elements. The number of rows in this block of zeros is equal to the cardinality of the subset of structurally dependent rows.
Similarly, the number of columns in this zero block is equal to the cardinality of the subset of structurally dependent columns. As we shall see, the problem of finding maximal and complete subsets of structurally dependent rows and columns is equal to finding a zero block with maximum perimeter.

Let u(i) be a zero-one variable to indicate that row i may be part of a maximal and complete subset of structurally dependent rows. Similarly, let v(j) denote a zero-one variable to indicate that column j may be part of a maximal and complete subset of structurally dependent columns.

Consider the problem:

$$\begin{array}{ll} \text{maximize} & \text{SUM}(\, i \,,\, u(i)) \;+\; \text{SUM}(\, j \,,\, u(j)) \\ \text{s.t.} & u(i) + v(j) \;\le\; 1 \quad \text{for all } (i,j) \text{ with } N(i,j) = 1. \\ & u(i) \ge 0 \;,\;\; v(j) \ge 0 \;. \end{array}$$

In the absence of zero rows or columns in N, this program corresponds with finding the maximum independent set of vertices in the underlying bipartite graph of the incidence matrix N (see [6]). Independent of this observation one can show that the above linear program has a unimodular contraint matrix, because i) every row contains at most two nonzero elements equal to one and ii) the columns of the matrix can be partitioned into two groups (one associated with u-variables and the other with v-variables) such that the nonzeros in every row are in different groups of columns (see [8]). This means that the linear program will have a zero-one solution as long as N does not have zero rows or zero columns, which we can assume without loss of generality.

It is straightforward to verify from the construction of the above linear program and the zero-one property of its solution, that all entries in N(i,j) for which both u(i) = 1 and v(j) = 1 , must be equal to zero. Thus the optimal solution describes a subblock of zeros in N with maximum perimeter.

Note that this subblock may be empty if either all u(i) or all v(j) are zero. In the case that all u(i) are zero the matrix does not contain a subset of structurally dependent rows, while in the case that all v(j) are zero the matrix does not contain a subset of structurally dependent columns (see Theorem 2.2 below).

The optimal solution will be at least as large as MAX[m,n], the maximum of the row size and column size of the matrix N, because setting all u(i) = 1 and v(j) = 0 is a feasible solution, and setting all u(i) = 0 and v(j) = 1 is also a feasible solution.

<u>Theorem 2.1.</u> The structurally dependent subsets of rows and columns (if any) identified by the solution of the above linear program are maximal and complete.

<u>Proof.</u> Consider the solution of the above linear program, and permute the rows and columns of N such that the first rows correspond with u(i) = 1 and the first columns correspond with v(j) = 0 . Then the matrix is partitioned such that the north-east block (possibly vacuous) consists entirely of zeros. The set of rows in the north-west block and the set of columns in the south-east block are both complete. If the south-east block, for instance, were not complete, then there are structurally dependent rows in this block. By appending the corresponding rows of N to the rows of the north-west block, we obtain a new partition of N whereby the block with zero elements has changed. This block has gained more rows than it lost columns, and therefore has a larger perimeter than before. This contradicts the optimality of the linear programming solution. Similarly, if the north-west block were not complete, then there are structurally dependent columns in this block. By appending the corresponding columns of N to the columns of the south-east block, we obtain a new partition of N whereby the block with zero elements has changed. This block has gained more columns than it lost rows, and therefore

has a larger perimeter than before. Again, this contradicts the optimality of the linear programming solution.
In addition, the set of rows in the north-west block and the set of columns in the south-east block are both maximal. Let r denote the cardinality of the set of rows in the north-west block, and c denote the cardinality of the set of columns in the south-east block. The nullity of the north-west block nr is r-(n-c), while the nullity of the south-east block nc is c-(m-r). This implies that nr = r+c-n , and nc = r+c-m. Since r+c is the maximum value of the objective function, and both the north-west and the north-east blocks are complete, it follows that nr is equal to the row nullity of the entire matrix and nc is equal to the column nullity of the entire matrix. This proves that the set of rows in the north-west block and the set of columns in the south-east block are both maximal.

<u>Theorem 2.2.</u> Whenever the objective function value for the optimal solution of the above linear program is equal to MAX[m,n] for the (mxn)-matrix N, then N does not contain a subset of structurally dependent rows if $n \geq m$, and N does not contain a subset of structurally dependent columns if $m \geq n$.

<u>Proof.</u> Assume that $n \geq m$, and that the optimal value of the objective function is equal to n. Assume also that there is a set of r structurally dependent rows in N. Then by selecting the r rows and the (n-r+nr) columns for which these rows have zero elements, we can construct a feasible solution of the above linear program by setting u(i) = 1 for the r rows and v(j) = 1 for the (n-r+nr) columns. This feasible solution produces an objective function value equal to r+(n-r+nr) = n+nr. Since nr > 0 by the assumption of structural dependence, the optimal value of the objective function is greater than n instead of the assumed value equal to n. A similar argument can be used for the nonexistence of a structurally dependent subset of columns for the case $m \geq n$.

<u>Corollary.</u> In the event that N is a square (nxn)-matrix an optimal objective function value equal to n implies that N is structurally nonsingular.

As we stated previously, the mathematical problem of finding maximal and complete subsets of structurally dependent rows and columns is equivalent to finding the vertices of a maximum independent set in the underlying bipartite graph, assuming no zero rows and columns in the incidence matrix. The dual of this problem, written below, determines the number of edges in a minimum edge covering for the bipartite graph.

minimize	SUM((i,j) , N(i,j) . x(i,j))		
s.t.	SUM(i , N(i,j) . x(i,j))	≥ 1	for all i
	SUM(j , N(i,j) . x(i,j))	≥ 1	for all j
	x(i,j) ≥ 0 .		

One interpretation which is consistent with that of a minium edge cover is that every row and column must have at least one distinct representative, and that the optimal solution is formed with the fewest representatives. We again assume that N has no zero rows or zero columns, otherwise the above problem does not have a feasible solution. As we shall see, the solution does indicate the existence of structurally dependent subblocks, but it does not identify these subblocks. Only the dual values perform this task.

If the value of the objective function is z = MAX[m,n], then the optimal solution of the above dual program contains only one representative for each row when m ≥ n , and only one representative for each column when n ≥ m.

<u>Theorem 2.3.</u> Assume that the value of the objective function of the above dual linear program is z. If n ≥ m , then z - MAX[m,n] represents the nullity of any maximal and complete subset of structurally dependent rows in N. If m ≥ n, then z - MAX[m,n] represents the nullity of any maximal and complete subset of structurally dependent columns in N.

<u>Proof.</u> Assume that n ≥ m, and that the optimal value of the objective function is z > n. Then it is straightforward to verify that the row rank of the matrix is not full. Consider any maximal and complete subset of r structurally dependent rows with nonzeros in (r-nr) columns. Then the smallest number of representatives is exactly equal to r. This follows from the completeness property. In that case the corresponding columns have nr extra representatives.
As was shown in the proof of Theorem 1.3, the remaining (n-r+nr) columns also form a maximal and complete set of dependent columns with nonzeros in exactly (m-r) rows. Again, by completeness, the smallest number of representatives is equal to (n-r+nr), and the corresponding rows with nonzero elements have (n-r+nr)-(m-r) = (n-m)+nr extra representatives. By construction, the smallest total number of representatives is r+(n-r+nr) = n+nr which is the value of z. Since n ≥ m , we have that (z-n) = nr represents the nullity of any maximal and complete subset of structurally dependent rows in N. A similar proof can be given for the case that m ≥ n , where (z-m) > 0 represents the nullity of any maximal and complete subset of columns.

<u>Corollary 1.</u> If $z = MAX[m,n]$ and $n > m$, then there is no subset of structurally dependent rows.

<u>Corollary 2.</u> If $z = MAX[m,n]$ and $m > n$, then there is no subset of structurally dependent columns.

<u>Corollary 3.</u> If $z = MAX[m,n]$ and $m = n$, then there does not exist a subset of structurally dependent rows or columns, and the matrix is structurally nonsingular. In that case the representatives determine (after row permutations) a complete transversal of N, or equivalently, a complete cardinality matching of the bipartite graph induced by N.

3. Minimal Structurally Dependent Sets.

The search for the smallest subset of structurally dependent rows or columns in a matrix N can be limited to examining only maximal and complete subsets of structural dependent rows or columns. This observation is a direct consequence of Theorem 1.3. Even though this observation implies that a portion of N can be ignored during the search, the remaining portion may still be large in size.

In some applications the maximal complete subsets of structurally dependent rows or columns turn out to be also the smallest. In general, however, this is not the case, and finding the smallest subset of structurally dependent rows or columns in a matrix is difficult. That is the main result of this section.

When a smallest subset of k structurally dependent rows exists, then its row rank is (k-1). The intersection of the k rows with the remaining (n-k+1) columns will then be a zero matrix. This observation provides an alternative characterization of the smallest subset of structurally dependent rows.

Let $NN(i,j) = (1-N(i,j))$ be the boolean complement of the incidence matrix N, and let BNN(R,C,E) be the bipartite graph constructed from NN. The symbols R and C denote the two disjoint sets of vertices of BNN corresponding to the set of rows and the set of columns of NN, respectively. The symbol E denotes the set of edges between the vertices of BNN, where each edge corresponds to a nonzero of the matrix NN.

From the above two paragraphs we may conclude that any structurally dependent subset of rows in N corresponds to a rectangular submatrix of zeros in N, which, in turn, corresponds to a rectangular submatrix of ones in NN, which, in turn, corresponds to a subgraph of the bipartite graph BNN with vertex sets R1 in R and C1 in C such that for r in R1 and c in C1 the corresponding edge (r,c) is in the induced edge set $E1 \subset E$.

Theorem 3.1. The problem of finding the smallest subset of structurally dependent rows or columns in an (mxn)-incidence matrix N is NP-complete.

Proof. The problem of finding the smallest subset of structurally dependent rows or columns in an (mxn)-incidence matrix N can be restated using the above notation. Consider the bipartite graph BNN(R,C,E) and select two integers K1 and K2 such that $K1 \leq |R|$ and $K2 \leq |C|$. Find two subsets R1 in R and C1 in C such that $|R1| = K1$ and $|C1| = K2$, and such that r in R1 and c in C1 implies that the edge (r,c) is in E. By choosing $K2 = n-K1+1$ and K1 as small as possible but still allowing a solution, we will identify a smallest set of structurally dependent rows in N. Similarly, by choosing $K1 = m-K2+1$ and K2 as small as possible but still allowing a solution, we will identify a smallest set of structurally dependent columns. Whenever n is uneven and the integers K1 and K2 are selected such that $K1 = K2$, then the above problem reduces to the so-called Balanced Complete Bipartite Subgraph problem which is NP-complete (see [6] on page 196). The assumption that n must be uneven is artificial, since we can always append one isolated vertex to the set C in the graph BNN(R,C,E). This implies that the problem of finding the smallest subset of structurally dependent rows or columns is also NP-complete.

In the next section we will describe a practical approach that tends to identify small structurally dependent subsets of rows or columns in N.

4. A Practical Approach Using the Transversal Finder.

As defined in section 1, the structural rank of a matrix is equal to the cardinality of its largest transversal after row and column permutations. Computational experience has shown that an efficient implementation of the algorithm of Hall, despite its higher asymptotic bound for the worst case, has been found to perform better than any other transversal finding algorithm (see [9] p. 185).

A version of the Hall algorithm with row interchanges is ideally suited to find minimal subsets of structurally dependent rows. Likewise, an implementation with column interchanges is ideally suited to find minimal subsets of structurally dependent columns. Let us look at the version with row interchanges.

Of interest is the situation where r rows have been found to contain a full transversal, and row (r+1) is being considered. The basic step that is performed each time is to trace an alternating augmenting path. Such a path originates from a nonzero in row (r+1) at position k, going to element (k,k), next visiting a nonzero in row k at position ℓ, going to element (ℓ,ℓ), next visiting a nonzero in row ℓ etc. The path cannot traverse any row or column more than once. If for some row we cannot find an off-diagonal nonzero on an unvisited column, then we delete that row from the path (but not from the list of visited positions) and go back to the previous row to look for off-diagonal nonzero elements in columns not yet visited. If there is still one such nonzero element, then the forward search starts again. The path ends either i) when a nonzero in the submatrix situated in rows 1 through r and columns (r+1) through n has been found, or ii) when the path becomes empty and we are back at the starting point. In case i) the transversal can be augmented. In case ii) we find that we have visited (r+1) rows (including the originating row) with nonzeros in exactly r columns.

The transversal algorithm, therefore, detects (via a list of visited positions) minimal structurally dependent subsets of rows with structural nullity equal to one. By retaining the one with lowest cardinality, we have found a candidate smallest subset of structurally dependent set of rows. It is not necessarily the globally smallest subset, but the smallest subset relative to a particular transversal. Changing the transversal may produce even a smaller minimal set of structurally dependent columns. We will illustrate this with an example.

Consider the following pair of (8x6) matrices (one being a permutation of the other), and examine their structural row dependencies.

	1	2	3	4	5	6
1	1					
2		1	1			
3			1	1		1
4				1	1	
5					1	
6						1
7		1			1	
8			1			

	1	2	3	4	5	6
1	1					
2		1	1			
8			1			
4				1	1	
5					1	
6						1
7		1			1	
3			1	1		1

Both matrices have a nonzero transversal of cardinality six. Applying the algorithm of Hall to rows 7 and 8 of the first matrix gives us two minimal structurally dependent sets of rows. Rows (7,2,3,4,5,6) have nonzero elements in columns (2,3,4,5,6), and rows (8,3,4,5,6) have nonzero elements in columns (3,4,5,6). The smallest set of rows has cardinality 5. Applying the algorithm of Hall to rows 7 and 3 of the second matrix gives us rows (7,2,8,5) with non-zero elements in columns (2,3,5), and rows (3,8,4,5,6) with nonzero elements in columns (3,4,5,6). Notice that the smallest set of structurally dependent rows has cardinality 4, which is one less than the smallest cardinality found in the first matrix.

The above example illustrates that when the nullity of a set of structurally dependent rows is greater than one, then an enumeration over all possible transversals is needed in order to find that subset of structurally dependent rows which has lowest cardinality. The NP-completeness result of Theorem 3.1 does not encourage us to find an efficient enumeration process. One heuristic, that has been employed, is to order the rows in terms of the number of nonzero elements that they contain. This heuristic has a tendency to shorten the alternating paths that can be found from any transversal row that serves as originating row. This is best illustrated by again using the above example.

	1	2	3	4	5	6
1	1					
2		1	1			
3			1	1		1
4				1	1	
5					1	
6						1
7		1			1	
8			1			

	1	5	6	3	2	4
1	1					
5		1				
6			1			
8				1		
2				1	1	
4		1				1
7		1			1	
3			1	1		1

Rows (1,5,6) of the first matrix as originating rows in the Hall algorithm do not lead to an alternating augmenting path to any other rows, but rows (2,3,4) are such that row 2 leads to rows (3,4,5,6), row 3 leads to rows (4,5,6), and row 4 leads to row 5. The alternating paths resulting from the first six originating rows of the second matrix are shorter in length. In this case there are four rows, namely 1,5,6 and 8, that do not lead to any other row, and, in addition, row 4 only leads to row 5, and row 2 only leads to row 8.

In this example the ordering of rows in the second matrix caused the Hall algorithm to identify the globally smallest subset of structurally dependent rows, namely rows (7,5,8,2). It is not too hard, however, to construct examples whereby this heuristic fails.

In practical applications concerning large economy-wide models that were generated using a modeling system [4], the above methodology proved to be sufficient in identifying small sets of structurally dependent rows and columns. This is in part due to the nature of the application which imposes natural restrictions on the number and types of equations generated. In other applications the above methodology is likely to be insufficient, and further heuristics need to be developed.

References.

[1] Bisschop, J. and Meeraus, A., "On the Development of a General Algebraic Modeling System in a Strategic Planning Environment", Mathematical Programming Study, Vol. 20 (1982) pp. 1-29.

[2] Bunch, J.R., and Rose, D.J., Sparse Matrix Computations, Academic Press Inc. (1976) pp. 275-280.

[3] Brooke, A., Drud, A., and Meeraus, A., "Modeling Systems and Nonlinear Programming in a Research Environment", in: Computer in Engineering 1985 Volume Three, Edited by R. Raghaven and S.M. Rohde, The American Society of Mechanical Engineers (1985) pp. 213-219.

[4] Drud, A., Grais, W., and Pyatt, G., "The Transaction Value Approach - A Systematic Method of Defining Economywide Models Based on Social Accounting Matrices", in : Dynamic Modelling and Control of National Economies 1983 , T. Basar and L.F. Pau, eds., Pergamon Press, New York (1983) pp. 241-248.

[5] Fourer, R., "Modeling Languages Versus Matrix Generators for Linear Programming", ACM Transactions on Mathematical Software, Vol 9, No 2 (1983) pp. 143-183.

[6] Gary, M.R., and Johnson, D.S., Computers and Intractability: A Guide to the Theory of NP-Completeness, W.H. Freeman and Company, 1979.

[7] Hall, P., "On Representatives of Subsets", Journal of the London Mathematical Society, 10 (1935) pp. 26-30.

[8] Lawler E., Combinatorial Optimization: Networks and Matroids, Holt, Rinehart and Winston, 1976.

[9] Pissanetzky, S., Sparse Matrix Technology, Academic Press,Inc., 1984.

BEST NETWORK FLOW BOUNDS FOR THE QUADRATIC KNAPSACK PROBLEM[1]

PAUL CHAILLOU
Renault, Paris, France

PIERRE HANSEN
RUTCOR, Rutgers University, USA

YVON MAHIEU
Solvay, Brussels, Belgium.

A Lagrangean relaxation of the quadratic knapsack problem is studied. It is shown, among other properties, that the best value of the Lagrangean multiplier, and hence the best bound for the original problem, can be determined in at most $n-1$ applications of a maximum flow algorithm to a network with $n+2$ vertices and $n+m$ arcs, where n and m denote the numbers of variables and of quadratic terms. A branch-and-bound algorithm using this result is presented and computational experience is reported on.

1. INTRODUCTION.

The *quadratic knapsack problem*, introduced by Gallo, Hammer and Simeone ([3]), can be expressed as follows:

$$\text{Maximize } f(x_1, x_2, \ldots, x_n) = \sum_{i=1}^{n}\sum_{j=1}^{n} c_{ij}x_ix_j \tag{1}$$

subject to

$$\sum_{j=1}^{n} a_jx_j \leq b \tag{2}$$

$$x_j \in \{0,1\} \quad \text{for } j = 1, 2, \ldots, n \tag{3}$$

and $b > 0$, $a_j \geq 0$, $c_{ij} \geq 0$ for $j = 1, 2, \ldots, n$. Note that (1) contains linear terms as $c_{ii}x_ix_i = c_{ii}x_i$. This problem has several applications cited in [3] e.g. location of airports to maximize freight, location of weather measurement stations with least correlation between measures,

---

[1]Paper presented at the NETFLOW 83 International Workshop, Pisa, Italy, 28-31 March 1983.

detection of a k-clique in a graph.

Problem (1)-(3) is NP-complete, as the case $c_{ij} = 0$ for $i \neq j$ is the usual knapsack problem. Gallo, Hammer and Simeone ([3]) use *upper planes* of (1), i.e. linear majorizing functions to obtain knapsack problems as relaxations. They explore several variants and compare the performances of the corresponding branch-and-bound algorithms.

The present paper is devoted to a Lagrangean relaxation approach to problem (1)-(3), and is organized as follows: properties of the Lagrangean function are studied in the next section; a polynomial algorithm to obtain the best value of the Lagrangean multiplier, and hence the best bound for the original problem is presented in section 3, together with heuristic and exact algorithms for the quadratic knapsack problem itself; computational experience is reported on in the last section.

2. PROPERTIES OF THE LAGRANGEAN FUNCTION.

Using a non-negative multiplier λ, the introduction of the constraint (2) into the objective function (1) yields the Lagrangean function

$$h(x_1, x_2, \ldots, x_n, \lambda) = \sum_{i=1}^{n} \sum_{j=1}^{n} c_{ij} x_i x_j + \lambda (b - \sum_{j=1}^{n} a_j x_j). \tag{4}$$

We are interested here in the properties of the function

$$\overline{f}(\lambda) = \max_{x_1, \ldots, x_n \in \{0,1\}} h(x_1, x_2, \ldots, x_n, \lambda). \tag{5}$$

Let $f^* = f(x_1^*, \ldots, x_n^*)$ denote the value of the optimal solution of (1)-(3). The first three following properties are particular cases of well-known results on Lagrangean relaxation (see Minoux [8, Chap. 6]). The proofs are therefore omitted.

Property 1. *$\overline{f}(\lambda)$ is piecewise linear in λ.* ∎

Property 2. $\min_{\lambda \geq 0} \overline{f}(\lambda) \geq f^*$. ∎

Property 3. *$\overline{f}(\lambda)$ is convex.* ∎

The *inclusion property* which is next proved allows to find the value of λ which minimizes $\overline{f}(\lambda)$ in polynomial time. A similar property has been shown to be true for hyperbolic 0-1 programs by Picard and Queyranne ([9]), with a similar proof.

Property 4. *Let* $0 \leq \lambda^1 < \lambda^2$, $\max\{h(x_1, \ldots, x_n, \lambda^1)\} = h(x_1^1, \ldots, x_n^1, \lambda^1)$ *and* $\max\{h(x_1, \ldots, x_n, \lambda^2)\} = h(x_1^2, \ldots, x_n^2, \lambda^2)$. *Then* $x_j^1 \geq x_j^2$ *for* $j = 1, 2, \ldots, n$.

Proof. Let $I = \{j;\ x_j^1 = 1,\ x_j^2 = 0\}$, $J = \{j;\ x_j^1 = 1,\ x_j^2 = 1\}$, and $K = \{j;\ x_j^1 = 0,\ x_j^2 = 1\}$. We must show that $K = \emptyset$. Let us consider two more Boolean n-vectors defined by

$$x_j^3 = \begin{cases} 1 & \text{if } j \in I \cup J \cup K \\ 0 & \text{if } j \notin I \cup J \cup K \end{cases} \quad \text{and} \quad x_j^4 = \begin{cases} 1 & \text{if } j \in J \\ 0 & \text{if } j \notin J. \end{cases}$$

From the definitions above and the optimality of $(x_1^1, \ldots, x_n^1)$, we have

$$h(x_1^1, \ldots, x_n^1, \lambda^1) = \sum_{i \in I \cup J} \sum_{j \in I \cup J} c_{ij} + \lambda^1 (b - \sum_{j \in I \cup J} a_j)$$

$$\geq h(x_1^3, \ldots, x_n^3, \lambda^1) = \sum_{i \in I \cup J} \sum_{j \in I \cup J} c_{ij} + \sum_{i \in K} \sum_{j \in K} c_{ij} +$$

$$\sum_{i \in I \cup J} \sum_{j \in K} c_{ij} + \sum_{i \in K} \sum_{j \in I \cup J} c_{ij} + \lambda^1 (b - \sum_{j \in I \cup J} a_j - \sum_{j \in K} a_j)$$

which implies

$$\lambda^1 \sum_{j \in K} a_j \geq \sum_{i \in I} \sum_{j \in K} c_{ij} + \sum_{i \in J} \sum_{j \in K} c_{ij} +$$

$$\sum_{i \in K} \sum_{j \in I} c_{ij} + \sum_{i \in K} \sum_{j \in J} c_{ij} + \sum_{i \in K} \sum_{j \in K} c_{ij}. \tag{6}$$

From the optimality of $(x_1^2, \ldots, x_n^2)$ it follows that

$$h(x_1^2, \ldots, x_n^2, \lambda^2) = \sum_{i \in J} \sum_{j \in J} c_{ij} + \sum_{i \in J} \sum_{j \in K} c_{ij} +$$

$$\sum_{i \in K} \sum_{j \in J} c_{ij} + \sum_{i \in K} \sum_{j \in K} c_{ij} + \lambda^2 (b - \sum_{j \in J} a_j - \sum_{j \in K} a_j)$$

$$\geq h(x_1^4, \ldots, x_n^4, \lambda^2) = \sum_{i \in J} \sum_{j \in J} c_{ij} + \lambda^2 (b - \sum_{j \in J} a_j),$$

hence

$$\sum_{i\in J}\sum_{j\in K} c_{ij} + \sum_{i\in K}\sum_{j\in J} c_{ij} + \sum_{i\in K}\sum_{j\in K} c_{ij} \geq \lambda^2 \sum_{j\in K} a_j \tag{7}$$

and, as $c_{ij} \geq 0$, from (6) and (7)

$$\lambda^1 \sum_{j\in K} a_j \geq \lambda^2 \sum_{j\in K} a_j$$

which, as $\lambda^1 < \lambda^2$, is possible only if $K = \emptyset$. ■

Corollary 4.1. *The number of linear segments of $\overline{f}(\lambda)$ is at most n.* ■

Property 5. *Let $\overline{f}(\lambda^*) = \min_{\lambda \geq 0} \overline{f}(\lambda)$ (taking the smallest value of λ in case of multiple optima). Then the optimal solution of (5) is infeasible for all $\lambda < \lambda^*$ and feasible for all $\lambda \geq \lambda^*$.*

Proof. The slope of $\overline{f}(\lambda)$ is negative to the left of λ^* hence, for each linear segment, the coefficient of λ (i.e. $b - \sum_{j=1}^{n} a_j x_j$) is negative, and the corresponding solution is infeasible. The case $\lambda \geq \lambda^*$ is treated similarly. ■

The following property uses the Boolean expression of the minimum-cut problem, due to Hammer ([4]), in a similar way as Picard and Ratliff ([10]). Let us associate with (5) a network $N = (V, U, \lambda)$ with a vertex set $V = \{s, 1, \ldots, n, p\}$, where s denotes the source and p the sink, and with an edge set $U = \{(s, i);\ i = 1, \ldots, n\} \cup \{(i, j);\ c_{ij} \neq 0\} \cup \{(i, p);\ i = 1, \ldots, n\}$. Moreover, let us give to the arcs the capacities

$$c_{sj}(\lambda) = \max(0, \sum_{i=1}^{n} c_{ji} - \lambda a_j), \quad c_{ij}(\lambda) = c_{ij} \text{ and } c_{jp}(\lambda) = \max(0, \lambda a_j - \sum_{i=1}^{n} c_{ji}).$$

Property 6. *We have $\overline{f}(\lambda) = \lambda b + \sum_{j=1}^{n} c_{sj}(\lambda) + \Psi(\lambda)$ where $\Psi(\lambda)$ denotes the maximum flow in N.*

Proof. From Ford and Fulkerson's theorem ([1]), we have

$$\Psi(\lambda) = \min\{\sum_{j=1}^{n} c_{sj}(\lambda)(1 - x_j) + \sum_{i=1}^{n}\sum_{j=1}^{n} c_{ij} x_i (1 - x_j) + \sum_{j=1}^{n} c_{jp}(\lambda) x_j\}$$

$$= \sum_{j=1}^{n} c_{sj}(\lambda) + \min\{\sum_{j=1}^{n} (c_{jp}(\lambda) - c_{sj}(\lambda) + \sum_{i=1}^{n} c_{ji}) x_j - \sum_{i=1}^{n}\sum_{j=1}^{n} c_{ij} x_i x_j\}$$

$$= \sum_{j=1}^{n} c_{sj}(\lambda) + \min\{\sum_{j=1}^{n} (\max(\lambda a_j - \sum_{i=1}^{n} c_{ji}, 0) + \min(\lambda a_j - \sum_{i=1}^{n} c_{ji}, 0) + \sum_{i=1}^{n} c_{ji})x_j - \sum_{i=1}^{n}\sum_{j=1}^{n} c_{ij}x_i x_j\}$$

$$= \sum_{j=1}^{n} c_{sj}(\lambda) + \min(\sum_{j=1}^{n} \lambda a_j x_j - \sum_{i=1}^{n}\sum_{j=1}^{n} c_{ij}x_i x_j)$$

$$= \sum_{j=1}^{n} c_{sj}(\lambda) - \overline{f}(\lambda) + \lambda b,$$ from which the result follows. ∎

Let us now consider some particular values of λ. Let

$$\lambda_j = \sum_{i=1}^{n} c_{ij}/a_j \text{ for } j = 1, \ldots, n \text{ and } \lambda_0 = \sum_{i=1}^{n}\sum_{j=1}^{n} c_{ij}/\sum_{j=1}^{n} a_j = \sum_{j=1}^{n} \lambda_j a_j/\sum_{j=1}^{n} a_j.$$

Property 7. $\overline{f}(\lambda) = \lambda b$ *for all* $\lambda > \max\{\lambda_j\}$.

Indeed when $\lambda > \max\{\lambda_j\}$, $c_{sj} = 0$ for all j and hence $\Psi(\lambda^*) = 0$, from which the result follows. ∎

Note also that $\overline{f}(0) = \sum_{i=1}^{n}\sum_{j=1}^{n} c_{ij}$.

Property 8. $\overline{f}(\lambda) \geq \lambda_0 b$ *for all* $\lambda \geq 0$.

Proof. $\overline{f}(\lambda)$ is greater than or equal to (5) for all Boolean n-vectors, in particular for the vectors $(0, 0, \ldots, 0)$ and $(1, 1, \ldots, 1)$. Hence

$$\overline{f}(\lambda) \geq \max(\lambda b, \sum_{i=1}^{n}\sum_{j=1}^{n} c_{ij} + \lambda(b - \sum_{j=1}^{n} a_j))$$

and the minimum of this expression is attained for $\lambda = \lambda_0$. ∎

The results of this section are summarized in Figure 1.

3. ALGORITHMS.

The following algorithm uses the results of the previous section to determine λ^* such that $\overline{f}(\lambda^*) = \min_{\lambda > 0} \overline{f}(\lambda)$.

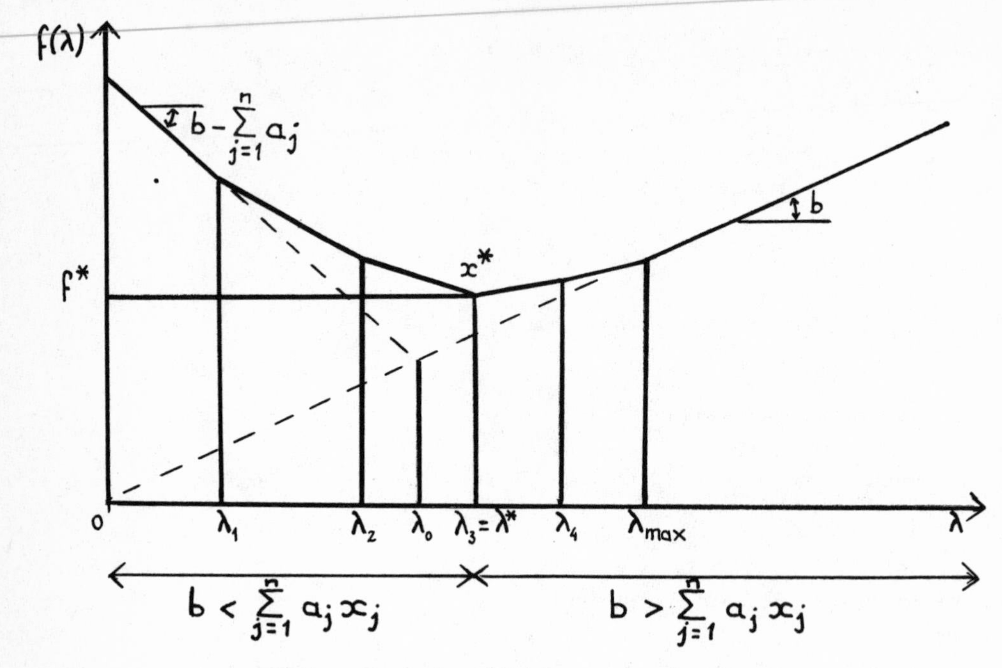

Figure 1. Properties of the Lagrangean function.

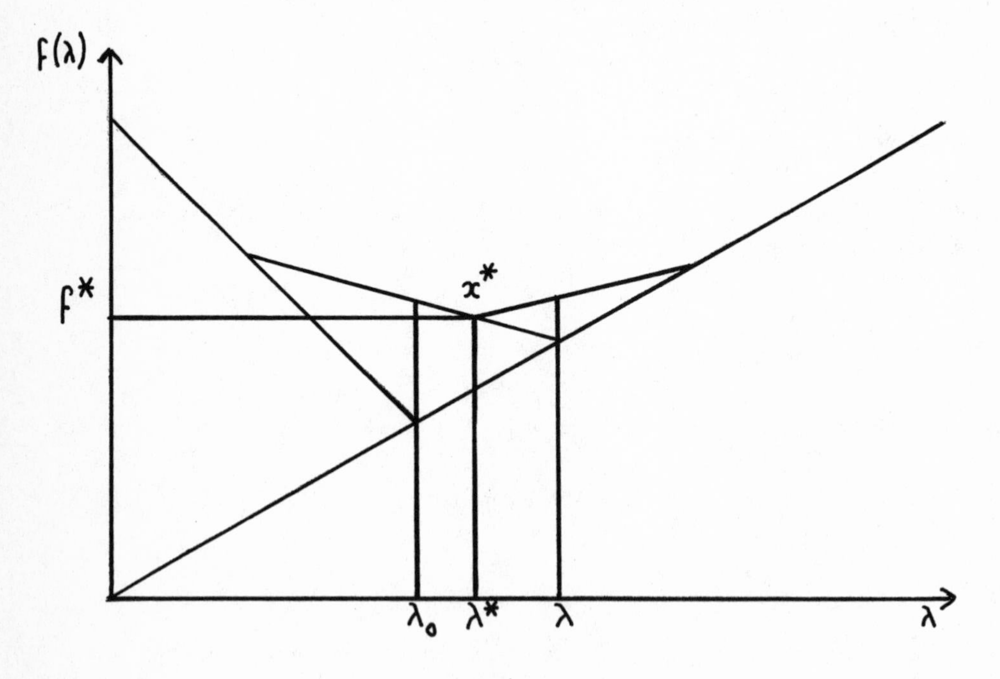

Figure 2. Obtention of λ^*.

ALGORITHM 1.

a) *Initialization.* Set $\alpha_1 := b - \sum_{j=1}^{n} a_j$ and $\beta_1 := \sum_{i=1}^{n}\sum_{j=1}^{n} c_{ij}$. Set $\alpha_2 := b$ and $\beta_2 = 0$.

b) *Maximum flow problem.* Compute $\lambda = (\beta_1 - \beta_2)/(\alpha_2 - \alpha_1)$. Construct the network $N = (V, U, \lambda)$ and determine a maximum flow between s and p. Let $\alpha_3\lambda+\beta_3$ denote the line in the $(\lambda, \overline{f}(\lambda))$-space corresponding to the optimal solution of (5), defined by Property 6.

c) *Optimality test.* If $\alpha_3\lambda+\beta_3 = \alpha_1\lambda+\beta_1$, then $\lambda^* = \lambda$, and so end. Otherwise if $\alpha_3 < 0$, set $\alpha_1 := \alpha_3$, $\beta_1 := \beta_3$ and go to b); if $\alpha_3 \geq 0$ set $\alpha_2 := \alpha_3$, $\beta_2 := \beta_3$ and go to b).

The application of Algorithm 1 to the function $\overline{f}(\lambda)$ of Figure 1 is illustrated in Figure 2; it is seen that λ^* is found in three iterations.

As two linear segments out of a maximum of n are considered at the first iteration and one more at each subsequent one, at most $n-1$ iterations are needed. As N has $n+2$ vertices and $n+m$ arcs, it follows that Algorithm 1 takes $O(n^4)$ operations in the worst case, if Karzanov's algorithm ([5]) or the "three Indians" one ([6]) is used in step b).

We now turn to the resolution of the quadratic knapsack problem itself, and first consider a heuristic algorithm of the "Attila" type.

ALGORITHM 2.

a) *Initialization.* Set $K_1 := \{1, 2, \ldots, n\}$, $K_0 := \emptyset$. Compute $q_j = \sum_{i=1}^{n} (c_{ij} + c_{ji})/a_j$ for $j = 1, 2, \ldots, n$ and $\sum_{j=1}^{n} a_j$.

b) If $\sum_{j \in K_1} a_j \leq b$, end with a heuristic solution X_h given by $x_j = 1$ for all j in K_1, $x_j = 0$ for all j in K_0.

c) *Setting a variable to 0.* Select $q_k = \min\{q_j;\ j \in K_1\}$. Set $K_1 := K_1 - \{k\}$, $K_0 := K_0 \cup \{k\}$. Update q_j for j in K_1 by setting $q_j := q_j - (c_{kj} + c_{jk})/a_j$. Return to step b).

If the quadratic terms with $c_{ij} \neq 0$ are given in linked lists, step a) requires $O(m)$ operations. The sum $\sum_{j \in K_1} a_j$ being updated, step b) is $O(1)$ and done n times at most, hence is $O(n)$ in all. If the q_j's are kept in a heap, updating the heap and selecting the smallest q_j can be done in $O(m \operatorname{Log} n)$ operations, which is also the worst-case complexity of Algorithm 2.

We now describe the exact branch-and-bound algorithm.

ALGORITHM 3.

a) *Heuristic solution.* Obtain a heuristic solution X_h by Algorithm 2. Set $X_{opt} := X_h$ and $f_{opt} := f(X_h)$.

b) *Lexicographic fixation of variables.*

b.1) Determine λ^* by Algorithm 1. Let X^* denote the corresponding optimal solution of (5). If

$\sum_{j=1}^{n} a_j x_j^* = b$, end, X^* being the optimal solution of the quadratic knapsack problem.

b.2) Set $K_1 := \emptyset$, $K_0 := \emptyset$, $K_2 := \{1, 2, \ldots, n\}$ i.e. consider all variables as free. Then, for all j in K_2 in turn:

b.3) Set $x_j := \bar{x}_j^*$ i.e. at the complement of its value in X^*. Then, if $x_j = 1$, set $x_k := 0$ for all k in K_2 such that $a_k > b - \sum_{l \in K_1} a_l - a_j$; if $x_j = 0$ set $x_k := 0$ for all k in K_2 such that $a_k > b - \sum_{l \in K_1} a_l$. Compute by Algorithm 1 the best Lagrangean bound $\overline{f}_j(\lambda_j^*)$ of the so-defined subproblem. If $\overline{f}_j(\lambda_j^*) \leq f_{opt}$ set $K_1 := K_1 \cup \{j\}$, $K_2 := K_2 - \{j\}$ if $x_j = 0$ and $K_0 := K_0 \cup \{j\}$, $K_2 := K_2 - \{j\}$ if $x_j = 1$.

c) *Branch-and-bound routine.*

c.1) Set $M = \emptyset$.

c.2) Compute $\underline{b} = b - \sum_{l \in K_1} a_l$. If $\underline{b} < 0$ go to c.7).

c.3) For all j in K_2, if $a_j > \underline{b}$ set $K_0 := K_0 \cup \{j\}$, $K_2 := K_2 - \{j\}$, add j to M by the right and underline it.

c.4) Compute by Algorithm 1 the best Lagrangean bound $\overline{f}(\lambda^*)$ for the current subproblem; let X^* denote the corresponding solution. If $\overline{f}(\lambda^*) \leq f_{opt}$ go to c.7).

c.5) If $\sum_{j=1}^{n} a_j x_j^* \leq b$, set $X_{opt} = X^*$, $f_{opt} = f(X^*)$; moreover if $\sum_{j=1}^{n} a_j x_j^* = b$ go to c.7).

c.6) Choose j in K_2 according to rule d) hereunder and fix x_j at the specified value. Add j to M by the right. Go to c.2).

c.7) Seek from right to left an index j in M that is not underlined. If no such index exists, end. Otherwise set free the variables with an index k to the right of j in M and erase those indices, set $x_j = \bar{x}_j$, underline j in M, update K_0, K_1 and K_2 accordingly and go to c.2).

d) *Branching.*

d.1) Determine for each j in K_2 the decrease in value of $\overline{f}(\lambda)$ when x_j^* is replaced by $\bar{x}_j^*$, the value of λ being that one corresponding to the last bound computed.

d.2) Choose j in K_2 such that the decrease computed in the previous step is the smallest possible. Give to x_j the value $\bar{x}_j^*$.

COMPUTATIONAL EXPERIENCE.

The algorithms of the previous section have been extensively tested on a fairly small IBM 360/65 computer ([7]). Four hundred problems, with 10 to 50 variables and between 25 and 100 % of the largest possible number of quadratic terms have been randomly generated. All coefficients come from uniform distributions on [1, 100] for the c_{ij}, on [1, 50] for the a_{ij} and on $[50, \sum_{j=1}^{n} a_j]$ for b. Each problem was allocated a maximum CPU time of 2700 seconds, which was not enough to complete resolution for some of the largest and densest problems. The results are summarized in Table 1.

It appears clearly that a large proportion (about two thirds) of the variables are fixed in the preliminary procedure (step b) of Algorithm 3. Also, the number of subproblems solved and the number of backtracking remain fairly moderate.

The values of the upper and lower bounds given by Algorithms 1 and 2 were compared to the optimal values of two further series of 30 problems with 50 and 40 variables and a maximum number of quadratic terms. The results are summarized in Table 2. It is seen that the heuristic often yields an optimal solution or near-optimal solution, while the uper bound is quite precise. Comparing these computational results with those of Gallo, Hammer and Simeone ([3]) we find that computer times are larger but, on a smaller computer, bounds are more precise and subproblems less numerous. Finally, we note that a few larger problems, with up to 150 variables were solved in several hours of computer time.

Key for Table 1: D = density of the c_{ij} matrix, n = number of variables, m = number of quadratic terms, N_{sp} = number of subproblems solved, t = computational time in CPU seconds, N_p = number of variables fixed in the preliminary procedure, N_b = number of backtrackings in the branch-and-bound procedure, N_{bb} = number of variables fixed in the branch-and-bound procedure, N_{spbb} = number of subproblems solved in the branch-and-bound procedure, S = percentage of problems solved in 2700 seconds.

Figures on odd and even lines are mean values and standard deviations respectively for 20 problems in each case.

D	n	m	N_{sp}	t	N_p	N_b	N_{bb}	N_{spbb}	S
100	50	1275	279	655.25	36	120	14	243	60
		0	265	397.33	8	136	8	271	
	40	820	352	607.83	24	165	16	328	78.26
		0	445	672.46	11	238	11	452	
	30	465	240	281.35	18	117	12	222	100
		0	309	409.82	9	173	9	317	
	20	210	98	55.56	11	42	9	87	100
		0	103	60.40	6	54	6	109	
	10	55	17	3.85	5	5	5	12	100
		0	12	3.06	3	7	3	4	
75	50	951	336	650.89	32	148	18	304	75
		13	300	519.85	14	151	14	305	
	40	619	256	443.11	24	111	16	232	90
		8	228	473.92	14	115	14	241	
	30	348	144	121.20	22	60	8	122	100
		7	230	147.02	7	115	7	235	
	20	157	123	63.50	9	59	11	114	100
		6	185	92.90	7	105	7	189	
	10	40	16	3.68	6	4	4	10	100
		3	9	2.04	2	5	2	10	
50	50	639	255	428.84	38	106	12	217	90.48
		20	309	277.10	8	157	8	315	
	40	412	171	277.47	28	68	12	143	100
		15	181	356.55	10	91	10	190	
	30	233	158	104.75	20	67	10	138	100
		11	250	118.05	9	125	9	255	
	20	105	64	29.46	13	24	7	51	100
		7	68	31.63	5	35	5	73	
	10	27	18	3.56	6	5	4	11	100
		4	14	3.13	3	8	3	17	
25	50	317	194	313.00	36	74	14	157	100
		15	209	347.38	11	104	11	219	
	40	204	158	145.48	28	61	12	130	100
		12	141	123.59	8	71	8	148	
	30	117	86	51.45	22	30	8	64	100
		8	105	47.80	8	52	8	109	
	20	51	31	11.29	16	7	4	14	100
		7	33	13.79	4	17	4	36	
	10	15	20	3.63	6	6	4	14	100
		4	15	3.44	3	9	3	19	

Table 1. Computational Results.

n	*UB − opt*	*opt − LB*	*UB − LB*
50	4.230	0.191	4.421
	7.701	0.332	7.773
40	3.985	0.051	4.037
	5.152	0.177	5.131

Table 2. Comparison of upper bounds, lower bounds and optimal values. Odd lines give mean values and even lines standard deviations for 30 problems (in %).

Note added on proofs: Gallo, Grigoriadis and Tarjan ([2]) have recently obtained an $O(n^3)$ algorithm for the parametric network flow problem. it can be used to find the best multiplier with a worst-case complexity an order of magnitude lower than that of Algorithm 1.

REFERENCES

[1] FORD L.R. and D.R. FULKERSON, *Flows in networks*, Princeton University Press, 1962.

[2] GALLO G., M. GRIGORIADIS and R.E. TARJAN, "A Fast Parametric Network Flow Algorithm", Research Report, Department of Computer Science, Rutgers University, N.J. (forthcoming).

[3] GALLO G., P.L. HAMMER and B. SIMEONE, "Quadratic Knapsack Problem", *Math. Prog. Study* ***12*** (1980), 132-149.

[4] HAMMER P.L., "Some Network Flow Problem Solved with Pseudo-Boolean Programming", *Operations Research* ***13*** (1965), 388-399.

[5] KARZANOV A.V., "Determining the Maximum Flow in a Network by the Method of Preflows", *Soviet Math. Dokl.* ***15*** (1974), 434-437.

[6] MALHOTRA V.M., M.P. KUMAR and S.N. MAHESHWARI, "An $O(V^3)$ Algorithm for Finding Maximum Flows in Networks", *Information Processing Letters* ***7*** (1978), 277-278.

[7] MAHIEU Y., Un Algorithme pour le problème du sac de campeur quadratique, Mémoire de fin d'études, Faculté Universitaire Catholique de Mons, Belgium, 1981.

[8] MINOUX M., *Programmation Mathématique*, 2 Volumes, Dunod, Paris, 1983.

[9] PICARD J.C. and M. QUEYRANNE, "Networks, Graphs and Some Non-Linear 0-1 Programming Problems", Techn. Rep. EP77-R-32, Ecole Polytechnique de Montréal, Canada, 1977.

[10] PICARD J.C. and H.D. RATLIFF, "Minimum Cuts and Related Problems", *Networks* ***5*** (1975), 357-370.

$(K_4\text{-}e)$-FREE PERFECT GRAPHS AND STAR CUTSETS

Michele Conforti
Department of Statistics and Operations Research
New York University

Abstract.

We show that a perfect graph not containing $(K_4\text{-}e)$ as an induced subgraph, and whose clique-node incidence matrix does not belong to a restricted class of totally unimodular matrices, has a star cutset. This result yields a new proof that the Strong Perfect Graph Conjecture is true for this class of graphs.

1. Introduction

The graphs G(V,E) we consider are simple, with nodeset V and edgeset E. Let $\alpha(G)$ be the maximum size of a stable set of G, and $\theta(G)$ the minimum number of cliques which cover V(G). A graph G is perfect if $\alpha(G') = \theta(G')$ for every induced subgraph G' of G. Berge [BER 1] formulated the following two conjectures:

1.1. A graph is perfect if and only if its complement is perfect.

1.2. A graph is perfect if and only if it does not contain an odd hole or an odd antihole.

An odd hole is a chordless cycle of length ≥ 5 and an antihole its complement.

Lovász [LOV] has proven the first conjecture. The second conjecture, which is stronger than 1.1, is still unsolved, but has been shown to hold for several classes of graphs, as planar graphs [TU 1], claw-free graphs [PR 1], $(K_4\text{-}e)$-free graphs [PR 2], to mention a few. An exhaustive collection of papers dealing with perfect graphs can be found in [BC].

In two recent papers, [CCM 1], [CCM 2] the problem of covering the family of cliques of size i (i-cliques) of a graph with cliques of size i - 1 has been studied. In particular, a graph is defined in [CCM 2] to be K_i-perfect if, for every subfamily F of i-cliques, the maximum number of i-cliques in F no two of which have i-1 nodes in common, is equal to the minimum number of (i-1)-cliques necessary to cover all the i-cliques of F. (A (i-1)-clique covers a i-clique if it is completely contained in the i-clique.) The main result of [CCM 2] is a characterization of K_i-perfect graphs, in terms of forbidden subgraphs. This result uses the fact that intersection graphs of K_i-perfect graphs do not contain as induced subgraph the graph $(K_4\text{-}e)$ in figure 1.3, which is referred to as a diamond in our treatment.

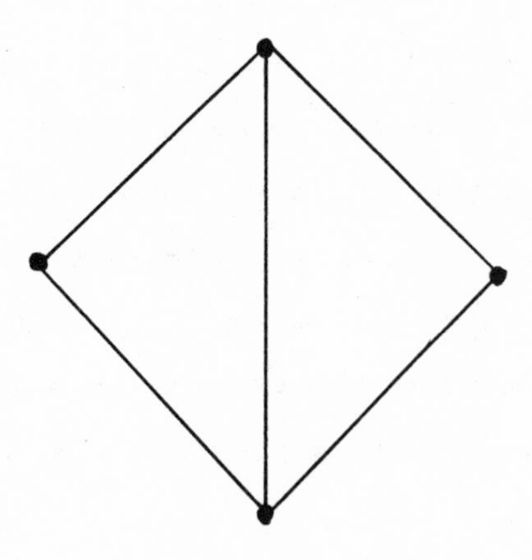

Figure 1.3

Furthermore, the characterization of K_i-perfect graphs uses the fact that a diamond-free graph is perfect if and only if it does not contain an odd hole [PR 2]. Hence the study of structural properties of diamond-free graphs is crucial in the study of K_i-perfect graphs.

In this paper we give a new proof of the fact that diamond-free graphs satisfy conjecture 1.2.

Our terminology uses the theory of hypergraphs. Hypergraphs are a generalization of graphs in the sense that edges can be adjacent to a nonempty set of nodes of any cardinality. Berge defines a hypergraph to be linear if any two edges have at most one node in common. This property

holds for any simple graph, hence these hypergraphs can be thought of as a first generalization of simple graphs. A hypergraph whose set of edges is the family of maximal cliques of a graph is linear if and only if the graph does not contain a diamond. The hypergraph defined on the family of maximal cliques of a diamond-free graph does not have odd cycles of size 3. A graph which does not contain a cycle of size 3 is perfect if and only if it does not contain an odd hole.
We show that this statement can be extended in the sense that a graph whose family of maximal cliques is a linear hypergraph is perfect if and only if the graph does not contain an odd hole, which is a restatement of the result of Pathasarathy and Ravindra [PR 2].

Our paper can also be seen as an attempt to use the definitions and the theory of hypergraphs in this domain. In a recent paper [CC], the theory of hypergraphs has been used to provide an algorithmic framework for the solution of the set packing problem in some classes of matrices.

2. Some basic hypergraph theory and results on diamond-free hypergraphs

In our treatment, we use theory and definitions of hypergraphs. We refer to Berge's [BER 2] classical textbook for an extensive treatment. Given a finite set $V = \{v\}$ and a family $E = \{e\}$ of subsets of V, the pair H(V,E) is a hypergraph if $V = \bigcup_{e \in E} e$. The set E(H) is the set of edges of H and V(H) is the set of nodes of H.

The set of edges incident with a node $v \in V(H)$ is indicated by E_v. The set V_e is the set of nodes contained in an edge of E(H), and the degree of e is the cardinality of V_e.

Graphically, the nodes of H are represented as points, edges with degree 1 as loops, edges with degree 2 as edges of a graph and edges with degree greater than 2 as circles containing the nodes of it. Thus any 0-1 matrix with no zero row or column can be represented as the edge-node incidence matrix of a hypergraph and vice versa. In particular, given a graph G, its hypergraph of the cliques is the hypergraph H(V,E) having the same nodeset as G, and E is the family of all maximal cliques of G. The incidence matrix of H is the clique matrix of G in the sense of [PAD]. A partial hypergraph H - E' of the hypergraph H(V,E), induced by $E-E' \subseteq E$ is obtained from H by removing all the edges in E' and the nodes which become isolated, or equivalently by removing some rows of the incidence matrix and the columns which become zero vectors. A subhypergraph H - V' induced by $V-V' \subseteq V$ is obtained by removing from H the nodes in V' and the edges which become empty, or equivalently, by removing some columns of the incidence matrix and the rows which become zero vectors.

If A is the edge-node incidence matrix of H, the dual hypergraph of H is the hypergraph $H^*(V^*,E^*)$, whose incidence matrix is the transpose of A. Therefore, there is a one-to-one correspondence between the elements of V and E^* and the elements of E and V^*. Also, the dual hypergraph of H^* is H itself.

A path P is a sequence $v_1,e_1,v_2 \ldots e_{n+1}$ of distinct nodes and edges, and $v_i,v_{i+1} \in V_{e_i}$ $\forall\, i = 1 \ldots n$. The members of P are the nodes and edges of P. The first and the last members of P are the ends of P. A member of P which is not an end is called intermediate.

Ends of P can be nodes or edges. If both ends are nodes, the path is a node-node path. If the first (last) end is a node and the last (first) end is an edge, P is a node-edge (edge-node) path. If both ends are edges, P is an edge-edge path. A cycle has the same definition as a node-node path, except that $v_1 \equiv v_{n+1}$. A cycle or a path is odd if the cardinality of its edge set is odd, even otherwise. A chord of a cycle C is an edge containing at least 3 nodes of C. A chord containing exactly three nodes of C is triangular. A chord is short if it contains 3 consecutive nodes of C. A hole is a chordless cycle of H. A cycle C is balanced if there exists an edge of E(C) which is a chord of C.

A hypergraph is perfect if it is the hypergraph of the cliques of a perfect graph. It is well known that every odd cycle of a perfect graph must have a chord. A hypergraph is P-critical, see [PAD], if H is not perfect, but H - {v} is perfect, $\forall$ $v \in V(H)$. A hypergraph which is not perfect contains a P-critical subhypergraph. A hypergraph is balanced if every odd cycle of it is balanced. Balanced hypergraphs are perfect, see [BER 2].

Let H(V,E) be the hypergraph of the cliques of a diamond-free (DF, for short graph G, then H has the properties listed below. The first and the last ones follow immediately from the DF property. The other ones are immediate consequences of the preceding ones.

<u>Property 2.1</u>. H is linear if and only if G is diamond free.

<u>Property 2.2</u>. Not more than one edge of H can join two nodes. Hence a cycle C is completely specified by its nodeset and a chord of C containing two consecutive nodes of V(C) belongs to E(C).

<u>Property 2.3</u>. The number of edges of H is bounded by the number of edges of G.

Property 2.4. No cycle of H contains a short chord.

Property 2.5. Every cycle of H has length ≥ 4.

Property 2.6. The dual of H is diamond free.

An odd cycle C is minimal with respect to its nodeset if no subset of nodes of V(C) induces a smaller odd cycle.

Lemma 2.7. Let H' be the subhypergraph of a DF hypergraph H not containing odd holes, induced by the nodes of an odd cycle C minimal with respect to its nodeset. H' is a balanced odd cycle, with a unique edge e^* of degree 3, a unique node v^* of degree 3 and formed by two even holes of length greater or equal to 4, (see figure 2.8).

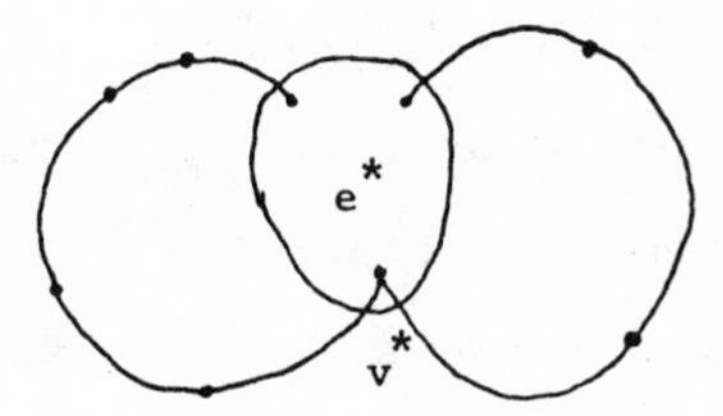

Figure 2.8

Proof: No edge of E(H) - E(C) contains exactly two nodes of V(C), since by property 2.1, these nodes have to be nonconsecutive in C, and therefore a smaller odd cycle exists.

Suppose now an edge e contains three nodes, v_i, v_j, v_ℓ, no two of them consecutive in C, see figure 2.9. Since e contains at most two consecutive nodes of V(C), we can assume that the three nodes v_i, v_j, v_ℓ

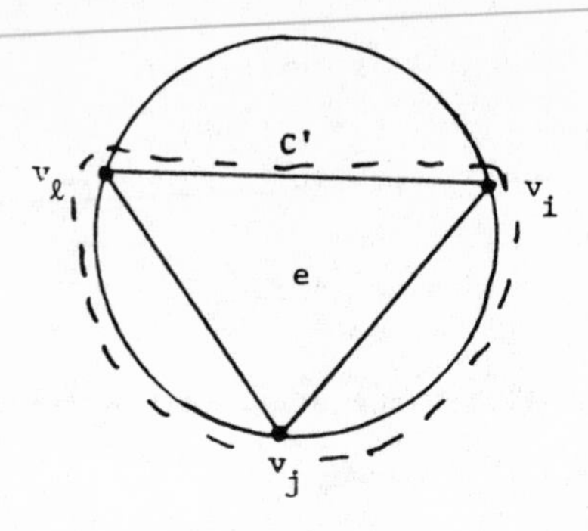

Figure 2.9

are chosen so that no node v_k between v_i and v_j, or v_j and v_ℓ, belongs to $V(C) \cap V_e$. The cycles between v_i, v_j and v_j, v_ℓ closed by e must be even, else a smaller odd cycle exists, but then the cycle $C' = v_j,\ldots,v_\ell,e,v_i,\ldots,v_j$ is a smaller induced odd cycle, since v_i and v_ℓ are nonconsecutive in C. Hence no chords of E(C) can contain four or more nodes of C and no chords of E(H) - E(C) can contain three or more nodes of C. This implies that the only existing chords are triangular and belong to E(C). Therefore it remains to be shown that no two or more such chords exist, since the fact that the two even holes have size ≥ 4 is a consequence of property 2.5.

Let $e_1 = (v_i,v_{i+1},v_k)$, $e_2 = (v_j,v_{j+1},v_\ell)$ be two such chords. We distinguish two cases.

Case 1. e_1,e_2 do not cross, that is, there exist two nodes, v_1,v_2, whose removal divides C into two paths, P_1,P_2 with ends v_1,v_2, and V_{e_1} is contained in P_1, V_{e_2} is contained in P_2, see figure 2.10. In this case, since the cycles $v_j,e_2,v_\ell,\ldots,v_k,e_1,v_{i+1},\ldots,v_j$ and $v_i,e_1,v_k,\ldots,v_i$ must be even, then $v_j,e_2,v_\ell,\ldots,v_k,\ldots,v_i,e_1,v_{i+1},\ldots,v_j$ is an odd cycle with less nodes, since v_{j+1} is not in it.

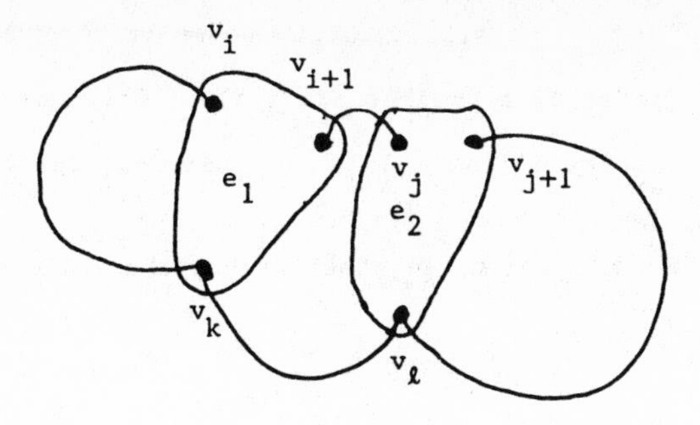

Figure 2.10

Case 2. e_1, e_2 cross. That is, no two nodes satisfying the above property exist, and suppose that nodes v_i, v_{i+1}, v_j, v_{j+1}, v_k, v_ℓ are in the above order in the cycle, see Figure 2.11.

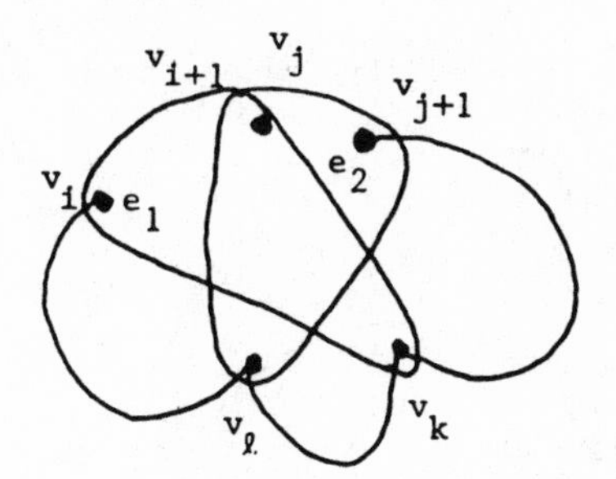

Figure 2.11

Label with + and - signs the nodes of V(C) so that the only consecutive nodes with the same label are v_i, v_{i+1}. Node v_k has label opposite to the label of v_i, else a smaller odd cycle exists. If nodes v_{j+1} and v_ℓ have the same label, then e_2 closes a smaller odd cycle. Otherwise, if nodes v_{j+1} and v_ℓ have opposite labels, then nodes v_j and v_ℓ have the same label. The cycle $v_j, e_2, v_\ell, \ldots, v_k, e_1, v_{i+1}, \ldots, v_j$ is odd, since all its nodes have alternating labels except for nodes v_j and v_ℓ. Furthermore this cycle is smaller than C, otherwise nodes v_i and v_ℓ, v_j+1 and v_k would coincide, but then edges e_1, e_2 would contradict property 2.1. ∎

An odd cycle is minimal if no subset of its nodes or no subset of its edges is the node set or the edgeset of a smaller odd cycle. Obviously, a minimal odd cycle must satisfy requirements of lemma 2.7. Furthermore, the following holds:

Lemma 2.12. If C is a minimal odd cycle, then it has the following two properties:

(2.13) $\qquad e_i, e_j \in E(C) \rightarrow V_{e_i} \cap V_{e_j} \backslash V(C) = \phi.$

(2.14) $\qquad v_i, v_j \in V(C) \rightarrow E_{v_i} \cap E_{v_j} \backslash E(C) = \phi.$

Proof: We prove property 2.13. Property 2.14 is a direct consequence of lemma 2.7.

Suppose e_i, e_j have a node $v \notin V(C)$ in common, and e_i, e_j contain the pairs $(v_i, v_{i+1})(v_j, v_{j+1})$ of consecutive nodes in C. Let P_1 and P_2 be the paths joining v_{i+1} to v_j and v_{j+1} to v_i in C (see figure 2.15).

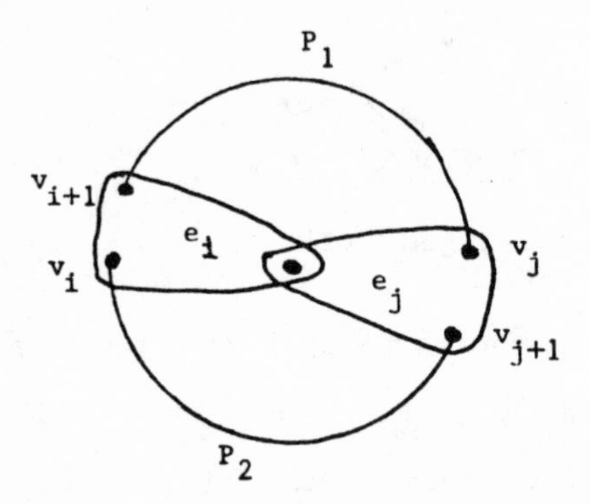

Figure 2.15

By property 2.1, we have that $V_{e_i} \cap V_{e_j} \cap V(C) = \phi$, in particular, e_i and e_j are not consecutive in C; hence P_1 and P_2 contain at least two nodes and have different parity, since C is odd; hence a smaller odd cycle exists. ■

3. The strong perfect graph conjecture is true for diamond-free graphs

In this paragraph we give a proof that if a graph is diamond free and has no odd hole, then it is perfect.

If S is a set, we indicate with $|S|$ its cardinality, and with $\|S\|$ its parity. $\|\bar{S}\|$ is the complement of its parity. Parities follow boolean laws of addition, hence we indicate with $\oplus$ the addition modulo 2. If C is a cycle containing a triangular chord $e \in E(C)$, then e contains two consecutive nodes in the cycle and a third node which is not consecutive to any of the two previous ones, else property 2.4 is contradicted. Such a node is defined to be a star of the cycle.

A cycle C of a DF hypergraph is said to be chorded if it has the following properties:

Property 3.1. The only chords of C are triangular and belong to E(C).

Property 3.2. C has at most two stars.

Property 3.3. No pair of consecutive nodes of a chord contains the star of another chord.

Lemma 3.4. Let T be the set of chords of a chorded cycle C. Then, either $\|C\| = \|T\|$, or else there exists an odd hole.

Proof: The proof is by induction on $|T|$, the cardinality of the set of chords of C. The lemma is obviously true for $|T| = 0$, $|T| = 1$.

Suppose C contains a chord e with consecutive nodes v_i, v_{i+1} and star v_k, such that, if we define the two cycles $C_1 = v_{i+1}, e, v_k, v_{k-1}, \ldots, v_{i+1}$ and $C_2 = v_i, e, v_k, v_{k+1}, \ldots, v_i$, then every chord of T different from e has its nodeset completely contained in $V(C_1)$ or $V(C_2)$ (see figure 3.5.).

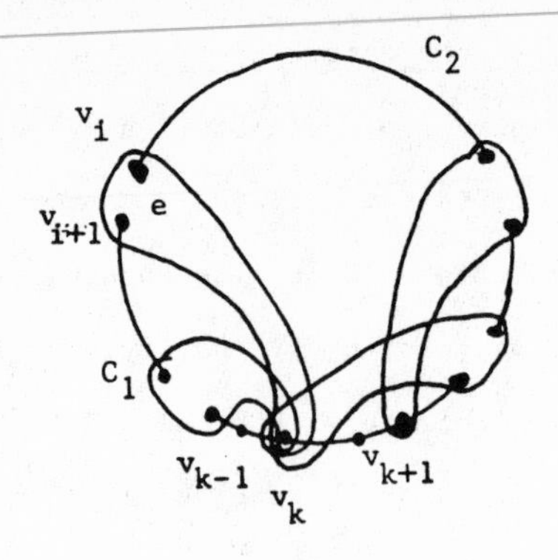

Figure 3.5

Let T_1 and T_2 be the set of chords of C_1 and C_2 respectively. By induction hypothesis $\|T_1\| = \|C_1\|$ and $\|T_2\| = \|C_2\|$, since C_1 and C_2 have at least one less chord than C.

But $\|T_1\| \oplus \|T_2\| = \|\bar{T}\|$ since T - {e} is partitioned into T_1 and T_2. Also $\|C_1\| \oplus \|C_2\| = \|\bar{C}\|$ since V(C) - $\{v_k\}$ is partitioned into $V(C_1)$, $V(C_2)$ and v_k belongs to both. Hence $\|\bar{T}\| = \|T_1\| \oplus \|T_2\| = \|C_1\| \oplus \|C_2\|$. Therefore $\|T\| = \|C\|$.

We now consider the case in which no chord, satisfying the above property exists. This implies that two chords $e_1 = \{v_i, v_{i+1}, v_k\}$ $e_2 = \{v_\ell, v_{\ell+1}, v_m\}$ having different stars, cross, see figure 3.6. (Crossing has the same meaning as in case 2 of lemma 2.7.)

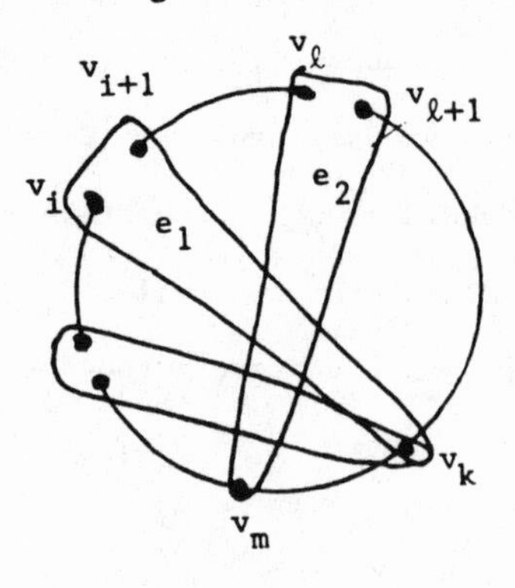

Figure 3.6

Let $C_1 = v_i,\ldots,v_m,e_2,v_{\ell+1},\ldots,v_k,e_1,v_i$, $C_2 = v_{i+1},\ldots,v_\ell,e_2,v_m,\ldots,v_k,$ e_1,v_{i+1}. Let T_1 and T_2 be the set of chords of C_1 and C_2. By the induction hypothesis, $\|T_1\| = \|C_1\|$ and $\|T_2\| = \|C_2\|$, since, by property 3.3, neither V_{e_1} nor V_{e_2} belong to $V(C_1)$ or $V(C_2)$, hence C_1 and C_2 have at least one less chord than C. Furthermore, $\|T_1\| \oplus \|T_2\| = \|T\|$, since the only stars of C are v_k and v_m, hence every chord of $T - \{e_1,e_2\}$ has its nodeset either in $V(C_1)$ or in $V(C_2)$ and, by the above argument, e_1 and e_2 do not belong to either $V(C_1)$ or $V(C_2)$. Also, $\|C\| = \|C_1\| \oplus \|C_2\|$ since $V(C)\backslash(v_k,v_m)$ is partitioned into $V(C_1)$ and $V(C_2)$ and v_k,v_m belong to both. Therefore

$$\|T\| = \|T_1\| \oplus \|T_2\| = \|C_1\| \oplus \|C_2\| = \|C\|. \qquad \blacksquare$$

We now consider a perfect diamond-free hypergraph containing a minimal odd cycle C, with v^* its unique mode of degree 3 and e^* its unique triangular chord of E(C), see figure 3.7.

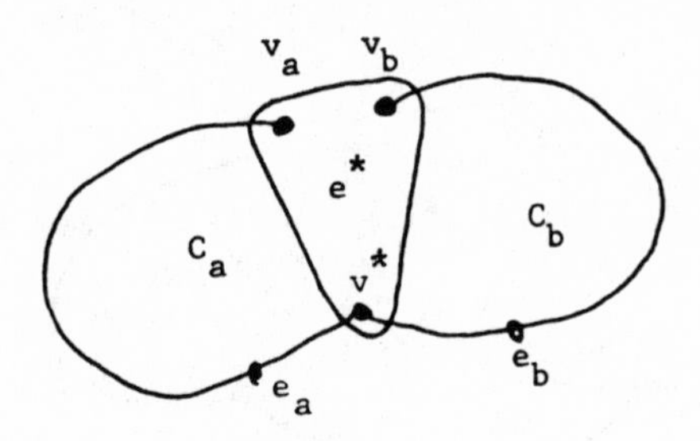

Figure 3.7

Nodes v_a,v_b are the two nodes of V(C) belonging to $V_{e*} - (v^*)$ and edges e_a,e_b are the two edges of E(C) adjacent to v^*.

Let C_a and C_b be the two even holes, having e^* and v^* in common, formed by the odd cycle. That is, $C_a = v^*,e_a,\ldots,v_a,e^*,v^*$ and $C_b = v^*,e_b,\ldots,v_b,e^*,v^*$.

We are now ready to state our key lemma. A path joining C_a to C_b is simple if no subset of its nodes forms a path connecting C_a to C_b.

Lemma 3.8. Every simple path joining C_a to C_b either contains a node in V_e^* or contains an edge of E_{v*}.

Proof: Suppose a path exists contradicting the above statement. Let P be the shortest such path. By lemma 2.12, if one end of P is a node, then the edge of P containing it can not contain any other node of C. If one end of P is an edge, then the node of P adjacent to it can not belong to any other edge of C. P can be a node-node, edge-node, or edge-edge path. Lemmas 3.9, 3.11 and 3.13 rule out these three possibilities. ■

Lemma 3.9. $P = v_1, e_1, v_i, \ldots, e_n, v_{n+1}$ cannot be a node-node path.

Proof: We obviously assume $v_1 \neq v_a$, $v_b \neq v_{n+1}$. Consider the subhypergraph induced by $V(P) \cup V(C)$, see figure 3.10.

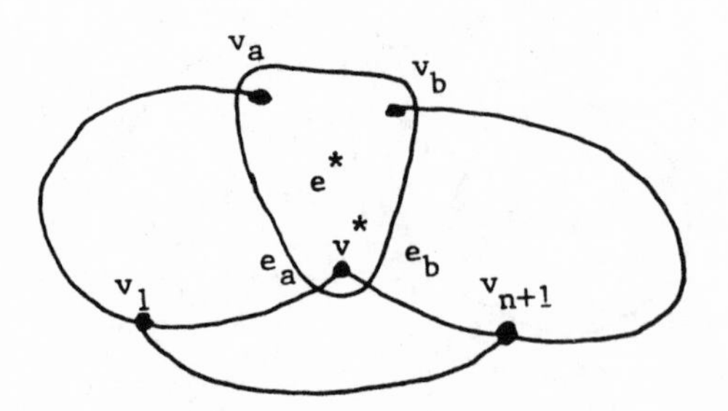

Figure 3.10

Consider the following four paths, formed by nodes of V(C) joining v_1 to v_{n+1}.

$$P_{e*} = v_1, \ldots, v_a, e^*, v_b, \ldots, v_{n+1}$$

$$P_{v*} = v_1, \ldots, e_a, v^*, e_b, \ldots, v_{n+1}$$

$$P_{v_a} = v_1,\ldots,v_a e^*,v^*,e_b,\ldots,v_{n+1}$$

$$P_{v_b} = v_1,\ldots,e_a,v^*,e^*,v_b,\ldots,v_{n+1}$$

Since C is an odd cycle, $\|P_{e*}\| = \|\overline{P_{v*}}\|$, and since C_a, C_b are even holes, we have that $\|\overline{P_{e*}}\| = \|P_{v*}\| = \|P_{v_a}\| = \|P_{v_b}\|$. No edge of C can contain an intermediate node of V(P), and an intermediate edge of P can contain only nodes v_a or v_b, but not both, else property 2.1 is contradicted. Let C_{e*}, C_{v*}, C_{v_a}, C_{v_b} be the cycles closed by P with P_{e*}, P_{v*}, P_{v_a}, P_{v_b} respectively. The only chords of these cycles are edges of P containing v_a or v_b, hence each of these cycles satisfies properties 3.1, 3.2 and 3.3 of lemma 3.4. C_{v*} is a chordless cycle, hence it must be an even hole. This means that both C_{v_a} and C_{v_b} are even, therefore the number of chords containing nodes v_a is even and the number of chords containing v_b is even too. But then the cycle C_{e*} is odd and has an even number of chords, namely the edges of P containing v_a and v_b. As a consequence of lemma 3.4, there exists an odd hole. ∎

<u>Lemma 3.11</u>. $P = e_1,v_1,e_2,\ldots,e_{n-1},v_n$ cannot be an edge-node path.

<u>Proof</u>: The proof is analogous to the one for lemma 3.9, hence it is sketched. We obviously assume $e_0 \neq e_a$ and $v_n \neq v_b$. Consider the subhypergraph induced by $V(P) \cup V(C)$, see figure 3.12.

Consider the following four paths:

$$P_{e*} = e_0,\ldots,v_a,e^*,v_b,\ldots,v_n$$

$$P_{v*} = e_0,\ldots,e_a,v^*,e_b,\ldots,v_n$$

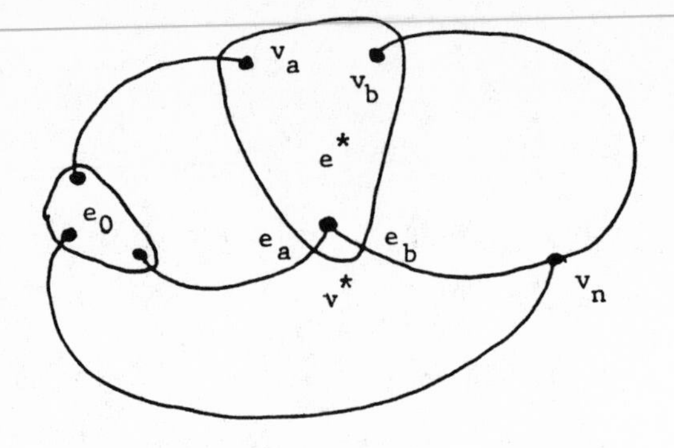

Figure 3.12

$$P_{v_a} = e_0,\ldots,v_a,e^*,v^*,e_b,\ldots,v_n$$

$$P_{v_b} = e_0,\ldots,e_a,v^*,e^*,v_b,\ldots,v_n$$

We have that $\|P_{e*}\| = \|P_{v*}\| = \|P_{v_b}\| = \|\overline{P_{v_a}}\|$. Let C_{e*}, C_{v*}, C_{v_a}, C_{v_b} be the cycles closed by P with P_{e*}, P_{v*}, P_{v_a}, P_{v_b} respectively. Then C_{v*}, C_{e*},C_{v_b} are even, implying that C_{v_a} is an odd cycle with an even number of chords, satisfying the properties, 3.1, 3.2 and 3.3 of lemma 3.4, hence an odd hole exists. ■

<u>Lemma 3.13</u>. P cannot be an edge-edge path.

<u>Proof</u>: The proof is the same as lemma 3.9 and is omitted. ■

<u>Lemma 3.14</u>. Let P be a path joining C_a to C_b. Then
$V(P)\cap V(E_{v*}) \neq \emptyset$.

<u>Proof</u>: Suppose P contradicts the above hypothesis. Let P' be the shortest path in the subhypergraph induced by $V(P)\cup V(C)$, joining C_a to C_b and not having nodes v_a or v_b as ends. P' exists since $|V(P)\cap V(E_{v*})| = \emptyset$, and it contradicts lemma 3.8.

■

A node v of a graph is a star cutset, see [CHVA], if the subgraph induced by removing all the nodes in $\{v \cup N(v)\}$ is disconnected, where N(v) is the set of nodes adjacent to v. Equivalently, a hypergraph of the cliques of a graph has a star cutset v if $H - V(E_v)$ is disconnected. Chvátal has shown that no P-critical hypergraph has a star cutset. However, lemma 3.14 says that v^* is a star cutset.

It is well known that a hypergraph having no odd cycles is totally unimodular (T.U. for short), see [BER 2]. That is, its edge-node incidence matrix is T.U. A T.U. hypergraph is perfect, hence for a hypergraph to qualify as imperfect or P-critical, it must contain an odd cycle. If it contains an odd hole, then the nodes of the smallest possible odd hole form a P-critical subhypergraph, since, as a consequence of property 2.5 all the odd cycles of H have length at least 5. If H contains an odd cycle but no odd hole, and H is DF, then it contains a star cutset and cannot be P-critical.

We have therefore shown the result of Parthasarathy and Ravindra [PR 2].

<u>Theorem 3.15</u>. If H is p-critical and linear, then H is an odd hole.

REFERENCES

[BER 1] Berge, C., "Farbung von Graphen, deren samtliche bzw deren ungerade kreise starr sind (Zusammenfassung)," Will. Z. Martin Luther Univ., Halle Wittenberg Math. Natur. Reihe, 114, 1961.

[BER 2] Berge, C., "Graphs and Hypergraphs," North Holland, 1973.

[BC] Berge, C., and Chvátal, V., "Topics on perfect graphs," Annals of Discrete Mathematics 21, North Holland, 1984.

[CHVA] Chvátal, V., Star cutsets and perfect graphs, Technical Report, SOCS, McGill University, 1983.

[CC] Conforti, M., and Cornuéjols, G., "An algorithmic framework for the matching problem in some hypergraphs," Working Paper 24-84-85, G.S.I.A., Carnegie Mellon University, To appear in Networks.

[CCM 1] Conforti, M., Corneil, D., and Mahjoub, A. R., "K_i-covers I: Complexity and polytopes," to appear in Discrete Mathematics.

[CCM 2] Conforti, M., Corneil, D., and Mahjoub, A. R., "K_i-covers II: K_i-perfect graphs," submitted for publication.

[LOV] Lovász, L., "Normal hypergraphs and the perfect graph conjecture," Discrete Mathematics 2, 253-267, 1972.

[PAD] Padberg, M., "Perfect zero-one matrices," Mathematical Programming 6, 180-196, 1974.

[PR 1] Parthasarathy, K. R., and Ravindra, G., "The strong perfect-graph conjecture is true for $K_{1,3}$-free graphs," J. Combin. Theory B 21, 212-223, 1976.

[PR 2] Parthasarathy, K. R., and Ravindra, G., "The validity of the strong perfect-graph conjecture for $(K_4 - e)$-free graphs," J. Combin. Theory B 26, 98-100, 1979.

[TU 1] Tucker, A., "The strong perfect graph conjecture for planar graphs," Canad. J. Math. 25, 103-114, 1973a.

[TU 2] •Tucker, A., Circular arc graphs: New uses and a new algorithm, in "Theory and Application of Graphs," Lecture Notes in Math 642, pp. 580-589, Springer-Verlag, 1978.

A Bound on the Roof-duality Gap

P.L. Hammer*

RUTCOR, Rutgers University, New Brunswick, NJ, 08903, U.S.A.

B. Kalantari

Department of Computer Science, Rutgers University, New Brunswick, NJ, 08903, U.S.A.

Abstract Zero-one maximization of a quadratic function $f(x)$ is NP-hard. In [4] several equivalent polynomially solvable relaxations are described whose common optimal value, w, yields an upper bound of $f*$, the zero-one maximum of $f(x)$. One of these relaxations is the maximization of a piecewise linear concave function, $R(x)$, over the full unit hypercube. Using this relaxation we obtain a bound on $(w-f*)$. In the special case where the off-diagonal elements of the Hessian matrix are nonnegative, we show that $R(x)$ coincides with the concave envelope.

Keywords: Quadratic Zero-One, Roof duality, Concave envelope.

1. Introduction

Consider

(P0) $max\ \{f(x) : x \in B\}$,

where, as in [4], we assume

$$f(x) = \sum_{i=1}^{n} q_{ii} x_i + \sum_{i=1}^{n} \sum_{j=i+1}^{n} q_{ij}\, x_i x_j,$$

and B $=$ $\{x:\ x_i=0,\ 1, \text{ for } i=1,\dots,n\}$. Let $\overline{B} = \{x : 0 \le x_i \le 1, \text{ for } i=1,\dots,n\}$, $M = \{(i,j) : 1 \le i < j \le n, q_{ij} \ne 0\}$, $P = \{(i,j) \in M : q_{ij} > 0\}$, and $N = \{(i,j) \in M : q_{ij} < 0\}$.

It is well-known that (P0) is NP-hard (see Garey et al. [2]). In [4] several equivalent polynomially solvable relaxations are described whose common optimal value, w, yields an upper bound of $f*$, the zero-one maximum of $f(x)$. One relaxation of (P0) defined in [4] is the continuous Rhys form (*crf*)

(P1) $max\ L(x,y)$

$$\begin{aligned}
\text{s.t. } & y_{ij} \le x_i,\ y_{ij} \le x_j, && \text{for all } (i,j) \in P,\\
& y_{ij} \le (1-x_i),\ y_{ij} \le x_j, && \text{for all } (i,j) \in N,\\
& 0 \le x_i \le 1, && \text{for } i=1,\dots,n,\\
& 0 \le y_{ij} \le 1, && \text{for } i,j=1,\dots,n,\ i<j,
\end{aligned}$$

---

* The partial support of NSF Grant No. ECS85-03212 and AFOSR Grant No. 0271 is gratefully acknowledged.

where,

$$L(x,y)=\sum_{(i,j)\in P} q_{ij}y_{ij}-\sum_{(i,j)\in N} q_{ij}y_{ij}+\sum_{i=1}^{n}(q_{ii}+\sum_{j|(j,i)\in N} q_{ji})x_i.$$

The optimal value w of (P1) is an upper bound of f^*, the optimal value of (P0).

For each $x\in\overline{\mathrm{B}}$, define

$$R(x)=\sum_{(i,j)\in P} q_{ij}\,min\{x_i,x_j\} - \sum_{(i,j)\in N} q_{ij}\,min\{1-x_i,x_j\}+\sum_{i=1}^{n}(q_{ii}+\sum_{j|(j,i)\in N} q_{ji})x_i.$$

Consider

(P2) $max\ \{R(x) : x\in\overline{\mathrm{B}}\}$.

In the next section we focus on the relaxation defined by (P2) and obtain a bound on the *gap* $(w-f^*)$. In Section 3, we show that in the special case where the off-diagonal elements of the Hessian matrix are nonnegative, $R(x)$ coincides with the concave envelope.

2. A Bound on the Roof-duality Gap

In this section we obtain a bound on the gap.

Theorem 2.1. [4] Let w' be the optimal value of (P2), then $w=w'$.

Proof. Let $z^*=(x^*,y^*)$ be an optimal solution of *crf*. Let $\overline{z}=(x^*,\overline{y})$, where

$$\overline{y}_{ij}=\begin{cases} min\{x_i^*,x_j^*\} & \text{, if } q_{ij}>0 \\ min\{1-x_i^*,x_j^*\} & \text{, if } q_{ij}<0 \end{cases}$$

Note that $\overline{z}$ is a feasible solution to *crf*, $R(x^*)=L(\overline{z})$, and that $L(z^*)\leq L(\overline{z})$. Thus, $L(z^*)=L(\overline{z})$ which implies $w\leq w'$. Conversely, if x^* is an optimal solution of $R(x)$, and $\overline{z}$ is defined as before, then $L(\overline{z})=R(x^*)$. ■

By converting the *crf* to a continuous relaxation of maximum weighted stable set problem, and utilizing a result due to Balinski [1], in [4] it was shown that there exists an optimal solution to *crf* with components 0, 1/2, or 1. From this and Theorem 2.1 we conclude

Corollary 2.1. There exists an optimal solution x^* to (P2) with $x_i^*=0, 1/2$, or 1, for all $i=1,\ldots,n$. ■

Lemma 2.1. Let $S=\{(i,j)\subset M : x_i^*=x_j^*=1/2\}$. Then,

$$w-f(x^*)=1/4\sum_{(i,j)\in S}|q_{ij}|.$$

Proof. Consider a term $q_{ij}x_ix_j$ in $f(x)$. The corresponding term in $R(x)$ is $q_{ij}min\{x_i,x_j\}$, if $q_{ij}>0$ and $|q_{ij}|(min\{1-x_i,x_j\}-x_j)$, if $q_{ij}<0$. The only time an error will be incurred is when $x_i^*=x_j^*=1/2$, with a corresponding error of $1/4|q_{ij}|$. ■

Lemma 2.2. Given x^*, one can obtain in polynomial-time a point $\bar{x} \in B$ such that

$$w - f(\bar{x}) \leq 1/4 \sum_{(i,j)\in S} |q_{ij}|.$$

Proof. Suppose $x_i^* = 1/2$, for some $i \in \{1,...,n\}$. By fixing all variables except for the i-th one, we obtain a linear function in that variable. Maximizing this function over the unit interval, we get a solution whose i-th component is either 0 or 1 and whose function value is at least $f(x^*)$. Repeating this process results in a point $\bar{x} \in B$ having the desired property. ■

It follows that

Theorem 2.2.

$$w - f^* \leq w - f(\bar{x}) \leq w - f(x^*) = 1/4 \sum_{(i,j)\in S} |q_{ij}|. \ ■$$

Remark. The above bound which also applies to $(f^* - f(\bar{x}))$, is independent of the linear term in $f(x)$. In some special cases, we can use this bound to measure the quality of $\bar{x}$ as an approximate solution. As an example let Q^+ and Q^- denote the sum of the positive and negative q_{ij}'s, respectively. Let $|Q| = Q^+ - Q^-$ and $Q = Q^+ + Q^- = f(1,1,...,1)$. Assume that $Q > 0$ and $\rho = -(Q^-/|Q|) < 1/2$. It can be shown that $(f^* - f(\bar{x}))/f^* \leq 1/4(1-2\rho)$. Thus, for ρ small, one expects $\bar{x}$ to be a very good approximate solution.

3. The Relationship between R(x) and the Concave Envelope

Let x^i, $i=1,...,2^n$ be the vertices of B. For each $x \in \bar{B}$, define $\Gamma(x) = \max_{\alpha} \sum_{i=1}^{2^n} \alpha_i f(x_i)$ subject to $\sum_{i=1}^{2^n} \alpha_i = 1$, $\alpha_i \geq 0$, and $\sum_{i=1}^{2^n} \alpha_i x^i = x$. The function $\Gamma(x)$ is concave, it agrees with $f(x)$ over $x \in B$ and its maximum over $\bar{B}$ coincides with the maximum of $f(x)$ over B. It can be shown that $\Gamma(x)$ is the best concave overestimator of $f(x)$ over $\bar{B}$, i.e., if $g(x)$ is any concave function over $\bar{B}$ with the property that $g(x) = f(x)$, for all $x \in B$, then $g(x) \geq \Gamma(x)$, for all $x \in \bar{B}$ (see Falk and Hoffman [3], and Rockafellar [6] for general definition of concave envelope).

Thus, $R(x) \geq \Gamma(x)$, for all $x \in \bar{B}$ and whenever $R(x)$ and $\Gamma(x)$ have the same maximum value over $\bar{B}$, (P0) may be solved in polynomial time. Rhys's work [5] implies that if $q_{ij} \geq 0$, for all $i,j=1,...,n$, $i<j$, then (P0) is polynomially solvable. This also follows from the unimodularity of crf (see [4]). In this case, $R(x)$ and $\Gamma(x)$ are in fact identical:

Theorem 3.1. If $q_{ij} > 0$ for all $(i,j) \in M$, then $R(x) = \Gamma(x)$.

Proof. Let $x=(x_1,...,x_n)\in \overline{\mathrm{B}}$. Without loss of generality assume $x_1 \geq x_2... \geq x_n$. Let $x^1=(1,...,1)$, $x^2=(1,...,1,0)$, ..., $x^n=(1,0,...,0)$, $x^{n+1}=(0,0,...,0)$. Let $\alpha_1=x_n$, and for i=2,...,n define $\alpha_i=x_{n+1-i}-x_{n+2-i}$. Let $\alpha_{n+1}=1-x_1$. Note that $\alpha_i\geq 0$, for all i=1,...,n+1, $\sum_{i=1}^{n+1}\alpha_i=1$, and that $x=\sum_{i=1}^{n+1}\alpha_i x^i$. Consider $\sum_{i=1}^{n+1}\alpha_i f(x^i)$. For each $i\leq j$, the coefficient of the q_{ij} term in this sum is equal to x_j. Since $x_i\geq x_j$ for $i<j$, these coefficients coincide with the corresponding ones in $R(x)$. Thus, $R(x)=\sum_{i=1}^{n+1}\alpha_i f(x^i)$. From the fact that $R(x)\geq \Gamma(x)$ and the definition of $\Gamma(x)$, it follows that $R(x)=\Gamma(x)$. ■

Acknowledgments. We would like to thank Professor Bruno Simeone for making valuable suggestions.

REFERENCES

1. M.L. Balinski, "Integer Programming: Methods, uses, computation", in: G.B. Dantzig and A.F. Veinott, Jr., eds., *Mathematics of the Decision Sciences*, Part I (American Mathematical Society, Providence 1968), 38-64.
2. M.R. Garey, D.S. Johnson and L. Stockmeyer, "Some simplified NP-complete graph problems", *Theoretical Computer Science* 1 (1976), 237-267.
3. J.E. Falk, and K.L Hoffman, "A successive underestimation method for concave minimization problems", *Mathematics of Operations Research* 1 (1976), 251-259.
4. P.L. Hammer, P. Hansen and B. Simeone, "Roof duality, complementation and persistency in quadratic 0-1 optimization", *Mathematical Programming* 28 (1984), 121-155.
5. J. Rhys, "A selection problem of shared fixed costs and networks", *Managements Science* 17 (1970), 200-207.
6. R.T. Rockafellar, *Convex Analysis* (Princeton University Press, Princeton 1970).

HYPERPATHS AND SHORTEST HYPERPATHS

Sang Nguyen[1]
Stefano Pallottino[2]

ABSTRACT

A generalization of the concept of a path in a directed graph, called a hyperpath, is the object of this paper. Fundamental properties of a path are generalized for a hyperpath. In particular, the shortest hyperpath problem is studied for a subclass of hyperpaths. Shortest hyperpath properties are derived from the classical one, and efficient algorithms are presented in detail.

Sommario: L'oggetto di questo lavoro è una generalizzazione del concetto di cammino in un grafo orientato: l'ipercammino. Vengono presentate le più importanti proprietà degli ipercammini come generalizzazione delle proprietà dei cammini. In particolare viene studiato il problema della ricerca dell'ipercammino minimo. Vengono mostrate le proprietà relative e vengono descritti diversi algoritmi per la determinazione degli ipercammini minimi, oltre che la loro complessità computazionale.

Introduction

In this paper we investigate a multipath concept on a directed graph – called a hyperpath. In broad terms, a hyperpath connecting two points is an acyclic graph with a single source and a single sink, paired with a given vector of real arc values. Such a concept plays a crucial role in transit networks with overlapping routes (see for example [6]).

Following a brief recall of the classical properties of a path, we derive a definition of a hyperpath from a relaxation of the definition of a path, and examine the principal properties of a hyperpath.

From this point on, we focus our attention on an important particular case where components of the hyperpath characteristic vector are defined locally in terms of arc attributes. The problem of determining shortest hyperpaths from every node to a specified destination node is then investigated. Potential efficient shortest hyperpath algorithms are developed and analysed.

---

[1] Département d'informatique et de recherche opérationnelle, Université de Montréal, Canada.

[2] Istituto per le Applicazioni del Calcolo "Mauro Picone", Consiglio Nazionale delle Ricerche, Roma, Italy.

2. Path and hyperpath formulations

2.1 A path formulation

Consider an oriented graph $G = (N, A)$. As usual a simple path p from node r to node s is a finite sequence of distinct arcs of the form:

$$A_p = \{(r, i_1), (i_1, i_2), \cdots ,(i_{k-1}, i_k), (i_k, s)\}. \tag{1}$$

Path p is elementary if it does not traverse any node more than once. An elementary path p is completely described by a boolean arc-vector $x^p = (x^p_{ij})$ where $x^p_{ij} = 1$, *if* p traverses (i,j), and 0, otherwise.

If we express every component x^p_{ij} as a product of two boolean variables y^p_i and π^p_{ij}:

$$x^p_{ij} = y^p_i \, \pi^p_{ij}, \ \forall (i,j) \in A, \tag{2}$$

then the node-vector y^p and the arc-vector π^p must satisfy:

$$y^p_r = 1 \tag{3}$$

$$\sum_{(j,i) \in BS(i)} y^p_j \, \pi^p_{ji} = \begin{cases} y^p_i, & \forall i \in N - \{r\} \\ 0, & \text{for } i = r \end{cases} \tag{4}$$

$$y^p_i \sum_{(i,j) \in FS(i)} \pi^p_{ij} = \begin{cases} y^p_i, & \forall i \in N - \{s\} \\ 0, & \text{for } i = s \end{cases} \tag{5}$$

$$\pi^p_{ij} \in \{0,1\} \quad \forall (i,j) \in A \tag{6}$$

where $BS(i)$ and $FS(i)$ denote, the sets of arcs entering and leaving node $i \in N$, respectively.

If a path arc-vector π is given, then a boolean node vector $y(\pi)$ can be determined dynamically from equations (3) and (4) by a topological search on the graph starting at node r and ending at node s. Such vector $y(\pi)$ is called a derived node-vector. A boolean vector $\pi = (\pi_{ij})$, such that the pair $(y(\pi), \pi)$ satisfies the system (3)-(6), completely specifies an elementary path p from r to s ($(r\text{–}s)$path). For this reason, such an arc-vector π will be called the characteristic vector of the defined path:

$$G_p = (N_p, A_p) \tag{7}$$

where

$$N_p = \{i \in N \mid y^p_i > 0\}, \quad A_p = \{(i,j) \in A \mid \pi^p_{ij} > 0\}$$

2.2 Path relaxation: a hyperpath

In the previous section, we described a rather elaborate definition of an elementary path in a graph. This allows us to formally present a hyperpath as a continuous relaxation of the boolean characteristic vector π^p.

Consider replacing equations (5) and (6) by:

$$\sum_{(i,j)\varepsilon FS(i)} \pi_{ij}^{p} = \begin{cases} 1, & \text{if } y_i^p > 0 \quad \forall i \,\varepsilon\, N - \{s\} \\ 0, & \textit{otherwise} \end{cases} \tag{8}$$

$$0 \le \pi_{ij}^{p} \le 1, \qquad \forall (i,j) \,\varepsilon\, A. \tag{9}$$

Note that (9) leads to similar relaxation for the derived node-vector $y^p(\pi^p)$ and x^p. We now have all the ingredients to introduce formally the concept of hyperpath.

Definition 1. A subgraph $G_h = (N_h, A_h)$ induced by the doublet $(y^h(\pi^h), \pi^h)$ satisfying (3)-(4) and (8)-(9) is a hyperpath h connecting r and s.

Obviously an $(r\text{–}s)$ path is also a hyperpath. It is straightforward to show that G_h has the following characteristics:

1) G_h is acyclic*;

2) G_h admits a unique source r and a unique sink s: $A_h \cap BS(r) = A_h \cap FS(s) = \emptyset$;

3) every node $i \,\varepsilon\, N_h$, and every arc $(i,j) \,\varepsilon\, A_h$, belong to at least one $(r\text{–}s)$ path in G_h.

From a network flow point of view, x_{ij}^h, y_i^h may be interpreted as the arc and node flows resulting from sending a single unit of flow from r to s, and π_{ij}^h is the proportion of the flow traversing node i on arc (i,j). On the other hand, for probabilistic applications, given the subgraph G_h, π_{ij}^h may be interpreted as the conditional probability of traversing arc (i,j) from node i:

$$\pi_{ij}^h = Prob\{ \text{ traversing } (i,j) \,\varepsilon\, A_h \mid i \,\varepsilon\, N_h\},$$

and consequently y_i^h and x_{ij}^h are the probability of traversing node i and arc (i,j), respectively.

2.3 A multipath description

Let P_h be the set of all paths p connecting r and s in graph G_h. Let us introduce the following multipath characterisation of the $(r\text{–}s)$ hyperpath (G_h, π^h). For every elementary path $p \,\varepsilon\, P_h$, define:

$$\omega_p^h = \prod_{(i,j)\varepsilon A} (\pi_{ij}^h)^{\delta_{ijp}} \tag{10}$$

where $\delta_{ijp} = 1$, if path p traverses arc (i,j), and 0, otherwise; 0^0 is defined as 1. Components of x^h and y^h can now be derived from that of ω^h :

* A cycle is understood here as a closed path.

$$y_i^h = \sum_{p \varepsilon P_h} \omega_p^h \, \delta_{ip}, \quad \forall i \; \varepsilon \; N \tag{11}$$

$$x_{ij}^h = \sum_{p \varepsilon P_h} \omega_p^h \, \delta_{ijp}, \quad \forall (i,j) \; \varepsilon \; A \tag{12}$$

where $\delta_{ip} = 1$, if p traverses node i, and 0, otherwise.

The path attribute ω_p^h may be interpreted either as the proportion of a unit flow on path p, or the probability of traversing path p.

Let $X_h \subset N_h$ such that node $r \; \varepsilon \; X_h$ and $s \not\varepsilon \; X_h$. Let $D(X_h)$ and $I(X_h)$ denote, respectively, the sets of arcs leaving and entering X_h. Define

$$\gamma(X_h) = \sum_{(i,j) \varepsilon D(X_h)} x_{ij}^h - \sum_{(i,j) \varepsilon I(X_h)} x_{ij}^h \tag{13}$$

as the net flow out of set X_h, it is easy to show that:

Proposition 1.

$$\gamma(X_h) = \sum_{p \varepsilon P_h} \omega_p^h = 1, \qquad \forall X_h \, . \tag{14}$$

2.4 The hyperpath cost

It is now assumed that associated with every arc $(i,j) \; \varepsilon \; A$ are two real values a_{ij} and b_{ij}, where a_{ij} represents the ordinary arc cost. For every node i, there is also a node traversing cost w_i^h, which varies among hyperpaths. Furthermore, it will be assumed that w_i^h is given by a real valued function of vectors b and π^h

$$w_i^h = g(b, \pi^h), \tag{15}$$

such that any two hyperpaths h' and h'' with identical arc-set $A_{h'} \cap FS(i) = A_{h''} \cap FS(i)$ have equal traversing costs at node i: $w_i^{h'} = w_i^{h''}$. This also implies that $w_s^h = 0$, for any $(r\text{–}s)$ hyperpath h.

The cost of a given hyperpath h is now defined as:

$$C_h \overset{def}{=} \sum_{(i,j) \varepsilon A} a_{ij} x_{ij}^h + \sum_{i \varepsilon N} w_{ih}^h y_i^h \tag{16}$$

Note that if h is an elementary path, then (16) reduces to

$$C_h = \sum_{(i,j) \varepsilon A} \delta_{ij_h} (a_{ij} + w_i^h), \tag{17}$$

and from the assumption made on function $g(b, \pi^h)$, $w_i^h = g(b_{ij})$ where j is the successor of i on path h. Consequently, (17) has the traditional form:

$$C_h = \sum_{(i,j)\varepsilon A} \delta_{ijh} a'_{ij} \tag{18}$$

where $a'_{ij} = a_{ij} + g(b_{ij})$.

For general hyperpaths, there is an alternate cost expression in terms of path costs:

Proposition 2.

$$C_h = \sum_{p\varepsilon P_h} c_p \omega_p^h \tag{19}$$

where

$$c_p = \sum_{(i,j)\varepsilon A} \delta_{ijp} a_{ij} + \sum_{i\varepsilon N} \delta_{ip} w_i^h \tag{20}$$

is the cost incurred on $(r–s)$ path p of G_h.

2.5 Concatenation of sub-hyperpaths

In the classical case, it is trivial to express a path p connecting r and s as a concatenation of two sub-paths p'_{ri} and p''_{is} for any node $i \varepsilon p$. And the additivity of the path costs holds:

$$C_p = C_{p'} + C_{p''}. \tag{21}$$

This equality is frequently used in the derivation of a dynamic approach for computing the path cost.

For the hyperpath case, such a decomposition is less trivial.

Definition 2. A hyperpath h' is a sub-hyperpath of h if and only if

$$\pi_{ij}^{h'} = \pi_{ij}^{h}, \quad \forall (i,j) \; \varepsilon \; A_{h'}. \tag{22}$$

It is easy to prove the following.

Proposition 3. For every node i of h, there exists a unique sub-hyperpath connecting i and s.

This property permits the design of a dynamic approach for computing the hyperpath cost, and consequently shortest hyperpaths. Let $f_h(i)$ denote the set of arcs leaving node i in hyperpath (G_h, π^h). Let $h(j)$ be the unique sub-hyperpath connecting node $j \; \varepsilon \; f_h(r)$ and s, and $C_h(j,s)$ its cost, then clearly:

$$\begin{aligned} C_h &= \sum_{(r,j)\varepsilon f_h(r)} \pi_{rj}^h (w_r^h + a_{rj} + C_h(j,s)), \\ &= w_r^h + \sum_{(r,j)\varepsilon FS(r)} \pi_{rj}^h (a_{rj} + C_h(j,s)). \end{aligned} \tag{23}$$

Applying the above formula to all sub-hyperpaths of h having node s as their sink, leads to the following dynamic program:

$$\begin{cases} C_h(s,s) = 0 \\ C_h(i,s) = w_i^h + \sum\limits_{(i,j)\varepsilon FS(i)} \pi_{ij}^h(a_{ij} + C_h(j,s)), \quad \forall i \in N_h - \{s\} \end{cases} \tag{24}$$

Since G_h is acyclic, it will be easy to scan all nodes $i \in N_h$ in a topological order, starting from node s, in computing the cost $C_h(r,s)$ of hyperpath h.

3. Shortest hyperpaths

If the characteristic vector π^h is exogeneously defined for every hyperpath h connecting r and s, finding a shortest hyperpath amounts to simply computing explicitly the cost of each hyperpath and choosing the one with the smallest cost. On the other hand, when the characteristic vector π^h of any hyperpath h is a function $\pi^h(a,b)$ of the arc attributes, then finding a shortest hyperpath is not a trivial task.

In the sequel, we will study the shortest hyperpath problem for a subclass of hyperpaths which has important applications in the traffic assignment problem on transit networks.

Consider the subclass of hyperpaths with characteristic vectors π^h and node traversing costs defined as follows:

$$\pi_{ij}^h \overset{def}{=} \begin{cases} b_{ij} / \sum\limits_{(i,k)\varepsilon f_h(i)} b_{i_k} & \forall \ (i,j) \in f_h(i) \\ 0, & otherwise \end{cases} \tag{25}$$

and

$$w_i^h = g(b, \pi^h) \overset{def}{=} 1 / \sum_{(i,j) \in f_h(i)} b_{ij} \tag{26}$$

where $b_{ij} > 0$ for every $(i,j) \in A$.

One can readily check that these definitions satisfy conditions (4) and (5) and the assumptions made on function $g(b, \pi^h)$. It is further assumed that arc costs $a_{ij} \geq 0$.

Since common arcs of sub-hyperpaths of a given hyperpath h have equal characteristic values, one can rewrite the previously defined dynamic program (24) as:

$$\begin{cases} C_h(s,s) = 0 \\ C_h(i,s) = (1 + \sum\limits_{(i,j) \in f_h(i)} b_{ij}(a_{ij} + C_h(j,s))) / \sum\limits_{(i,j)\varepsilon f_h(i)} b_{ij} \end{cases} \tag{27}$$

which is a prelude to the program for computing minimum hyperpath costs.

Let $C^*(i, s)$ denote the cost of a shortest hyperpath connecting nodes i and s. From the preceding analysis, we can now establish the following:

Proposition 4 (generalized Bellman's equations).

The shortest cost vector C^* is the unique solution of the following system of equations:

$$C^*(s, s) = 0 \tag{28}$$

$$C^*(i, s) = \min_{\Gamma_i \subseteq FS(i)} \left\{ \frac{1 + \sum_{(i,j)\varepsilon\Gamma_i} b_{ij}(a_{ij} + C^*(j,s))}{\sum_{(i,j)\varepsilon\Gamma_i} b_{ij}} \right\}, \qquad \forall i \ \varepsilon\ N-\{s\}.$$

The proof of the above generalized Bellman's equations ([1]) follows closely that for the standard case.

Let

$$B(\Gamma_i) = \sum_{(i,j)\varepsilon\Gamma_i} b_{ij} \tag{29}$$

where $\Gamma_i \subseteq FS(i)$ is any subset of arcs with tail node i. Consider two hyperpaths h_{is} and h'_{is} with identical common sub-hyperpaths, such that $f_h(i) \supset f_{h'}(i)$. Let $\Gamma_i = f_h(i)$ and $\Gamma'_i = f_{h'}(i)$, and $C(i,s)$ and $C'(i,s)$ denote the cost of h_{is} and that of h'_{is} respectively. For these hyperpaths, we can write equation (27) as

$$B(\Gamma_i)C(i,s) = 1 + \sum_{(i,j)\varepsilon\Gamma_i} b_{ij}(a_{ij} + C(j,s))$$

and

$$B(\Gamma'_i)\ C'(i,s) = 1 + \sum_{(i,j)\varepsilon\Gamma'_i} b_{ij}(a_{ij} + C(j,s)).$$

Rewriting Γ_i as $\Gamma'_i \cup (\Gamma_i - \Gamma'_i)$, it follows that

$$B(\Gamma_i)\ C(i,s) = B(\Gamma'_i)\ C'(i,s) + \sum_{(i,j)\varepsilon\Gamma_i - \Gamma'_i} b_{ij}\ (a_{ij} + C(j,s)) \tag{30}$$

and finally

$$B(\Gamma_i)(C(i,s) - C'(i,s)) = \sum_{(i,j)\varepsilon\Gamma_i - \Gamma'_i} b_{ij}(a_{ij} + C(j,s) - C'(i,s)). \tag{31}$$

From symmetry, the following equation also holds:

$$B(\Gamma'_i)(C(i,s) - C'(i,s)) = \sum_{(i,j)\varepsilon\Gamma_i - \Gamma'_i} b_{ij}(a_{ij} + C(j,s) - C(i,s)). \tag{32}$$

The above equations imply that the cost of hyperpath h_{is} is strictly less than that of h'_{is}, only if the right hand sides of both (31) and (32) are strictly negative. In particular,

Proposition 5. If a single arc (i,j) is added to the set Γ'_i to form $\Gamma_i = \Gamma'_i \cup \{(i,j)\}$, then one of the following statements holds:

1) $a_{ij} + C(j,s) < C(i,s) < C'(i,s)$
2) $a_{ij} + C(j,s) = C(i,s) = C'(i,s)$
3) $a_{ij} + C(j,s) > C(i,s) > C'(i,s)$.

A direct implication of this property is the following crucial characterisation of the arc set $f_{h^*}(i) \subseteq FS(i)$ of a shortest hyperpath h^*_{is}.

Proposition 6. Let $f_{h^*}(i) \subseteq FS(i)$ be the set of arcs leaving node i of a shortest hyperpath h^*_{is}, then:

$$C^*(i,s) \geq a_{ij} + C^*(j,s), \qquad \forall (i,j) \in f_{h^*}(i)$$

and

$$C^*(i,s) \leq a_{ij} + C^*(j,s), \qquad \forall (i,j) \notin f_{h^*}(i).$$

Recall that a directed tree, rooted at node s, is an oriented graph on which every node $j \neq s$ is connected to the roots by a unique path. Similarly,

Definition 3. A hypertree rooted at s is an oriented graph, on which every node distinct from s is connected to s by a unique hyperpath.

A shortest hypertree, rooted at s, is therefore a hypertree formed by shortest hyperpaths.

4. Shortest hypertree algorithms

We now have all the ingredients to state a general procedure for finding a shortest hypertree, in a finite number of iterations, and then suggest several efficient implementations.

General shortest hypertree procedure

initialisation. Determine a hypertree T_s, rooted at s ; Let $C(i,s)$ denote the cost of the unique hyperpath h_{is} on T_s, and Γ_i the set of arcs of T_s with common tail node i.

general step. While there exists a node i such that

$$C(i,s) > C'(i,s) \tag{33}$$

where

$$C'(i,s) = \min_{\Gamma'_i \subseteq FS(i)} \left\{ \frac{1 + \sum_{(i,j) \varepsilon \Gamma'_i} b_{ij}(a_{ij} + C(j,s))}{B(\Gamma'_i)} \right\},$$

update the cost $C(i,s)$ and arc set Γ_i according to:

$$C(i,s) = C'(i,s) \tag{34}$$

$$\Gamma_i = arg \min_{\Gamma'_i \subseteq FS(i)} \left\{ \frac{1 + \sum_{(i,j) \varepsilon \Gamma'_i} b_{ij}(a_{ij}+C(j,s))}{B(\Gamma'_i)} \right\}. \tag{35}$$

4.1 A label-correcting algorithm

To avoid unnecessary operations at each iteration of the general step, all nodes $i \, \varepsilon \, N$ that may satisfy condition (33) are collected in a candidate list Q. If no initial hypertree T_s is available, the following default setting may be used:

- $Q = N - \{s\}$;
- $\Gamma_s = \varnothing, \quad C(s,s) = 0$;
- $\Gamma_i = \{(i,s)\}$, $C(i,s) = a_{is} + \frac{1}{b_{is}}, \quad \forall (i,s) \, \varepsilon \, BS(s)$;
- $\Gamma_i = \varnothing, \quad C(i,s) = +\infty, \quad \forall (i,s) \not\varepsilon \, BS(s)$.

Whenever a candidate node j is selected and removed from list Q – operation QOUT – its arc-set Γ_j and label $C(j,s)$ are updated according to (35) and (34). Every node i, such that

$$a_{ij} + C(j,s) < C(i,s)\ ,\quad (i,j)\ \varepsilon\ BS(j) \tag{36}$$

is then inserted into Q, if it is not already there – operation QIN. The following algorithm for determing a shortest hypertree (SHT), stated in a high level Pascal-like language, implements the above ideas.

```
Procedure SHT(s) :

begin
  comment: initialisation;
  C(s,s) := 0;
  Q := N-{s};
  for each  i ε N  do
    begin
      Γ_i := ∅;
      C(i,s) :=  + ∞
    end;
  for each (i,s) ε BS(s)  do
    begin
      Γ_i := {(i,s)};
      C(i,s) := a_is + 1/b_is
    end;
  comment: fetch node  j ε Q, and find optimal  Γ_j, C(j,s);
  while  Q ≠ ∅  do
    begin
      QOUT(j, Q);
      FIND(Γ_j,FS(j),C,j);
      comment: scanning for potential candidates;
      for each (i,j) ε BS(j) do
        if  a_ij + C(j,s) < C(i,s)  then  QIN(i,Q)
    end
end  SHT;
```

The procedure FIND determines the optimal arc-set Γ_j with respect to the current labels $C(k,s)$, $(k,s)\ \varepsilon\ FS(j)$ and updates $C(j,s)$. This task can be performed with an efficient algorithm, based on proposition 6, in $0(m_j \log m_j)$ time, where $m_j = |FS(j)|$. This algorithm works as follows. First, sort the arcs of the forward star: $FS(j) = \{(j,k)\ \varepsilon\ A\}$ in non-decreasing order of values $[a_{jk} + C(k,s)]$, and then insert one arc at a time into Γ_j, until no further improvement can be obtained for $C(j,s)$.

```
Procedure  FIND (Γ_j, FS(j), C, j):
begin
comment: sort the arcs in  FS(j) in non-decreasing order of  a_jk + C(k,s);
SORT (FS(j), a, C);
 C(j,s) := a_jk1 + C(k_1, s) + 1/b_jk1;
 Γ_j := {(j, k_1)};
 B(Γ_j) := b_jk1;
 α := 2;
comment:  scan the sequence of sorted arcs ;
 while  α ≤ m_j  and  C(j,s) > a_jkα + C(k_α, s)  do
   begin
     comment:  add arc (j, k_α)  to  Γ_j  and update C(j,s);
     Γ_j := Γ_j ∪ {(j, k_α)};
```

$$C(j,s) := C(j,s) - (C(j,s) - a_{jk_\alpha} - C(k_\alpha, s)) \frac{b_{jk_\alpha}}{B(\Gamma_j)} ;$$

$\alpha := \alpha + 1$

end

end FIND;

FIND can be executed in $0(m_j \log m_j)$ time using a heap-sort. Note that determining the optimal arc-set Γ_j may also be formulated as a boolean hyperbolic program for which there is already a greedy solution algorithm (see for example [4]).

To evaluate the complexity of SHT, let us assume that nodes in Q are selected in lexicographic order. After every complete node selection cycle, at least one node $i \in N$ has received a permanent minimum label $C(i,s)$. This follows directly from the absence of negative cycles in G and thus limits the maximum number of node selection cycles to $n = |N|$.

In every node selection cycle, three distinct operations are performed:

i) the selection of nodes in lexicographic order, which requires $0(n)$ operations,

ii) the determination of optimal arc-sets, which requires $0(\sum_{i \in N} m_i \log m_i) \leq 0(m \log n)$ operations,

iii) the scanning of the arcs in backward star $BS(i)$, $i \in N$, which require $0(m)$ operations.

Since QIN and QOUT have constant complexity, algorithm SHT runs in $0(mn \log n)$ time. In the shortest path nomenclature, the above SHT algorithm is a label-correcting one. Note that for shortest path problems, label-correcting algorithms run in $0(mn)$ time, therefore, the extra work required by SHT is entirely due to the sorting of the forward star arc-sets. However, for expository purposes, we have chosen to describe a procedure for determining the optimal sets Γ_i, $i \in N$, which does not exploit the ordering of arcs in the previous sets Γ_i . It is possible, as suggested in [5], to design updating procedures which fully exploit this ordering of arcs, and obviate the complete sorting of the forward star arc-sets.

4.2 A label-setting algorithm

Consider node i, with its current label $C(i,s)$ and corresponding arc-set Γ_i . From proposition 5, if there is an arc $(i,j) \in FS(i)$ such that

$$a_{ij} + C(j,s) < C(i,s), \tag{37}$$

then adding (i,j) to Γ_i will produce an improved label $C(i,s)$. The value $(a_{ij} + C(j,s))$ may be considered as an arc label, and thus suggests an arc-selection based algorithm, which produces a shortest hypertree after at most m arc selections.

Consider the following selection rule. At each iteration, arc (i,j) with the lowest arc label $(a_{ij} + C(j,s))$ among all unscanned arcs with head node $j \varepsilon Q$ is selected:

$$a_{ij} + C(j,s) = \min_{j' \varepsilon Q} \min_{(i',j') \varepsilon BS(j')} \{a_{i'j'} + C(j',s)\}. \tag{38}$$

This rule ensures that label $C(j,s)$ has reached its minimum value; since

$$C(j,s) \leq C(k,s) + a_{jk} \qquad \forall (j,k) \not\varepsilon \Gamma_j$$

Note that one can initialise the procedure with empty sets Γ_i, $i \varepsilon N$, and whenever (37) occurs, with arc (i,j) selected according to (38), then arc (i,j) becomes a permanent member of arc-set Γ_i.

To implement the above selection rule efficiently, one must consider a pre-ordering of the arcs of every backward star $BS(i)$, $i \varepsilon N$, in non-decreasing order of arc attributes a_{ji}. Such pre-ordering is based on an idea suggested in [2] (see also [3], algorithm SPT1). With such pre-ordering of arcs, it suffices to maintain, for every node $j \varepsilon Q$, a pointer to the first unscanned arc $(i,j) \varepsilon BS(j)$ in the pre-ordering. When node j is selected, its arc pointer is updated as long as an unscanned arc (i,j) remains, otherwise j is removed from list Q .

The following algorithm, denoted DSHT (Dantzig shortest hypertree), implements these ideas.

Procedure DSHT(s):

begin
  comment: initialisation;
  PREORDER (N,BS);

  for each $i \varepsilon N$ **do**
    begin
      $\Gamma_i := \varnothing$;
      $B(\Gamma_i) := 0$;
      $C(i,s) := +\infty$
    end;

  $Q := \{s\}$;
  $C(s,s) := 0$;

  while $Q \neq \varnothing$ **do**
    begin
      comment: select (i,j) with minimum arc label;

      QOUT $(j, Q, (i,j))$;

      comment: check Bellman's condition;

```
if C(i,s) ≥ C(j,s) + a_ij then
     comment: label updating;
  if Γ_i = ∅ then
      begin
         comment: inserting first arc;
         Γ_i := {(i,j)} ;
         B(Γ_i) := b_ij;
         C(i,s) := C(j,s) + a_ij + 1/b_ij
      end
  else
     begin
         comment: updating Γ_i and C(i,s);
         Γ_i := Γ_i ∪ {(i,j)};
         B(Γ_i) := B(Γ_i) + b_ij;
         C(i,s) := C(i,s)-(C(i,s)-C(j,s)-a_ij)b_ij/B(Γ_i);
         QIN(i,Q)
     end
  end
end DSHT;
```

Procedure PREORDER sorts the arcs in every backward star set in non-decreasing order of arc attributes a_{ij} in $0(m \log n)$ time. Procedure QOUT determines the minimum label arc (i,j) and updates the arc pointer of node j or removes j from Q. Finally, procedure QIN inserts node i into Q and sets the arc pointer of node i to the first arc in the pre-sorted backward star set $BS(i)$.

If list Q is implemented as an n-elements binary heap, then QOUT and QIN run in $0(\log n)$ time. Since at most m arc selections are required, the complexity of DSHT is $0(m \log n)$, which is equal to that of a binary heap algorithm for shortest path calculations.

Conclusion

This paper investigates a multipath concept on a directed graph, called here a hyperpath. It shows that the principal properties of paths can be retrieved for hyperpaths. In particular, efficient shortest hyperpath algorithms can be designed from the theory developped for shortest paths. It is believed that the hyperpath concept may advantageously replace the standard path concept in many applications.

References

[1] R. Bellmann, "On a routing problem", *Quaterly of Applied Mathematics 16* (1958) 87-90.

[2] G.B. Dantzig, *Linear programming and extensions* (Princeton University Press, Princeton, NJ, 1963).

[3] G. Gallo and S. Pallottino, "Shortest path methods: a unifying approach", *Mathematical Programming Study* 26 (1986) 38-64.

[4] P.L. Hammer and S. Rudeanu, *Boolean Methods in Operations Research* (Springer-Verlag, NY, 1968).

[5] S. Nguyen and S. Pallottino, "Assegnamento dei passeggeri ad un sistema di linee urbane: determinazione degli ipercamini minimi", Quaderno IAC 6, 1985.

[6] S. Nguyen and S. Pallottino, "Equilibrium traffic assignment for large scale transit networks", Quaderno IAC 14, 1985.

A CHARACTERIZATION OF CENTROIDAL GRAPHS

Wiktor Piotrowski* and Maciej M. Sysło
Institute of Computer Science
University of Wrocław
Przesmyckiego 20
51151 Wrocław, Poland

ABSTRACT

We give necessary and sufficient conditions for a graph to be a centroid of an - otherone as well as of itself. Then, we apply these results to some particular classes of graphs: chordal, Halin, series-parallel and outerplanar.

1. INTRODUCTION

A graph $G = (V(G),E(G))$ consists of a finite set of *vertices* $V(G)$ and a set of *edges* $E(G)$, where $E(G)$ is a family of 2-element subsets of vertices. Graph-theoretic terms not defined can be found in [3].

If T is a tree and $u \in V(T)$, then *the branch weight* of u, denoted by $bw(u)$, is the largest number of vertices in a component of $T-\{u\}$. Evidently,

(1) $$bw(u) = \max \{|S|: u \notin S \text{ and } S \in C\},$$

where C is the family of subsets $S \subseteq V(T)$ such that the subgraph $T[S]$ induced by S is connected. *The branch weight centroid of* T is the set of vertices for which the function bw has the minimal possible value.

The concept of a branch weight centroid has been extended in [10] so that it can be defined for an arbitrary finite set X with a distinguished family C of "convex" subsets of X. In particular, the centroid of a graph was defined for X to be the vertex set of G and the chordless path convexity.

Let G be a connected graph. A path P in G is *chordless* if the only pairs of vertices in P that are adjacent in G are consecutive along P. A set S of vertices is

*Research supported by the Government Grant CPBP 02.17

monophonically convex (or, simply, *convex*) if S contains every vertex on every chordless path between vertices in S.

The *weight* of a vertex v in G is defined as

$$w(v) = \max \{|S|: v \notin S \text{ and } S \in C\}, \qquad (2)$$

where C is the family of convex subsets of $V(G)$. The *centroid* of a graph G, denoted by Cent(G), is the subgraph of G induced by vertices with the smallest weight.

In the previous paper [10], the following generalization of the well-known Jordan's Theorem [8] was proved.

THEOREM 1. *The centroid of a graph G is a convex set. Moreover, if G is a connected chordal graph, then the centroid of G is a complete subgraph of G.*

Our considerations are related to the maximin facility location problems. Suppose we want to find a location for a control point in a network. Let the structure of a network be such that the information between two vertices of the network goes along all minimal (with respect to inclusion) paths. Observe that "convex" means now "closed under information flow". The vertices of a centroid are good candidates for control points because they minimize the cardinality of maximal non-controlled set. Other work related to the maximin facility location and other extensions of the concept of branch weight centroid can be found in [11,12]. For detailed discussion about convexity in graphs we refer the reader to [2,4-7].

In this paper we characterize those graphs which may occur as centroids and we give necessary and sufficient conditions for a graph to be self-centroidal. Then in Section 3 we describe centroids in some special families of graphs.

2. CENTROIDAL GRAPHS

Let $G = (V(G),E(G))$ be a connected graph. A *clique* is a set of pairwise adjacent vertices. By a *clique separator* we mean a clique whose removal disconnects the graph.

PROPOSITION 1. *Suppose S is a convex subset of $V(G)$ and let A be an arbitrary component of the graph G-S. Then, the set*

$$S_A = \{x \in S: \exists s \in A \ \{x,s\} \in E(G)\}$$

forms a clique.

P r o o f. Since for $|S_A| = 1$ the above statement is true, suppose that x and y are distinct elements of S_A. There are elements s and t in A adjacent to the vertices x and y, respectively.

If $s = t$, then (x,s,y) is a path of G. Since $s \notin S$, this path contains a chord. But the only possible chord is $\{x,y\}$.

If $s \neq t$, there is a path $(s = v_1, v_2, \ldots, v_k = t)$ in A, since A is connected. Moreover, in the path $(x, s = v_1, \ldots, v_k = t, y)$ there is a chordless path P with the end vertices x and y. On the other hand, since $v_i \notin S$, we have $v_i \notin P$ for all i. This means that $P = \{x,y\}$, what completes the proof. □

By a *copoint* (or semispace, see [9]) at a vertex $v \in V(G)$ we mean any maximal convex subset of the set $V(G)-\{v\}$. Obviously,

(3) $$w(v) = \max \{|S|: S \text{ is a copoint at } v\}.$$

PROPOSITION 2. *If S is a copoint at v, then there is a clique K such that*

(4) $$S = K \cup A_1 \cup \ldots \cup A_p,$$

where $A_0, \ldots, A_p$ are the components of the graph $G-K$ and $v \in A_0$.

P r o o f. Suppose that S is a copoint at v. Let $B_0, \ldots, B_m$ be the components of the graph $G-S$ and let v be in B_0. According to Proposition 1, the set S_{B_0} is a clique. Let $K = S_{B_0}$ and let $A_0, \ldots, A_p$ be the components of $G-K$ with $v \in A_0$.

We will show that the set $S' = K \cup A_1 \cup \ldots \cup A_p$ is convex and that $S \subseteq S'$. By the definition of a copoint, this means $S = S'$.

To show the first statement, suppose that P is a chordless path between two vertices of S'. We must show that $P \subseteq S'$. If P contains a vertex x not in S', it must be in A_0. Let s be the least vertex of S' on P before x, and t be the first vertex of S' on P after x. Then s and t are in K. Hence $\{s,t\}$ is a chord of P, contrary to P being chordless.

Now, let x be an arbitrary element of S and suppose that x does not belong to S', that is $x \in A_0$. Let P be a path in A_0 with x and v as its end vertices. Let y be the last point of S encountered along this path. Then y is an element of S_{B_0}, which is equal to K. Finally, y is in the intersection of K and A_0, what is impossible. □

For an arbitrary set $U \subseteq V(G)$ and a vertex $v \notin U$, let us denote by $C_v(G-U)$ the component of the graph $G-U$ which contains v. Let further $c: V(G) \to N$ be a function defined as follows:

(5) $$c(v) = \min \{|C_v(G-K)|: K \text{ is a clique of } G \text{ such that } v \notin K\}.$$

PROPOSITION 3. *For every vertex v, $c(v)+w(v) = |V(G)|$.*

P r o o f. According to (3), there is a copoint S such that $w(v) = |S|$. By Proposition 2, $S = K \cup A_1 \cup \ldots \cup A_p$, where K is a clique of G and $v \notin K$. Thus $A_0 = C_v(G-K)$ and $c(v) \leq |A_0| = |V(G)| - w(v)$.

On the other hand, there is a clique K' such that $c(v) = |C_v(G-K')|$. Let $A_0', \ldots, A_k'$ be components of the graph $G-K'$ with $v \in A_0'$. Observe that $c(v) = |V(G)| - |K' \cup A_1' \cup \ldots \cup A_k'|$. However, in the proof of Proposition 2 we have already observed that sets of the form (4) are convex. Thus, by (2), $|K' \cup A_1' \cup \ldots \cup A_k'| \leq w(v)$ and, consequently, $c(v) \geq |V(G)| - w(v)$. Finally, $c(v) = |V(G)| - w(v)$. □

COROLLARY 1. *The centroid of a graph G consists of those vertices at which the function c takes its maximum.* □

COROLLARY 2. *If a graph G contains no clique separator, then*

$$w(v) = \max \{|K| : K \text{ is a clique and } v \notin K\}.$$

P r o o f. Observe that in this case

$$c(v) = \min \{|V-K| : K \text{ is a clique and } v \notin K\}. \square$$

Recall that $\omega(G)$ denotes the maximal cardinality of a clique in G.

THEOREM 2. *If a graph G contains no clique separator, then the centroid of G forms a clique or is equal to the whole graph G.*

P r o o f. Let $S = \bigcap \{K : K \text{ is a clique and } |K| = \omega(G)\}$. There are two possible cases: S is empty or not.

If $S = \emptyset$ then for every vertex $v \in V$ there is a clique K such that $v \notin K$ and $|K| = \omega(G)$. Thus, by Corollary 2, for every vertex v we have $w(v) = \omega(G)$ and this means that $\mathrm{Cent}(G) = G$.

If $S \neq \emptyset$ then $\mathrm{Cent}(G) = S$. Indeed, $w(v) = \omega(G)-1$ or $w(v) = \omega(G)$, accordingly as v is an element of S or not. Since S is an intersection of cliques, it is a clique too. □

PROPOSITION 4. *Let $c(x) = c(s)$ for two distinct vertices x, s of a connected graph H, and assume that K and K' are cliques of H with $x \notin K$, $s \in K$ and $c(s) = |C_s(H-K')|$. Then x and s are in distinct components of the graph $H-K'$ or the intersetion $K' \cap C_x(H-K)$ is non-empty.*

P r o o f. Suppose to the contrary that x and s are in the same component of $H-K'$ and that there is no point in $K' \cap C_x(H-K)$. Observe that in this case $C_x(H-K') = C_s(H-K')$. Moreover, the set $C_x(H-K)$ is a proper subset of $C_x(H-K')$. Thus

$$c(x) \leq |C_x(H-K)| < |C_x(H-K')| = |C_s(H-K')| = c(s),$$

what contradicts the assumption $c(x) = c(s)$. □

Let us call a graph G *centroidal* if there is a connected graph H such that $\mathrm{Cent}(H) = G$, and *self-centroidal* if $\mathrm{Cent}(G) = G$.

THEOREM 3. *If a graph is centroidal, then it contains no clique separator.*

Proof. Let G be centroidal and let H be such that $\mathrm{Cent}(H) = G$. Suppose that K is a clique separator of G. We claim that if vertices x and y are in distinct components of the graph G-K, then they are in distinct components of H-K. Otherwise there is a chordless path P with the end vertices x and y such that $P \cap K = \emptyset$. By Theorem 1, G is convex. Hence, P is contained in $V(G)$, contradicting the choice of x and y.

Now, we can choose two vertices x and y in distinct components of G-K in such a way that there are elements s and t in K adjacent to x and y respectively. We have already noticed that x and y must be in distinct components of the graph H-K.

Let K' be a clique in H with $c(s) = |C_s(H-K')|$. Observe that K' has a non-empty intersection only with one of the components $C_x(H-K)$ and $C_y(H-K)$. On the other hand, by applying Proposition 4 to the vertices x and s, we obtain that there is an element $x' \in K' \cap C_x(H-K)$. Thus, $C_y(H-K)$ must be disjoint from K'. Again by Proposition 4, we obtain that also y and s are in distinct components of the graph H-K'. In particular, we have proved the following statement:

(6) If x and y are in distinct components of G-K, then they have no common neighbour in K.

Moreover, the path (s,t,y) must contain an element of K'. Hence $t \in K'$. Since $x' \in C_x(H-K)$, we can choose a chordless path $(x,v_1,\ldots,v_k,x')$. Let x'' be the first element on the path $(x,v_1,\ldots,v_k,x',t)$ that is adjacent to vertex t. Then the path $(x,v_1,\ldots,x'',t)$ is chordless and, moreover, its endpoints are in the convex set $V(G)$. Thus $x'' \in V(G)$.

Finally, we obtain vertices x'' and y in distinct components of G-K with the common neighbour t in the clique K, contrary to (6). □

By Theorem 3, we can confine our consideration to the class of connected graphs without clique separators and different from complete graphs, since they are self--centroidal.

COROLLARY 3. *Let G be a non-complete graph without clique separators. Then, G is self-centroidal if and only if the intersection of largest cliques in G is an empty set.*

Proof. It is an immediate consequence of the proof of Theorem 2. □

THEOREM 4. *Let G be a non-complete graph without clique separators. Then, the graph G is centroidal if and only if the intersection of inclusion-maximal cliques in G is the empty set.*

Proof. Necessity. Let H be such a graph that $\mathrm{Cent}(H) = G$ and suppose to the contrary that there is an element s belonging to all maximal cliques of G. Let K' be a clique such that $c(s) = |C_s(H-K')|$. The clique K' intersects at most one component of the graph $H-G$.

Case 1. The clique K' is disjoint with all components of $H-G$. In this case $K' \subseteq V(G)$ and there is a maximal clique in G, say K, containing K'. Observe that $s \in K$ and, moreover, since G is not complete, there is an element $x \in V(G)-K$. According to Proposition 4 and to the relation $K' \cap C_x(H-K) = \emptyset$, the vertices x and s are in distinct components of the graph $H-K'$. Hence K' is a clique separator of G, what is impossible.

Case 2. K' intersects a component A of the graph $H-G$. By Proposition 1, since $V(G)$ as a centroid is convex, $V(G)_A$ forms a clique in G. Let K be a maximal clique including $V(G)_A$. Then $s \in K$ and, moreover, since G is not complete, there is an element $x \in V(C)-K$. Observe that x and y cannot be in distinct components of $H-K'$, because otherwise $K' \cap V(G)$ would be a clique separator of G. By Proposition 4, there is an element $z \in K'$ such that z and x are in the same component of $H-K$. Thus, there is a path $(z, v_1, \ldots, v_k, x)$ which is disjoint with K. Since the vertex z does not belong to K and $K' \subseteq A \cup K$, we have $z \in A$. Hence, the first vertex u of the path not belonging to A has to be in $V(G)_A$. Thus $u \in K$ and this contradicts the choice of the path.

Sufficiency. Let G be a graph in which the intersection of all maximal cliques is empty. We will construct a graph H in the following way: for every maximal clique K of the graph G add $\omega(G)-|K|$ new vertices adjacent to all vertices in K. We show that $\mathrm{Cent}(H) = G$.

If $H = G$, then by Corollary 3, we have $\mathrm{Cent}(G) = G$. So, we can assume that H has at least one new vertex v. Observe that $c(v) = 1$ for every such a vertex v.

We claim that $c(v) = |V(H)|-\omega(G)$ for every vertex v of the graph G. From the construction of the graph H it follows that $\omega(H) = \omega(G)$. Thus, if K is not a clique separator of H and $v \notin K$, then

$$|C_v(H-K)| = |V(H)|-|K| \geq |V(H)|-\omega(G).$$

On the other hand, if K is a clique separator of H not comprising v, then $K' = K \cap V(G)$ is a maximal clique in G. However, $C_v(H-K) = C_v(H-K')$ and, simultaneously, K' separates exactly $\omega(K')-|K'|$ new vertices adjacent to K' from v. Thus,

$$|C_v(H-K)| = |C_v(H-K')| = |V(H)|-\omega(G).$$

Finally, according to (5), $c(v) = |V(H)|-\omega(G)$, which completes the proof of our claim.

To complete the proof let us observe that since G is not complete and H-G is non-empty, then we have $|H(V)|-\omega(G) > 1$. □

3. CENTROIDS IN SOME FAMILIES OF GRAPHS

In this section we apply the results of the previous section to the study of centroids in some particular classes of graphs. Let us note that a centroid of a graph is always a connected subgraph and, by Theorem 3, it contains no separating vertex.

Complete graphs. For every vertex v in K_n, we have $w(v) = n-1$, thus $\text{Cent}(K_n) = K_n$.

Cycles. A cycle C_n has no clique separators and the intersection of all maximal cardinality cliques is the empty set. Thus, by Corollary 3, we have $\text{Cent}(C_n) = C_n$.

Trees. The well-known Jordan's Theorem [8] states that centroids of trees are isomorphic to K_1 or to K_2. Observe, that this theorem follows immediately from our results. Indeed, the graphs K_1 and K_2 are the only subgraphs of a tree which are connected and without separating vertices.

Chordal graphs. A graph is *chordal* if it contains no cycle of length greater than 3 as an induced subgraph. A vertex is *simplicial* if its neighborhood induces a clique. It is well-known [1] that a graph G is chordal if and only if every induced subgraph of G has a simplicial vertex.

By Theorem 1, a centroid of a chordal graph is a clique. This was proved in [10] as a corollary to some general results on chordless path convexity in chordal graphs. We here include a simple, immediate proof of this fact.

If G is a chordal graph, then the subgraph $\text{Cent}(G)$ contains a simplicial vertex v. Let $N(v)$ be the neighborhood of v and let us consider the graph $\text{Cent}(G)-N(v)$. Since $\text{Cent}(G)$ does not contain a clique separator, this graph is connected. On the other hand, $\{v\}$ is its component. Thus, $\text{Cent}(G) = N(v)\cup\{v\}$.

Halin graphs. Halin graphs were introduced to provide examples of edge minimal planar 3-connected graphs. The usual method of constructing a *Halin graph* can be described as follows: take a tree T, in which each nonleaf vertex has degree at least three; embed the tree in the plane and then connect the leaves with a cycle in such a way that the planar embedding remains. We denote such a graph by $H(T)$ and call T an *interior tree* of $H(T)$. Observe that a Halin graph does not contain a clique separator. Thus we can apply Theorem 2.

If an interior tree T is a star $K_{1,n}$, then the Halin graph $H(T)$ is isomorphic to the wheel W_n $(n \geq 3)$. Since W_3 is a complete graph, it is selfcentroidal. If $n \geq 4$ then the intersection of maximal cardinality cliques (namely, 3-element cliques) of W_n consists of a unique element - an interior vertex of the wheel. Thus, $\mathrm{Cent}(W_n) = K_1$ for $n \geq 4$.

Now, let T have at least two nonleaves. Observe that in this case there are two disjoint triangles in $H(T)$ and, moreover, the graph $H(T)$ does not contain a clique of size greater than three. Thus, by Corollary 3, $H(T)$ is selfcentroidal.

Thus, if H is a Halin graph, then $\mathrm{Cent}(H) = H$ or $\mathrm{Cent}(H) = K_1$.

Series-parallel graphs. The class of *series-parallel graphs*, denoted by SPG, is defined recursively by the following rules of composition:

(1) An edge $\{u,v\}$ is series-parallel with *terminals* u and v. We call this graph a *primitive graph*.

(2) If G_1 and G_2 are series-parallel graphs with terminals u_1, v_1 and u_2, v_2, respectively, then:

 (a) The graph obtained by identifying u_1 and u_2 is a series-parallel graph, with v_1 and v_2 as its terminals. This graph is the *series composition* of G_1 and G_2.

 (b) The graph obtained by identifying u_1 and u_2 and also v_1 and v_2 is a series-parallel graph, the *parallel composition* of G_1 and G_2. This graph has u_1 $(= u_2)$ and v_1 $(= v_2)$ as its terminals.

For a convenience, we assume that K_1 is also in SPG.

We now prove the following statement:

(7) A graph H is a centroid of a series-parallel graph G if and only if H is a series-parallel graph with no clique separator.

By Theorem 3, centroid of G does not contain clique separators. Moreover, since the graphs K_1, K_2 and K_3 are series-parallel, assume that $\mathrm{Cent}(G)$ has at least four vertices. In this case, since the SPG class is closed under taking 2-connected induced subgraphs, a centroid is a series-parallel graph.

To show the converse, let us assume that H is a series-parallel graph with no clique separator. We show that H is selfcentroidal. Notice, that every triangle in a series-parallel graph is produced by a parallel composition with adjacent terminals. Thus we have $\omega(H) = 2$, since otherwise H would contain a triangle and terminals of this triangle would form a clique separator. Finally, maximal cardinality cliques of H have an empty intersection. Hence, by Corollary 3, this completes the proof of (7).

It is not hard to see that all non-primitive series-parallel graphs G without clique separators arise in the following way:

(8) Let G be obtained by a sequence of series or parallel compositions $p_1, \ldots, p_n$ $(n > 1)$. Then p_n must be a parallel composition and if p_i is a parallel composition for $i < n$, then both graphs entering into this composition are non-primitive. Moreover, only if p_{n-1} is a series composition, one of the graphs entering into p_n can be primitive.

Outerplanar graphs. An *outerplanar graph* can be viewed as a special kind of a series-parallel graph, namely one in which at least one of the graphs entering into a parallel composition is a primitive graph.

Let G be an outerplanar graph. It is an immediate consequence of (8) that if Cent(G) is obtained by a sequence $p_1, \ldots, p_n$ $(n > 1)$ of compositions, then p_i is a series composition for all $i < n$. Thus Cent(G) is a parallel composition of two paths, that is, Cent(G) is a cycle. Hence a centroid of an outerplanar graph is K_1, K_2 or C_n. If G is a *maximal outerplanar graph*, that is G is simultaneously outerplanar and chordal graph, then a centroid of G is a clique. Thus, in this case, a centroid is K_1, K_2 or K_3.

4. CONCLUSIONS

We have proved some characterizations of centroidal graphs and introduced some useful techniques for investigating centroids in graphs. We have also applied these results to some particular classes of graphs: chordal, Halin, series-parallel, and outerplanar.

Another direction of research appears when in the above definition of convexity in graphs we replace "chordless path" by "shortest path" or "simple path" (see [6] for details). Then it may be interesting to find similar characterizations of centroids for these, perhaps more natural, notions of convexity.

REFERENCES

[1] G.A. DIRAC, On rigid circuit graphs, Abh. Math. Seminar Univ. Hamburg 25 (1961), 71-76.

[2] M. FARBER, R.E. JAMISON, Convexity in graphs and hypergraphs, Res. Rep. Corr 83-46, Faculty of Math., University of Waterloo.

[3] M.C. GOLUMBIC, *Algorithmic Graph Theory and Perfect Graphs*, Academic Press, New York 1980.

[4] R.E. JAMISON-WALDNER, A perspective on abstract convexity: classifying alignments by varieties, in: D.C. Kay and B. Drech (eds.), *Convexity and Related Combinatorial Geometry*, M. Dekker, New York 1982, pp. 113-150.

[5] R.E. JAMISON-WALDNER, Copoints in antimatroids, *Congr. Numer.* 29 (1980), 535-544.

[6] R.E. JAMISON, P.H. EDELMAN, The theory of convex geometries, *Geom. Dedicata.* 19 (1985), 247-270.

[7] E.E. JAMISON, R. NOWAKOWSKI, A Helly theorem for convexity in graphs, *Discrete Math.* 51 (1984), 35-39.

[8] C. JORDAN, Sur les assemblages de lignes, *J. reine und angew. Math.* 70 (1869), 185-190.

[9] G. KÖTHE, *Topologische Lineare Räume* I, Berlin, Springer-Verlag, 1960, pp. 188-193.

[10] W. PIOTROWSKI, A generalization of branch weight centroids, to appear.

[11] P.J. SLATER, Maximin facility location, *J. Res. Nat. Bur. Standards* B 79 (1975), 107-115.

[12] P.J. SLATER, Accretion centers: a generalization of branch weight centroids, *Discrete Appl. Math.* 3 (1981), 187-192.

TOPOLOGICAL NETWORK SYNTHESIS

PAWEL WINTER

Department of Computer Science, University of Copenhagen

Universitetsparken 1, DK-2100 Copenhagen O, Denmark

ABSTRACT. We consider several families of deterministic network optimization problems (NOPs) of particular importance for the design (synthesis) of real-life transportation, communication, and distribution networks. These families of NOPs include determination of optimal spanning and Steiner trees, multiconnected networks, distance bounded networks, and capacitated networks. Problems belonging to these families are formulated in an unified manner, and exact algorithms, heuristics, as well as algorithms for special cases are surveyed. Commonalities within each family as well as across family boundaries are identified. A wide range of open problems is given.

1. INTRODUCTION. Our society abounds in large networks for traffic and transportation (urbanization regional planning, shipping), communication (computer networks, broadcasting integrated circuits), and distribution of energy (oil, gas electricity water). The cost complexity, and resource and environment constraints make decision problems concerning rational design (synthesis) and effective utilization (analysis) of such systems both important and difficult. Fortunately, a wide range of operations research techniques (together with concepts from graph theory, probability theory, statistics, and queuing theory), and the continuing progress of computer science, offer methods that permit to deal with such problems.

Synthesis and analysis problems or their subproblems can be formulated in terms of certain prototypical (abstract) network optimization problems (NOPs) [35]. Among the most well-known NOPs are the shortest path problem, the minimum cost spanning tree problem, the travelling salesman problem, and the network design problem. Also multiple objective synthesis and analysis problems can in many cases be reformulated as (single objective) NOPs. NOPs are not exclusively applicable to synthesis and analysis of real-life networks but also to project planning, production and inventory control, optimal capacity scheduling and many others.

We will consider prototypical NOPs which are of particular importance for the synthesis problems Due to the inherent combinatorial nature of most (but not all) synthesis problems. we will emphasize NOPs which can be approached by the techniques of integer programming and combinatorial optimization. A complete knowledge of the parameters associated with edges and vertices is assumed. Such NOPs are called deterministic. In probabilistic NOPs [24], some parameters are available only as random quantities with known or unknown probability distribution functions.

Among deterministic NOPs of particular importance for the synthesis problems, the most common group is concerned with the determination of an optimal topology (i.e., a subset of edges such that the subnetwork has some particular structure, and is in some way optimal). Typical representatives are the minimum (cost) spanning tree problem and the minimum cost k-connected subnetwork problem. Another group is concerned with the selection of optimal edge capacities so that certain flow requirements can be fulfilled. The most well-known and the most general representative is the network design problem.

Facility location problems in which one or more vertices are to be selected to minimize some appropriately chosen centrality measure may be considered as the third group of NOPs of particular importance for the synthesis of real-life networks. In this paper we will not consider location problems. Comprehensive surveys on the facility location problems can be found in [53.54.76,77].

This paper is organized as follows. In Section 2 we discuss spanning and Steiner tree NOPs. Section 3 deals with the design of multiconnected networks. Section 4 is concerned with the design of small diameter networks. Section 5 discusses the network design problem. Conclusions and suggestions for further research are given in Section 6.

All graph-theoretic concepts not defined in the sequel follow [42], with the only exception that they are in a natural way extended to networks. The reader is assumed to be acquainted with both linear programming (LP) and integer programming (IP) [33].

2. TREES. Connectivity is a common structural requirement when designing networks: there must be at least one path between every pair of vertices. Usually, a dominating objective is to keep the construction costs as low as possible. Since the cost in most situations is the sum of (nonnegative) construction costs of involved edges, the network must be a tree. For a comprehensive survey of tree problems, see [82].

2.1. SPANNING TREES. In this section we discuss problems in which trees are required to span all vertices. The most well-known problem within this class is the minimum spanning tree (MST) problem. It is formulated as follows.

GIVEN: A network $G=(V,E,c)$ with n vertices, m edges, and edge-cost function $c:E\rightarrow R_+$.
FIND: A minimum cost subtree of G spanning V.

Several polynomial time algorithms are available. They are all based on the fact that a minimum cost edge in any edge cutset $(W,\overline{W})$ belongs to at least one MST.

In Kruskal's algorithm [55], the edges are sorted in nondecreasing order of their costs. They are then added in that order to the (initially edge-less) solution, provided that no cycle is created.

In Prim's algorithm [67], the tree is expanded one vertex at a time. Initially the tree consists of an arbitrarily chosen vertex. A vertex added is the one connected to the tree by a minimum cost edge.

Kruskal's algorithm requires O(mlogn) time, while Prim's algorithm requires $O(n^2)$ time. O(mloglogn) algorithms using sophisticated data structures to determine edges to be added are also available [17,93].

Other spanning tree NOPs have been considered. Despite differences in their formulations, there are some common techniques available. The most important, so called edge exchange, is based on repetitive replacements of some edges while preserving the spanning tree property. The usefulness of this approach is due to the fact that an arbitrary spanning tree can be obtained from any other spanning tree by a finite number of edge exchanges.

Edge exchanges have been applied to the degree-constrained spanning tree problem [27], k-th least cost spanning tree problem [26,49], and the enumeration of all spanning trees [28,88]. In particular, the enumeration algorithms based on edge exchanges are useful for NP-complete spanning tree problems. Also various heuristics for such spanning tree problems (as well as for other NOPs) are based on edge exchanges.

Another approach to various minimum cost spanning tree problems (as well as many other problems) is based on branch-and-bound procedures in which the lower bounds are obtained using Lagrangian relaxation (LR) methods combined with efficient approximation procedures for obtaining the Lagrange multipliers (LMs). The most well-known spanning tree problem approached successfully by the LR methods is the degree-constrained minimum cost spanning tree problem described by Gavish [34]. The reader is referred to this paper for a more detailed discussion (including a brief description of the subgradient optimization method).

Some minimum cost spanning tree problems (as well as many other problems) can be formulated as mixed integer programming problems. Often there is a natural partition between continuous and integer variables; for fixed integer variables the resulting problem reduces to some easily solvable network flow problem. Benders decomposition algorithm is a general algorithm for mixed integer programming problems. Basically, it solves the problem for some fixed integer variables, and uses the solution to determine a better set of integer variables. Among spanning tree problems approached by the Benders decomposition algorithm, one could mention the capacitated spanning tree problem [16,50.34], and the Telpak problem [34].

Among other important spanning tree problems are: min-max spanning tree problem

[12], minimum ratio spanning tree problem [14], minimum communication spanning tree problem [44], minimum cost-reliability ratio spanning tree problem [15].

2.2. STEINER TREES. Sometimes a network does only need to span some subset Z of p vertices. The remaining vertices $S=V\setminus Z$ are included in the solution only if they reduce the cost. If the solution is merely required to be connected and the edge-costs are positive, the problem is known as the Steiner (tree) problem in networks (SPN), and the optimal solution is called the Steiner minimum tree (SMT). For a comprehensive survey on the SPN and related problems, see [86].

The SMT may sometimes be required to satisfy some side constraints (e.g., bounded number of leaves, bounded diameter). No specialized algorithms nor heuristics for such problems are available. An investigation of their computational complexity can be found in [13].

2.2.1. EXACT ALGORITHMS. The SPN was originally formulated by Hakimi [39]. He also suggested two very simple (and inefficient) algorithms that enumerate trees spanning Z, and choose the SMT among them.

A dynamic programming algorithm [21] is based on the following observation: Given the SMT, split any of its vertices (say, i) into k=deg(i) copies, each adjacent to one edge. Each tree is the SMT for i and its Z-vertices. The algorithm determines SMTs of larger subnetworks by combining SMTs of smaller subnetworks.

An implicit enumeration algorithm [72] is based on inclusion/exclusion of edges. The original method of determining lower bounds, based on simple tree properties, was too weak. Various IP formulations of the SPN have been considered, and several relaxation techniques were suggested to determine better lower bounds. One way of formulating the SPN is as follows.

$$(2.1) \qquad \min \sum_{(i,j)\in E} c_{ij} x_{ij}$$

s t.

$$(2.2) \qquad \sum_{\substack{(i,j)\in E \\ i\in W,\, j\in \overline{W}}} x_{ij} \geq 1 \qquad \forall W\subset V,\ W\cap Z\neq\emptyset,\ \overline{W}\cap Z\neq\emptyset$$

$$(2.3) \qquad x_{ij}\in\{0,1\} \qquad \forall (i,j)\in E$$

The LP relaxation (LPR) of (2.1–2.3) yields a lower bound. The LPR is solved with a subset of constraints (2.2). If the solution is not optimal, a violated constraint is identified and added to the problem. The process is repeated until an optimal solution to the LPR is found. Essentially the approach is identical to the cutting plane algorithm for the set covering problem [33].

Another IP formulation of the SPN is as follows:

(2.4) $$\min \sum_{(i,j)\in E} c_{ij}x_{ij}$$

s.t.

(2.5) $$x_{ij} \geq f_{kij} + f_{kji} \qquad \forall (i,j)\in E,\ \forall k\in Z_1$$

(2.6) $$\sum_{h\in V} f_{kih} - \sum_{j\in N} f_{kji} = \begin{cases} 1, & i=1 \\ -1, & i=k \\ 0, & i\neq 1,k \end{cases} \qquad \forall k\in Z_1$$

(2.7) $$x_{ij}\in\{0,1\} \qquad \forall (i,j)\in E$$

(2.8) $$f_{kij} \geq 0 \qquad \forall i,j:(i,j)\in E, \forall k\in Z_1$$

where 1 is a Z-vertex and $Z_1=Z\setminus\{1\}$.

Beasley [4] suggested two Lagrangian relaxations (LRs) obtained by introducing constraints (2.5) (resp. (2.6)) into the objective function (2.4). In both cases, the LRs for a given set of Lagrangian multipliers reduce to very simple problems (e.g., shortest path problem). The subgradient optimization algorithm is used to obtain better multipliers.

Wong [91] developed a (dual ascent) heuristic for a nearly optimal solution to the dual of the LPR of (2.4-2.8), yielding a lower bound for (2.4-2.8).

The SPN can be formulated as yet another IP model [5]: Add to G an additional vertex 0, and connect it by zero-cost edges to an arbitrarily chosen Z-vertex 1 and to all S-vertices. The problem of finding the MST in the expanded network $G'=(V\cup 0,E',c')$ such that all S-vertices adjacent to 0 are leaves, solves the SPN in G (delete the edge (0,1); the component containing Z-vertices is the SMT). This degree-constrained MST problem can be formulated as the following IP model.

(2.9) $$\min \sum_{(i,j)\in E} c_{ij}x_{ij}$$

s.t.

(2.10) $$\{(i,j) \mid x_{ij} = 1\} \text{ is a spanning tree of } G'$$

(2.11) $$x_{0k} + x_{kq} \leq 1 \qquad \forall k\in V\setminus Z, \forall (k,q)\in E$$

(2.12) $x_{ij} \in \{0,1\}$ $\forall (i,j)\in E'$

where (2.10) can be expressed mathematically, but is here given in verbal form for transparency. The lower bound is obtained by relaxing (2.11) in the Lagrangian fashion. The LR for a given set of Lagrangian multipliers reduces to the MST problem. The subgradient optimization method is used to obtain better multipliers.

In general, it is very difficult to say anything definite about the performance of the algorithms for the SPN. However, the lower bounds obtained by the LR of the degree-constrained formulation and by the dual ascent heuristic are both easy to determine and the solutions obtained are (nearly) optimal in most cases. Furthermore, their costs seem to be very close to the costs of SMTs (often they produce all-integer solutions before any branching is carried out).

2.2.2. HEURISTICS. Since the SPN is NP-complete [48], it is of practical importance to develop heuristics for low-cost trees spanning Z.

A heuristic [75,92] related to the Prim's MST algorithm expands a tree T_Z one Z-vertex at a time. Initially T_Z consists of an arbitrarily chosen Z-vertex. A Z-vertex i added is the one closest to T_Z. T_Z is augmented by the edges on the minimum cost path between T_Z and i. The worst-case error ratio between the cost of T_Z and the SMT is tightly bounded by 2-2/p.

Another heuristic [51] starts with a distance network $D_Z(G)$ for all Z-vertices. Determine its MST T. Let H denote a subnetwork of G containing the edges of paths in T. Determine the MST T_Z in H. Remove from T_Z S-vertices of degree 1 (one at a time). The worst-case error ratio between the cost of T_Z and the SMT is tightly bounded by 2-2/p.

A third heuristic [68], to some degree related to the Kruskal's MST algorithm, examines a list $L=\{T_1,T_2,\dots,T_k\}$ of trees. Initially L consists of isolated Z-vertices. During each iteration two trees are joined together. A special average tree distance function f is used to join trees: Let i be a vertex in G. Relabel trees in L according to their nondecreasing distance from i yielding a list L(i). Let

$$(2.13) \qquad f(i) = \min_{2\leq r\leq k} \{\sum_{j=1}^{r} d(i,T_j)/(r-1)\}$$

Let i^* denote a vertex minimizing f. Join the first to trees T_1^* and T_2^* in $L(i^*)$ by a minimum cost path (passing through i if $i^*\in T_1^*$. In a related heuristic [64], not just one pair but several trees may be joined at once during each iteration. The worst-case error ratio of this heuristic is tightly bounded by 2-2/p.

The distance network heuristic is inferior to the other two [69]. None of the latter can be said to be the best. The differences between the costs of obtained solutions were not large. The error ratio for all three heuristics was far below 2-2/p.

2.2.3. SPECIAL CASES. The SPN remains NP-complete even if edge costs are uniform, G is planar or bipartite [32]. Thus, it is of practical and theoretical importance to identify polynomially solvable cases of the SPN.

A network is said to be a k-tree, $k \geq 1$, if and only if it satisfies one of the following conditions:

(i) It is the complete network on k vertices, K_k,
(ii) It has a vertex i of degree k such that the subnetwork induced by i together with its neighbours is a k-clique.

A linear time (for fixed k) algorithm for the SPN in k-trees [79,2] iteratively eliminates vertices of degree k. Appropriate minimum cost subnetworks induced by the eliminated vertices are determined. When the k-tree is reduced to a single k-clique, these subnetworks are combined to determine the SMT for the original 2-tree. A linear time algorithm that completes any network to a 2-tree (if possible) by adding very high cost edges also is available [79]. Thus, the SPN algorithm is applicable for all partial subnetworks of 2-trees. These are exactly series-parallel networks. A polynomial time algorithm that completes any network to a 3-tree is also available [3]. No such polynomial completion algorithm is known for $k \geq 4$.

Winter [83] developed a linear time algorithm for the SPN in Halin networks, i.e., networks that can be embedded in the plane such that a subnetwork obtained by deleting the exterior face is a tree. As for k-trees, the algorithm is based on vertex eliminations. However, usually several vertices are eliminated at once. Winter [89] also developed a linear time algorithm that completes any network to a Halin network (if possible).

Erickson et al. [22] obtained an $O(n^2p^2+n^3)$ time algorithm for the SPN defined on Z-planar networks, i.e., planar networks that can be embedded in the plane such that all Z-vertices are on the exterior face. This algorithm is essentially a simplified version of the dynamic programming algorithm [21] for the general case.

White et al. [80] developed a polynomial time algorithm for the SPN in strongly chordal graphs. A graph is called chordal if every cycle of four or more edges has a chord. It is called strongly chordal if it is chordal and every even cycle with six or more edges has an odd chord (i.e., a chord that divides a cycle into two paths, each containing odd number of edges). The algorithm itself is not as important as the fact that the SPN in chordal graphs is NP-complete. Furthermore, the SPN in strongly chordal networks is NP-complete even if the edge-costs are strictly triangular (i.e., when the cost of an edge is strictly less than the cost of any path between the end-vertices of the edge).

2.3. TREES IN METRIC SPACES. Vertices to be spanned by a tree can be embedded in some metric space. Euclidean and rectilinear metric spaces are among the most well-known examples. If the edges are permitted to intersect at given vertices only, tree problems in metric spaces can be formulated as tree problems in networks. However, there are algorithms that exploit metric properties of edge costs. The reader is referred to [66] for a comprehensive discussion of efficient "metric" algorithms for the Euclidean MST.

If the edges are permitted to intersect not only in given vertices but also anywhere in the space, difficult Steiner problems arise for the majority of metric spaces. For a survey on the Steiner problem in various metric spaces, the reader is referred to [73]. For a detailed discussion of the Euclidean Steiner problem, see [84].

3. MULTICONNECTED NETWORKS. When designing communication, transportation, and distribution networks, two antagonistic goals are often present: minimization of total cost and maximization of reliability. While the cost is often expressed as the sum of the edge-costs, the reliability is expressed by the connectivity of the network.

Trees satisfy the primary goal of minimizing the total cost while being connected. However, they are extremely vulnerable if operational failures of vertices and/or edges due to scheduled maintenance, error, overload, or any other kind of disturbance or destruction are likely to occur. Only one vertex or edge failure causes a tree network to fail in its main objective of enabling communication between all pairs of vertices.

When vertex or edge connectivity is used as the reliability measure, the following synthesis problems arise:

GIVEN: A network $G=(V,E,c)$, a subset Z of p vertices, and an $p*p$ integer matrix $R=(r_{ij})$.

FIND: A subnetwork G_R such that there are at least r_{ij} vertex (edge) disjoint paths for every pair of vertices $i,j\in Z$, and the cost of G_R is minimized.

As it will be seen the special "spanning" case with Z=V received considerable attention. The more general "Steiner" case ($Z\subseteq V$) has been explicitly formulated for the first time by Krarup [52] as the generalized Steiner problem (GSP).

3.1. EXACT ALGORITHMS. The spanning edge version of the GSP can be formulated as the SPN in (2.1-2.3) with the right-hand side of (2.2) replaced by $r_W=\max\{r_{ij}|i\in W,j\in \bar{W}\}$, where $\bar{W}=V\setminus W$. Christofides and Whitlock [18] suggested the following algorithm.

Let X' denote the optimal solution vector to the linear relaxation of (2.1-2.3) obtained by disregarding all constraints (2.2) with $|W|>1$. Interpret X' as edge capacities in G. Suppose that there is a cutset $(W',\overline{W'})$ such that its (2.2) constraint is violated. Add this constraint to the relaxation and determine new X' using dual simplex method. Eventually, X' will be the optimal solution to the LPR of (2.1-2.3). If this X' is not all-integer, branch-and-bound is used.

In order to find violated cutsets at each iteration, the multiterminal flow algorithm [36] is used. It requires n-1 invocations of the maximum flow algorithm on successively reduced networks.

According to the computational experience [18], problem instances with "well over a hundred vertices" can be solved (no precise computational results are given).

To solve the spanning vertex version of the problem, solve the edge version first. If the solution contains a vertex disconnecting set W such that $|W|<r_{ij}$ for some pair of vertices $i,j \in V\setminus W$, then the constraint

$$(3.1) \qquad \sum_{\substack{(i,j)\in E \\ i\in S,\, j\in T}} x_{ij} \geq 1$$

where S and T are the components obtained by deleting W from the solution, is added and the entire process is repeated. This algorithm is not as effective as its edge counterpart. However, no time estimates are available.

The edge and vertex approaches are also applicable to directed networks. However, the computational effort increases since the multiterminal flow algorithm cannot be used. Thus, essentially n(n-1) maximum flow computations (instead of just n-1) are necessary.

No exact algorithm for the Steiner edge (vertex) version has been developed as yet. However, it should be quite obvious how to modify the above algorithms to cover the general GSP case.

3.2. HEURISTICS. The GSP is NP-complete since its special case Z=V is NP-complete both for the edge and the vertex version [32].

Steiglitz et al. [74] suggested a heuristic for the spanning vertex (and hence edge) version. Let $G_R=(V,E_R,c)$ denote a feasible solution. A greedy approach is used to obtain an initial feasible solution. Let p, q, s, t, denote four distinct vertices such that $(p,q)\cup(s,t)\in E_R$, while $(p,s)\cup(q,t)\notin E_R$. A pairwise edge exchange in G_R replaces edges (p,q) and (s,t) by edges (p,s) and (q,t). It is said to be valid if it reduces the cost and preserves feasibility. The heuristic applies valid pairwise edge exchanges to some initial feasible solution as long as possible. The feasibility of the new network can be checked by the algorithm of Frisch [25].

If the requirement matrix R is uniform (i.e., $r_{ij}=r$ for all $i,j\in V$), then vertices must be incident to at least r edges. Thus, (minimum cost) r-regular networks could be useful when solving both vertex and edge versions of GSP (especially when V=Z). However, this problem is NP-complete even if edge-costs are uniform [65].

3.3. SPECIAL CASES. Since the design of multiconnected networks is NP-complete, much effort has been concentrated on solving various special cases. Furthermore, connectivity issues have been extensively studied within graph theory. Consequently, there are some results concerning synthesis of multiconnected graphs. These results will not be discussed here. The reader is referred to [81,18].

Recently, Monma et al. [61] considered the 2-edge connected spanning GSP when the vertices are embedded in a metric space. The optimal solution is bound to be 2-vertex connected. All vertices are of degree at most 3. Also, the cost of an optimal travelling salesman cycle is not greater than 4/3 times the cost of an optimal 2-edge connected solution. Thus, solving the travelling salesman problem either exactly or by means of one of several available heuristics, yields a reasonable feasible solution. Similar results were obtained for the 2-edge-connected GSPs.

For some special cases of the GSP, linear time algorithms have been developed. If the underlying network is outerplanar, Winter [85] obtained linear time algorithms for 2-vertex (resp. edge) connected GSPs. Subsequently these algorithms were generalized to series-parallel networks [87]. If the underlying network is a Halin network, Winter [89] obtained linear time algorithms for 2- and 3-vertex (resp. edge) connected GSPs. It remains an open question whether both versions of GSP can be solved in polynomial time when the underlying network is Z-planar and R is uniform with all entries being equal to 2.

3.4. OTHER MEASURES OF CONNECTEDNESS. So far we considered problems of designing networks that are reliable with respect to the classical vertex and edge connectivity. But these reliability measures are not always adequate. A network can for example be regarded as functioning correctly as long as at least half of its vertices remains connected, or as long as there are at most k, $1 \leq k \leq n-1$, components. In this subsection we briefly describe some of these alternative reliability measures.

Vertex (resp. edge) k-cantle connectivity is the minimum size of any vertex (resp. edge) set whose removal causes G to split into more than k components, $1 \leq k \leq n$ [9]. The problem of determining vertex (resp. edge) k-cantle connectivity received very limited attention. Also design of low cost networks with specified vertex (resp. edge) k-cantle connectivity remains to be investigated.

Vertex (resp. edge) q-degree connectivity is the minimum size of any vertex (resp. edge) set whose removal causes G to become disconnected with q vertices isolated from the remaining vertices [10].

Vertex (resp. edge) strength of a vertex i is equal to the maximum vertex (resp. edge) connectivity of any subnetwork containing i. Vertex (resp. edge) strength of a network is equal to the maximum strength of any vertex (resp. edge) in the network [60].

For a more complete survey of these as well as other reliability measures, the reader is referred to [8].

4. DISTANCE BOUNDED NETWORKS. In the design of communication networks there is sometimes a requirement that the number of relay stations or distance betweeen sites should be bounded. The reason for such a requirement is the noise existing in communication lines.

4.1. REALIZATION OF DISTANCE MATRICES. Given a network $G=(V,E,l)$, distances between its vertices can be tabulated in an $n*n$ distance matrix $D=(d_{ij})$. Note that the cost function c in G has been replaced by the length function l. A natural problem that arises in this context is whether for a given matrix D there exists a network G with D as its distance matrix.

G realizing D exists if and only if $d_{ii}=0$, $d_{ij}=d_{ji}$, and $d_{ij}+d_{jk}\geq d_{ik}$ for all i, j, k [40]. By dropping the condition $d_{ij}=d_{ji}$, a necessary and sufficient condition for existence of a directed realization is obtained.

Usually there will be several different realizations of D. It is natural to look for one with the smallest possible total length. Let G be an arbitrary undirected realization of D. Remove from G (one at a time) redundant edges (i.e., edges whose removal does not affect distances between vertices). A network realizing D without redundant edges is unique and therefore of minimum total length [40].

Sometimes a network realizing D may be required to be a tree. Unfortunately, no characterization of distance matrices realizable by trees (or any other types of networks) is known. On the other hand, there are algorithms generating tree realizations if they exist [7].

It can be advantageous to introduce additional, S-vertices. The general problem of determining minimum length Steiner network realizing D is unsolved. A heuristic based on triangle transformations has been suggested [40]. The heuristic starts with a complete network K_n with edge lengths equal to the required distances between their end-vertices. Then a triangle Δijk is replaced by a star with a new S-vertex s connected to i, j, k by edges of appropriate length. This is repeated as many times as possible.

4.2. MINIMUM DIAMETER NETWORKS. The minimum diameter problem in a network $G=(V,E,c,l)$, where c and l are edge-cost and edge-length functions respectively, asks for a subnetwork with its total cost not exceeding some fixed budget B, and with its diameter (i.e., maximum distance between any pair of vertices) minimized.

The minimum diameter problem is NP-complete even if G is planar with maximum degree 3, and the budget restricting the choice to spanning trees. Furthermore, it is very unlikely that a polynomial heuristic with the worst-case error ratio less than 2 exists. Similar results are valid for the minimum radius problem where the radius is defined as the minimum over maximal distances from each vertex to all other vertices [63].

To the best of our knowledge, neither exact algorithms nor heuristics for the minimum diameter (radius) problem are available. However a special case received considerable attention. It concerns design of regular minimum diameter networks. Such networks are needed in connection with the design of microprocessors consisting of an enormous number of components. Due to cost and complexity considerations, components should have identical characteristics. In particular, the number of interconnections associated with each component should be uniform.

There are many partial results that provide construction of regular minimum diameter graphs for some particular number of vertices n and degree r. These constructions are based on so called Moore trees and Singleton forests. However, no general technique is available. Recently, Schumacher [70] developed an $O(n^2)$ algorithm generating regular graphs with n vertices and degree r which have diameters at most twice as large as the theoretical minimum. Furthermore, the graphs obtained are r-vertex connected.

Toueg and Steiglitz [78] developed a heuristic for the regular minimum diameter problem in an arbitrary graph. As the heuristic for the special case of the GSP (Section 3.2), it is also based on valid pairwise edge exchanges which reduce either the diameter or the number of vertex pairs at maximal distance. The heuristic generates regular graphs of diameter close to the minimum. It is also very stable (independent of initial solutions), and it can be easily modified to cover the directed case and the case where edge lengths are not limited to unity. However, due to considerable computational effort needed to find valid pairwise edge exchanges, the heuristic is impractical for graphs with more than 150 vertices. Another limitation is the lack of knowledge about possible symmetries in the obtained solutions.

A closely related (r,d) graph problem of interconnecting the maximum number of vertices, while keeping the diameter and degree below some fixed numbers because of technical constraints received considerable attention. As an additional constraint, the graph could be required to be r-vertex or r-edge connected. For a comprehensive survey, see [6].

Another interesting problem (which received only very little attention) is: Given a graph G, and three integers d, d', and q. Determine (if possible) a subgraph of G with the minimum number of edges such that its diameter is at most d, and removal of any q edges increases the diameter to at most d'. In addition, the degrees of vertices in the subgraph could be required to be bounded by r. A related problem is to find the largest (with respect to the number of vertices) network of diameter d, degree at most r, and such that removal of any q edges increases the diameter to at most d'. For a survey of few available results, see [6].

4.3. MINIMUM AVERAGE DISTANCE NETWORKS. Rather than looking for a budget-constrained network with minimum diameter (radius), one could ask for a budget-constrained network minimizing the average distance (or equivalently minimizing the sum of distances between all pairs of vertices). It will be seen in Section 5 that this problem can be formulated as a special case of the network design problem. Contrary to the problems discussed so far, the techniques used to solve the minimum average distance network problem are essentially the same as for the network design problem, and will be discussed in the next section.

5. NETWORK DESIGN PROBLEM. The network design problem (NDP) deals with choosing a subset of (perhaps capacitated) edges, and assigning to them flows in such a manner that requirements between all pairs of vertices can be simultaneously fulfilled and the total cost incurred is minimal. In some situations the cost is a function of edge flows while the design costs are incorporated as budget constraints. Under the assumption that the cost function is linear, this problem is often referred to as the budget design problem. If the cost function is linear and involves both design and flow costs, then the problem is called the fixed charge design problem.

A comprehensive survey on the NDP has been given by Magnanti and Wong [58]. It also covers the case when cost functions are nonlinear.

5.1. SPECIAL PROBLEMS. The importance of the NDP is not merely due to its obvious applicability in the design of various kinds of transportation networks. There is a large variety of NOPs that are special cases of the NDP. Still other NOPs turn out to be very closely related to the NDP.

Consider the fixed charge design problem with no flow costs. If the requirement pattern is complete (i.e., there is a requirement of flow between every pair of vertices), this is the MST problem (Section 2.1). If the requirement pattern is complete only with respect to a subset of vertices, this is the SPN (Section 2.2). If the network is directed and the requirement pattern is complete, this is the minimum weight equivalent subnetwork problem [59]. If however the requirement pattern is incomplete such that there is only one source while all the remaining vertices serve as sinks, this is the minimum arborescence problem. If there is one source but only some of the remaining vertices serve as sinks, this is the Steiner arborescence problem.

The fixed charge design problem with no design costs reduces to the minimum cost path problem (for all pairs of vertices with non-zero requirements). If the edges are in addition capacitated, the problem reduces to the multicommodity flow problem.

The fixed charge design problem with a complete requirement pattern, uniform flow costs and large, uniform design costs reduces to the travelling salesman problem.

Consider the minimum average distance network problem (Section 4.3). Interpret the length of an edge as the cost of shipping one unit of flow through it. Then the budget design problem with complete unitary flow requirements is equivalent to the minimum average distance network problem.

Consider a version of the fixed charge design problem with no flow costs, unitary edge capacities, and integral flow requirements. Furthermore, suppose that flow requirements no longer need to be satisfied simultaneously. Clearly, this is the edge version of the GSP (Section 3). Vertex-splitting can be used to obtain vertex version.

Finally, many facility location problems can be formulated as NDPs. This is achieved by means of the well-known vertex-splitting technique. Each vertex i that is a candidate for a facility is splitted in two vertices i_1 and i_2 connected by a (directed) arc (i_1,i_2) of capacity equal to the capacity of the potential facility. All arcs previously directed into (resp. from) i are directed into i_1 (resp. from i_2). If the arc (i_1,i_2) is chosen in the instance of NDP, then the vertex i is chosen in the corresponding instance of the facility location problem.

5.2. BUDGET DESIGN PROBLEM. In this section we discuss various methods to solve the (uncapacitated) budget design problem. They can in most cases be adapted to solve the capacitated version as well as the fixed charge design problem.

The (uncapacitated) budget design problem has potential uses in designing air, rail, and highway transportation systems. Although such systems are usually much more complex, solving the budget design problem can be valuable in screening potential configurations for a more detailed study.

The problem is NP-complete even if all flow and design costs are unitary. Furthermore, it remains NP-complete if the budget restricts the solution to be a tree [47].

5.2.1. EXACT ALGORITHMS. The budget design problem in an undirected network can be formulated as the following mixed IP problem:

$$(5.1) \qquad \min \sum_{s,t \in V} \sum_{\substack{i,j \in V \\ (i,j) \in E}} f_{ij}^{st} y_{ij}^{st}$$

s.t.

$$(5.2) \qquad \sum_{p \in V} y_{ip}^{st} - \sum_{q \in V} y_{qi}^{st} = \begin{cases} r_{st} & \text{if } i=s \\ -r_{st} & \text{if } i=t \\ 0 & \text{otherwise} \end{cases} \qquad \forall s,t \in V$$

(5.3) $$\sum_{s,t \in V} y_{ij}^{st} + y_{ji}^{st} \leq C_{ij}x_{ij} \qquad \forall (i,j) \in E$$

(5.4) $$\sum_{(i,j) \in E} c_{ij}x_{ij} \leq B$$

(5.5) $$y_{ij}^{st} \geq 0 \qquad \forall i,j \in V\colon (i,j) \in E,\ \forall s,t \in V$$

(5.6) $$x_{ij} \in \{0,1\} \qquad \forall (i,j) \in E$$

where B is the budget, C_{ij} is the capacity of the edge (i,j), c_{ij} is the design cost of the edge (i,j), and f_{ij}^{st} is the cost of sending one unit of flow from s to t along the edge (i,j) in direction from i to j.

Apart from the above "conservation of flow" formulation of the NDP, also "cut-flow" formulations have been investigated [62].

Several branch-and-bound algorithms for the uncapacitated budget design problem based on the inclusion/exclusion of edges have been suggested. Let IN and OUT denote subsets of edges that must be respectively included and excluded from the solution. Define $F=E\setminus(IN \cup OUT)$. Let respectively $c(G)$ and $f(G)$ denote the total design cost and the total flow cost of G.

$f(G-OUT)$ is a lower bound for (5.1-5.6) [11]. If $c(G-OUT) \leq B$, then G-OUT is a feasible and hence optimal solution; the set of solutions determined by IN and OUT can be fathomed. If $c(G-OUT)>B$, then at least one edge $(i,j) \in F$ has to become a member of OUT later on during the branching process. Let I_{ij} denote the increment of the length of the "shortest" path (in terms of flow costs) from i to j when the edge (i,j) is removed from G-OUT. The optimal solution to the following continuous knapsack problem (CKP):

(5.7) $$\min \quad f(G-OUT) + \sum_{(i,j) \in F} (1-x_{ij})I_{ij}$$

subject to

(5.8) $$\sum_{(i,j) \in IN} c_{ij} + \sum_{(i,j) \in F} c_{ij}x_{ij} \leq B$$

(5.9) $$0 \leq x_{ij} \leq 1 \qquad \forall (i,j) \in F$$

yields a lower bound for (5.1-5.6) for given IN and OUT [43]. The CKP can be solved by the simple one pass algorithm [19]. The optimal solution to CKP has at most one non-integer element which can be used as a branching variable [43]. Dionne and Florian [20] suggested a more effective strategy: a variable x_{ij}, $(i,j) \in F$, for which I_{ij}/c_{ij} is minimized is chosen as the branching variable. They also suggested a relatively efficient way of determining I_{ij}. Other lower bounds can be found in

[29,30,57].

Also Benders decomposition algorithm for the mixed IP problem [33] has been applied successfully to various special cases of NDP. This algorithm chooses some subnetwork (i.e., it fixes x-variables), finds the optimal flow pattern on this subnetwork , and uses it to obtain a better subnetwork. This approach was adapted to the fixed charge design problem [56]. A survey on the methods for generation of Benders and other cuts as well as their relation to lower bounds for the branch-and-bound has been given in [58].

As it was seen in connection with the SPN, good lower bounds were obtained by both Lagrangian relaxation [4,5], and dual ascent procedure [91]. However, for more general NDPs, these methods are not as easily applied due to the larger number of constraints.

Another approach which has been fruitful for special cases of the NDP is the facet generation technique. It was succesfully applied to the travelling salesman problem [37,38]. Study of facets for more general cases of the NDP can be useful in improving lower bounds based on linear programming relaxations.

5.2.2. HEURISTICS. Although considerable improvements on the efficiency of exact algorithms have been obtained in recent years, problem instances with 50-100 vertices are virtually unsolvable due to enormous computational effort involved. Consequently, several heuristics have been suggested.

The greedy "add" heuristic starts with the MST (with respect to design costs), and adds edges one at a time; each time the edge anticipated to cause the greatest flow cost decrease while preserving feasibility is chosen. If the cost of the MST exceeds the budget, then there is no solution at all. The edges can be added in nondecreasing order of their flow costs. Alternatively, edges can be added in nondecreasing order of their design costs. Yet another possibility is to add edges in nonincreasing order of ratios I_{ij}/c_{ij}. The relative error ratio of the greedy "add" heuristic with respect to the optimal solution is between 1 and 7% [20]. Unfortunately, its worst-case error ratio grows linearly with the number of vertices [90].

The greedy "delete" heuristic starts with (usually) infeasible solution containing all edges. Edges are deleted one at a time; each time the edge anticipated to cause the smallest flow cost increase is chosen. This process continues until a feasible solution is obtained. Again, various removal orderings are possible.

Scott [71] developed a heuristic (subsequently refined by Dionne and Florian [20]) which combines the addition and deletion of edges. Let Q_{ij} denote the increment in the total flow cost when (i,j) is removed from G-OUT. Remove edges from the original network until the budget constraint is satisfied. At each iteration an edge with the smallest Q_{ij}/c_{ij} is removed. Then the edges are added in e.g., nonincreasing order of Q_{ij}/c_{ij} as long as the budget constraint remains satisfied. This heuristic is slower than the greedy "add" heuristic. However, its relative error ratio is usually less than 1%. In many cases the solution obtained is in fact optimal. Unfortunately, the worst-case error ratio for this heuristic grows linearly with the number of vertices [90]. In fact, the existence of a polynomial time heuristic with the worst-case error ratio less than $n^{1-\varepsilon}$, for any fixed , $0<\varepsilon<1$, would imply P=NP.

Perhaps the best heuristic with respect to the relative error is due to Dionne and Florian [20]. It is based on their branch-and-bound algorithm. Instead of I_{ij} when computing lower bounds in (5.7-5.9), Q_{ij} is used. Although this does not guarantee to produce a lower bound, computational experience indicates that the probability of excluding the optimal solution is very low. Thus, this heuristic (slower than those described previously but faster than the branch-and-bound algorithm with correct lower bounds) has very small relative error ($\leq$0.05%) and produces in almost all cases optimal solutions.

6. CONCLUDING REMARKS AND SUGGESTIONS FOR FURTHER RESEARCH. In this paper we have surveyed a number of NOPs of particular importance for the synthesis of real-life networks. The increasing number of papers within the network optimization gives credence to the statement that this branch of operations research is growing in importance. On the other hand, network optimization (and problems applicable to synthesis of real-life networks in particular) can be considered as relatively young. In fact, this survey shows that there are still interesting unsolved NOPs, as well as algorithms and/or heuristics to other NOPs need to be improved before they can be applied in connection with real-life problems. In this section we briefly discuss these open problems.

The MST problem (Section 2.1) is perhaps the only NOP which can be said to be completely solved. However, other spanning tree problems received considerably less attention. A more unified approach to these problems would be of interest. For example, more or less unified use of Lagrangian relaxations (where side constraints are introduced into objective functions) when these NOPs are formulated as integer programming problems should be investigated. Some results have been obtained in [34]. Also use of edge exchange techniques in heuristics for this kind of problems should be studied.

Several algorithms and heuristics for the SPN (Section 2.2) have been suggested. However, a convincing study of their performance is still not available. It is almost impossible to say anything definite about which algorithm or heuristic will perform best for particular instance type (dense versus sparse networks, few versus many Z-vertices, small versus large edge-cost variance). Several polynomial time heuristics for the SPN have been shown to have the worst-case error ratio tending to 2 as the number of Z-vertices grows. It is an open problem whether existence of a polynomial heuristic with better worst-case error ratio would imply NP=P. Algorithms and heuristics for special cases (e.g., when edge-costs satisfy the triangle inequality) remain to be developed. Finally, Steiner tree problems with side constraints received very little attention.

Contrary to the analysis of multiconnected networks, the design of (both spanning and Steiner) networks (Section 3) did not receive much attention. Both exact algorithms and heuristics (especially for the more general, Steiner case) would be of great interest. Another important group of problems which received no attention at all is concerned with the design of multiconnected Steiner networks in Euclidean and rectilinear planes. Finally, other measures of connectedness (Section 3.4) remains to be investigated. Algorithms and heuristics measuring degree of connectedness of existing networks (analysis), and designing low-cost networks satisfying connectivity requirements with respect to these new measures are needed.

Design of distance bounded networks (Section 4) has not been studied as extensively as the design of multiconnected networks (with the exception of the minimum average distance network problem which is a special case of the network design problem). Consequently there is an obvious need of developing more efficient exact algorithms and heuristics. Furthermore, the problem of designing low-cost networks satisfying both distance and (vertex or edge) connectivity requirements deserves to be investigated. In particular, good heuristic would be of interest.

Network design problems (NDPs) (Section 5) received considerable attention. In particular, good lower bounds for the branch-and-bound algorithms were obtained. As indicated in [58], future research will provide even better bounds, thus enlarging the scope of NDPs that can be solved optimally. Also use of Benders cuts will be further investigated. The same is the case with respect to facet generating constraints which can strengthen the LP relaxations of the IP formulations of NDPs. Investigation of average case performance of (existing and new) heuristics for the NDP and its variants will become more widespread.

REFERENCES

[1] Aneja Y.P. "An integer linear programming approach to the Steiner problem in graphs" Networks 10 (1980) 167-178.

[2] Arnborg S. and Proskurowski A. "Linear time algorithms for NP-hard problems on graphs embedded in k-trees" Tech. Report TRITA-NA-8404, Dept. of Numerical Analysis and Computing Science, The Royal Inst. of Techno logy, Stockholm, Sweden (1984).

[3] Arnborg S. and Proskurowski A. "Characterization and recognition of partial k-trees" Tech. Report TRITA-NA-8402, Dept. of Numerical Analysis and Computing Science, The Royal Inst. of Technology, Stockholm, Sweden (1984).

[4] Beasley J.E. "An algorithm for the Steiner problem in graphs" Networks 14 (1984) 147-159.

[5] Beasley J.E. "An SST-based algorithm for the Steiner problem in graphs" Technical Report, Dept. of Man. Sciences, Imperial College, London (1985).

[6] Bermond J.C., Bond J., Paoli M. and Peyrat C. "Graphs and interconnection networks: diameter and vulnerability" in Lloyd (ed.) "Surveys in Combinatorics" London Math. Soc. Lect. Notes 82 (1983) 1-30.

[7] Boesch F.T. "Properties of the distance matrix of a graph", Quart. Appl. Math. 26 (1969) 607-610.

[8] Boesch F.T. "Graph-theoretic models for network reliability studies" Tech. Report 8010, Dept. of Electr. Eng and Comp. Sci., Stevens Inst. of Technology, Hoboken, New Jersey (1980).

[9] Boesch F.T.and Chen S. "A generalization of line connectivity and optimally invulnerable graphs" SIAM J. Appl. Math. 34 (1978) 657-665.

[10] Boesch F.T. and Felzer A. "On the invulnerability of the regular complete k-partite graphs" SIAM J. Appl. Math. 20 (1971) 176-182.

[11] Boyce D.E., Farhi A. and Weischedel R. "Optimal network problem: A branch and bound algorithm" Environment Planning 5 (1973) 519-533.

[12] Camerini P.M. "The min-max spanning tree problem and some extensions" Inf. Proc. Letters 7 (1978) 10-13.

[13] Camerini P.M., Galbiati G. and Maffioli F. "On the complexity of Steiner-like problems" Proc. 7th Ann. Allerton Conf. on Comm., Control and Computing (1979) 969-977.

[14] Chandrasekaran R. "Minimal ratio spanning trees" Networks 7 (1977) 335-342.

[15] Chandrasekaran R., Aneja Y.P. and Nair K.P.K. "Minimal cost-reliability ratio spanning tree" in Hansen P. (ed.) "Studies on Graphs and Discrete Programming" North-Holland (1981) 53-60.

[16] Chandy K.M. and Lo T. "The capacitated minimum spanning tree" Networks (1973) 173-181.

[17] Cheriton D. and Tarjan R.E. "Finding minimum spanning trees" SIAM J. Comput. 5 (1976) 724-742.

[18] Christofides N. and Whitlock C. "Network synthesis with connectivity constraints" in Brans J.P. (ed.) "Operational Research '81" North-Holland, London (1981) 705-723.

[19] Dantzig G.B. "Discrete-variable extremum problems" Operations Research 5 (1957) 266-277.

[20] Dionne R. and Florian M. "Exact and approximate algorithms for optimal network design" Networks 9 (1979) 37-59.

[21] Dreyfus S.E. and Wagner R.A. "The Steiner problem in graphs" Networks 1 (1971) 195-207.

[22] Erickson R.E., Monma C.L. and Veinott A.F.Jr. "Minimum-concave-cost network flows" to appear in Math. of OR.

[23] Frank H. and Frisch I.T. "Analysis and design of survivable networks" IEEE Trans. on Comm. Techn. COM-18 (1970) 501-519.

[24] Frank H. and Frisch I.T. "Communication, Transmission, and Transportation Networks" Addison-Wesley (1971).

[25] Frisch I.T. "Flow variation in multiple min-cut calculations" J. Franklin Inst. 287 (1969) 61-72.

[26] Gabow H.N. "Two algorithms for generating weighted spanning trees in order" SIAM J. Comput. 6 (1977) 139-150.

[27] Gabow H.N. "A good algorithm for smallest spanning trees with a degree constraint" Networks 8 (1978) 201-208.

[28] Gabow H.N. and Myers E.W. "Finding all spanning trees of directed and nondirected graphs" SIAM J. Comput. 7 (1978) 280-287.

[29] Gallo G. "A new branch-and-bound algorithm for the network design problem"

Report L81-01, Instituto di Elaborazione della Informazione, Pisa, Italy (1981).

[30] Gallo G. "Lower planes for the network design problem" Networks 13 (1983) 411-426.

[31] Garey M.R. and Johnson D.S. "The rectilinear Steiner problem is NP-complete" SIAM J. Appl. Math. 32 (1977) 826-834.

[32] Garey M.R. and Johnson D.S. "Computers and Intractability: A Guide to the Theory of NP-completeness" Freeman, San Francisco (1979).

[33] Garfinkel R.S. and Nemhauser G.L. "Integer Programming" J. Wiley, N.Y. (1972).

[34] Gavish B. "Topological design of centralized computer networks - formulations and algorithms" Networks 12 (1982) 355-377.

[35] Golden B.L. and Magnanti T.L. "Deterministic network optimization: A bibliography" Networks 7 (1977) 149-183.

[36] Gomory R.E. and Hu T.C. "Multi-terminal network flows" J. SIAM 9 (1961) 551-570.

[37] Grotschel M. and Padberg M.W. "On the symmetric travelling salesman problem: I. Inequalities" Mathematical Programming 16 (1979) 265-280.

[38] Grotschel M. and Padberg M.W. "On the symmetric travelling salesman problem: II. Lifting theorems and facets" Mathematical Programming 16 (1979) 281-302.

[39] Hakimi S.L. "Steiner's problem in graphs and its implications" Networks 1 (1971) 113-133.

[40] Hakimi S.L. and Yau S.S. "Distance matrix of a graph and its realizability" Quart. Appl. Math. 22 (1965) 305-317.

[41] Hanan M. "Net wiring for large scale integrated circuits" IBM Res. Report RC1375 (1965).

[42] Harary F. "Graph Theory" Addison-Wesley, Reading (1969).

[43] Hoang H.H. "A computational approach to the selection of an optimal network" Management Science 19 (1973) 488-498.

[44] Hu T.C. "Optimum communication spanning problem" SIAM J. Comput. 3 (1974) 188-195.

[45] Hwang F.K. "On Steiner minimal trees with rectilinear distance" SIAM J. Appl. Math. 30 (1976) 104-114.

[46] Johnson D.S., Lenstra J.K. and Rinnooy Kan A.H.G. "The complexity of the network design problem" Networks 8 (1978) 279-285.

[47] Karp R.M. "Reducibility among combinatorial problems" in Miller and Thatcher (eds.) "Complexity of Computer Computations" Plenum Press, N.Y. (1972) 85-103.

[48] Katoh N., Ibaraki T. and Mine H. "An algorithm for finding k minimum spanning trees" SIAM J. Comput. 10 (1981) 247-255.

[49] Kershenbaum A. "Computing capacitated minimal spanning trees efficiently" Networks 4 (1974) 299-310.

[50] Kou L., Markowsky G. and Berman L. "A fast algorithm for Steiner trees" Acta Informatica 15 (1981) 141-145.

[51] Krarup J. "The generalized Steiner problem" unpublished note (1978).

[52] Krarup J. and Pruzan P. "Selected families of location problems" Ann. Discrete Math. 5 (1979) 327-387.

[53] Krarup J. and Pruzan P. "The simple plant location problem: survey and synthesis" European J. Operational Research 12 (1983) 36-81.

[54] Kruskal J.B. "On the shortest spanning subtrees of a graph and the travelling salesman problem" Proc. Amer. Math. Soc. 7 (1956) 48-50.

[55] Los M. and Lardinos C. "Combinatorial programming, statistical optimization and the optimal transportation network problem" Transportation Research 16B (1980) 89-124.

[56] Magnanti T.L., Mireault P. and Wong R.T. "Tailoring Benders decomposition for network design" Technical Report OR 125-83, Operations Research Center, MIT (1983).

[57] Magnanti T.L. and Wong R.T. "Accelerating Benders decomposition: Algorithmic enhancement and model selection criteria" Operations Research 29 (1981) 464-484.

[58] Magnanti T.L. and Wong R.T. "Network design and transportation planning: Models and algorithms" Transportation Science 18 (1984) 1-55.

[59] Martello S. and Toth P. "Finding minimum equivalent graph of a digraph" Networks 12 (1982) 89-100.

[60] Matula D.W. "k-components, clusters, and slicings in graphs" SIAM J. Appl. Math. 22 (1972) 459-480.

[61] Monma C.L., Munson B.S. and Pulleyblank W.R. "Minimum-weight two-connected spanning networks" unpublished paper.

[62] Petersen B.E. "A cut-flow procedure for transportation network optimization" Networks 10 (1980) 33-43.

[63] Plesnik J. "The complexity of designing a network with minimum diameter" Networks 11 (1981) 77-85.

[64] Plesnik J. "A bound for the Steiner tree problem in graphs" Math. Slovaca 31 (1981) 155-163.

[65] Plesnik J. "A note on the complexity of finding regular subgraphs" Discrete Math. 49 (1984) 161-167.

[66] Preparata F.P. and Shamos M.I. "Computational Geometry: An Introduction" Springer-Verlag, Berlin (1985).

[67] Prim R.C. "Shortest connection networks and some generalizations" Bell System Tech. J. 36 (1957) 1389-1401.

[68] Rayward-Smith V.J. "The computation of nearly minimal Steiner trees in graphs" Int. J. Math. Educ. Sci. Technol. 14 (1983) 15-23.

[69] Rayward-Smith V.J. and Clare A. "On finding Steiner vertices" Networks 16 (1986) 283-294.

[70] Schumacher U. "An algorithm for construction of a k-connected graph with minimum number of edges and quasiminimal diameter" Networks 14 (1984) 63-74.

[71] Scott A.J. "The optimal network problem: some computational procedures" Transportation Research 3 (1969) 201-210.

[72] Shore M.L., Foulds L.R. and Gibbons P.B. "An algorithm for the Steiner problem in graphs" Networks 12 (1982) 323-333.

[73] Smith J.M. "Generalized Steiner network problems in engineering design" in Gero J.S. (ed.) "Design Optimization" Academic Press (1985) 119-161.

[74] Steiglitz K., Weiner P. and Kleitman D.J. "The design of minimum cost survivable networks" IEEE Trans. on Circuit Theory CT-16 (1969) 455-460.

[75] Takahashi H. and Matsuyama A. "An approximate solution for the Steiner problem in graphs" Math. Japonica 24 (1980) 573-577.

[76] Tansel B.C., Francis R.L. and Lowe T.L. "Location on networks: A survey. Part I: The p-center and p-median problems" Management Science 29 (1983) 482-497.

[77] Tansel B.C., Francis R.L. and Lowe T.L. "Location on networks: A survey. Part

II: Exploiting tree network structure" Management Science 29 (1983) 498-511.

[78] Toueg S. and Steiglitz K. "The design of small diameter networks by local search" IEEE Trans. Comput. C-28 (1979) 537-542.

[79] Wald J.A. and Colbourn C.J. "Steiner trees, partial 2-trees, and minimum IFI networks" Networks 13 (1983) 159-167.

[80] White K., Farber M. and Pulleyblank W. "Steiner trees, connected domination and strongly chordal graphs" Networks 15 (1985) 109-124.

[81] Wilkov R.S. "Analysis and design of reliable computer networks" IEEE Trans. Commun. COM-20 (1972) 660-678.

[82] Winter P. "Network Synthesis: Spanning Trees" Tech. Report, Inst. of Datalogy, Univ. of Copenhagen, Denmark (1983).

[83] Winter P. "Steiner problem in Halin networks" to appear in Discrete Applied Mathematics.

[84] Winter P. "An algorithm for the Steiner problem in the Euclidean plane" Networks 15 (1985) 323-345.

[85] Winter P. "Generalized Steiner problem in outerplanar networks" BIT 25 (1985) 485-496.

[86] Winter P. "Steiner problem in networks - a survey" to appear in Networks.

[87] Winter P. "Generalized Steiner problem in series parallel networks" to appear in J. of Algorithms.

[88] Winter P. "An algorithm for the enumeration of spanning trees" BIT 26 (1986) 44-62.

[89] Winter P. "Generalized Steiner problem in Halin networks" presented at XII Math. Progr. Symp., Boston (1985).

[90] Wong R.T. "Worst-case analysis of network design problem heuristics" SIAM J. Alg. Disc. Meth. 1 (1980) 51-63.

[91] Wong R.T. "A dual ascent approach for Steiner tree problems on a directed graph" Math. Programming 28 (1984) 271-287.

[92] Wu Y.F., Widmeyer P. and Wong C.K. "A faster approximation algorithm for the Steiner problem in graphs" Acta Informatica 23 (1986) 223-229.

[93] Yao A.C. "An $O(|E| \log\log |V|)$ algorithm for finding minimum spanning trees" Inf. Proc. Letters 4 (1975) 21-23.

C.I.M.E. Session on "Combinatorial Optimization"

List of Participants

D. ACKETA, Institute of Mathematics, Dr. Ilije Djuricica 4, 21000 Novi Sad, Yugoslavia

G. ANDREATTA, Facoltà di Scienze Statistiche, Via del Santo 22, 35123 Padova, Italy

E. APARO, Via Duccio Galimberti 27, 00100 Roma, Italy

C. ARBIB, Via Tofana 2, 00141 Roma, Italy

A. BARLOTTI, Istituto Matematico Università, Viale Morgagni 67/A, 50134 Firenze, Italy

J. BISSCHOP, Westerhoutpark 28, 2012 JN Haarlem, The Netherlands

P.M. CAMERINI, Dipartimento di Elettronica, Politecnico di Milano, Piazza Leonardo da Vinci 32, 20133 Milano, Italy

C. CHEGIREDDY, Department of ISE, University of Florida, Gainesville, Fl. 32601, USA

R. COOK, Department of Pure Mathematics, Hicks Bldg, Hounsfield Rd., Sheffield S3 7RH, U.K.

B. DORHOUT, Twente University of Technology, P.O.Box 217, 7500 AE Enschede, The Netherlands

S. DRAGUTIN, Department of Mathematics, University of Zagreb, P.O.Box 187, 41000 Zagreb, Yugoslavia

A. GALLUCCIO, Via R. Forster 81, 00143 Roma, Italy

L. GUIDOTTI LENZI, Dipartimento di Matematica, Università, Piazza Porta S. Donato 5, 40127 Bologna, Italy

P.L. HAMMER, Operations Research Center, Hill Center for the Math. Sciences, Rutgers University, Pusch Campus, New Brunswick, New Jersey 08903, USA

O. HOLLAND, Institut für Operations Research, Nassestr. 2, 5300 Bonn 1, West Germany

E.L. JOHNSON, IBM, Thomas J. Watson Research Center, Yorktown Height, N.Y.10598, USA

B. KORTE, Inst. for Operations Research, Rheinische Friedrich-Wilhelms Universität, Nassestr. 2, 5300 Bonn 1, West Germany

E.L. LAWLER, Electr. Eng. and Computer Sciences, University of California, Computer Science Division, Berkeley, Cal. 94720, USA

M. LUCERTINI, L.go Tevere R. Sanzio 5, 00153 Roma, Italy

F. MAFFIOLI, Dipartimento di Elettronica, Politecnico di Milano, Piazza Leonardo da Vinci 32, 20133 Milano, Italy

P. MANGANI, Istituto Matematico, Università, Viale Morgagni 67/A, 50134 Firenze, Italy

G. NICOLETTI, Dipartimento di Matematica, Università, Piazza Porta S. Donato 5, 40127 Bologna, Italy

P. NOBILI, Via Medaglie d'Oro 106, 00136 Roma, Italy

P. OLIVO, Via Vallescura 45, 40136 Bologna, Italy

S. PALLOTTINO, IAC-CNR, Via del Policlinico 137, 00161 Roma, Italy

J. PELCZEWSKI, Institute of Mathematics, University of Łodz, Ul. S. Banacha 22, 90-238 Łodz, Poland

V. PERISIC, Pantovcak 156, 41000 Zagreb, Yugoslavia

W. PIOTROWSKI, Institute of Computer Science, University of Wrocław, Ul. Przesmyckiego 20 51-151 Wrocław, Poland

G. PIRILLO, Via Fra' Bartolomeo 70, 50047 Prato, Firenze, Italy

R. SCAPELLATO, Università degli Studi, 43100 Parma, Italy

B. SIMEONE, Dipartimento di Statistica, Probabilità e Statistiche Applicate, Università di Roma "La Sapienza", P.le Aldo Moro 5, 00185 Roma, Italy

L. STEFANINI, Via Pallino 38, 61029 Urbino, Pesaro, Italy

A. STEGER, Institut für Operations Research, Nassestr. 2, 5300 Bonn 1, West Germany

F. TARDELLA, IEI-CNR, Via S. Maria 46, 56100 Pisa, Italy

M. VICHI, Via Filippo Fiorentini 106, 00159 Roma, Italy

B. VISCOLANI, Istituto di Analisi e Meccanica, Via Belzoni 7, 35131 Padova, Italy

P. WINTER, Department of Computer Science, University of Copenhagen, Universitetsparken 1, DK-2100 Copenhagen O, Denmark

FONDAZIONE C.I.M.E.
CENTRO INTERNAZIONALE MATEMATICO ESTIVO
INTERNATIONAL MATHEMATICAL SUMMER CENTER

"Logic and Computer Science"

is the subject of the First 1988 C.I.M.E. Session.

The Session, sponsored by the Consiglio Nazionale delle Ricerche and the Ministero della Pubblica Istruzione, will take place under the scientific direction of Prof. PIERGIORGIO ODIFREDDI (Università di Torino) at Villa «La Querceta», Montecatini Terme (Pistoia), Italy, ***from June 20 to June 28, 1988.***

Courses

a) ***Overview of Computational Complexity Theory.*** (6 lectures in English).
Prof. Juris HARTMANIS (Cornell University, Ithaca).

Outline

Computational complexity theory is the study of the quantitative laws that govern computing. During the last twenty-five years, complexity theory has grown into a rich mathematical theory and today, it is one of the most active research areas in computer science. Among the most challenging open problems in complexity theory is the problem of understanding what is and is not feasible computable and more generally, a thorough understanding of the structure of the feasible computations. The best known of these problems is the classic P = ? NP problem. It is interesting to note that these problems, which were formulated in computer science, are actually basic problems about fundamental quantitative nature of mathematics. In essence, the P = ? NP problem is a question of how much harder is it to derive (computationally) a proof of a theorem than to check the validity of a proof.

The lectures on Computational Complexity will rewiew the basic concepts and techniques of complexity, summarize the earlier results and then review the more recent results about the structure of feasible computations.

b) ***Non-Traditional Logics for Computation.*** (6 lectures in English).
Prof. Anil NERODE (Cornell University, Ithaca).

Outline

A primer of non-traditional logics for non-experts. Many non-traditional logics are receiving wide attention in computer science because of potential importance in specification, development, and verification of programs and systems.

Prerequisites: some knowledge of undergraduate algebra, topology, computer science, and classical predicate logic. Otherwise self-contained.

I. Classical propositional and first order logic. Its models and Herbrand universes. Proof procedures of Gentzen natural deduction, Beth-Smullyan tableaux and resolution and unification. Discussion of automated classical propositional and first order reasoning.

II. Classical propositional and first order intuitionist logic. Its models such as Kripke models, Beth models, continuous function models, cpo models, Heyting valued models, categorical models. Models with discrete equality for Kroneckerian constructive algebra versus models with apartness for Brouwer's intuitionist analysis and Bishop's constructive analysis. Proof procedures of Gentzen, Fitting tableaux, natural deduction, resolution and unification for intuitionist logics. Discussion of automated intuitionist propositional and first order reasoning. Kleene's realizability for intuitionistic arithmetic as the archetype for lambda calculus realizability and the extraction of programs from proofs.

III. Modal first order logics, classical and intuitionist. The correspondence theory. Modal logics arising in computing such as algorithmic and dynamic logic, and temporal logics for description of concurrent processes and programs.

IV. Many sorted first order logics of all of the above sorts, model and proof theory and automation. Their use in algebraic specification theory, description of communicating processes, etc.

V. Logics for computing based on finite models, such as Gurevich's models for Pascal.

VI. Buchi's monadic second order theory of one successor and Rabin's second order monadic theory of two successors as languages for computing.

VII. Higher order intuitionist logic, de Bruijn and AUTOMATH, Constable and NUPRL, etc. Relation to rewrite rules and lambda calculus via the Curry-Howard isomorphism.

VIII. Other logics, such as logics of knowledge, probabilistic logics and uncertain reasoning logics.

c) ***Program verification.*** (6 lectures in English).
Prof. Richard PLATEK (Odissey Research Association, Ithaca).

Outline

In this lecture series we will review some of the fundamental logical theorems which form the basis of program verification. We will consider proofs of both sequential and concurrent programs. Some of the topics to be considered are:

I. Flowchart Verification: Floyd's verification condition theorem; its relationship to second order logic, infinitary logic, and PROLOG. Partial and total correctness; weakest precondition, strongest postcondition liberal and strict. Extensions of the method of invariants to include concurrency.

II. Hoare Logic: The verification of structured programs. Rules for constructs such as procedure call, recursion, pointers, loop exit statements, etc. Relative completeness results; incompleteness results. Extension of ordinary logic to a logic of partial terms in order to deal with undefined expressions (reading an uninitialized variable, indexing an array out of bounds, etc.).

III. Structured specification languages: An examination of Anna, a specification/assertion language for Ada.

IV. Concurrency: The Owicki-Gries approach; Hoare logic for CSP; the use of temporal logic.

There will also be a review of existing automated program verification systems.

d) ***Logic and Computer Science.*** (6 lectures in English).
Prof. Gerald SACKS (Harvard University, Cambridge, Mass).

Outline

- Logical foundations of prolog.
- Backtracking, cuts and operators.
- Prolog procedures.
- Database manipulation.
- Definite clause grammars and parsing.
- Classical recursion theory.

Basic references

- C. MARCUS, Prolog Programming, Addison-Wesley, 1986.
- I. BRATKO, Prolog Programming for Artificial Intelligence, Addison-Wesley, 1986.

e) ***Type Theory and Functional Programming.*** (6 lectures in English).
Prof. Andre SCEDROV (University of Pennsylvania, Philadelphia).

Outline

— Lambda Calculus (Typed, untyped, Church-Rosser Theorem, normal form; domains, fixed-points).
— Introduction to the ML language. (Syntax, typing rules, polymorphism, principal types, type-checking,...).
— Natural Deduction and the Proposition-as-Types Principle. (First-order minimal propositional calculus, normalization, second-order minimal propositional calculus, normalization, extensions to predicate calculi and arithmetic).
— Polymorphic Lambda Calculus. (Girard-Reynolds second-order lambda calculus, definable types: polyboole, polyint and other examples; expressiveness in terms of representable recursive functions, ...).
— Semantics. (Environment models, cartesian closed categories, realizability, coherent spaces, ...).
— Calculus of Constructions. (Syntax, typical ambiguity, examples, semantics, normalization).

Seminars

A number of seminars and special lectures will be offered during the Session.

FONDAZIONE C.I.M.E.
CENTRO INTERNAZIONALE MATEMATICO ESTIVO
INTERNATIONAL MATHEMATICAL SUMMER CENTER

"Global Geometry and Mathematical Physics"

is the subject of the Second 1988 C.I.M.E. Session.

The Session, sponsored by the Consiglio Nazionale delle Ricerche and the Ministero della Pubblica Istruzione, will take place under the scientific direction of Prof. MAURO FRANCAVIGLIA (Università di Torino), and Prof. FRANCESCO GHERARDELLI (Università di Firenze) at Villa «La Querceta», Montecatini (Pistoia), Italy, ***from July 4 to July 12, 1988.***

Courses

a) ***String Theory and Riemann Surfaces.*** (6 lectures in English).
Prof. L. ALVAREZ-GAUME (University of Boston, USA).

Contents

In these lectures we will review the new developments of string theory and its connections with the theory of Riemann surfaces, super-Riemann surfaces and their moduli spaces. The point of view taken will be to start with the conformal field theory formulation of string theory and then develop in detail the operator formalism for strings on higher genus surfaces. A tentative plane of the lectures is:

Lecture 1 - Introduction to string theory and conformal field theory (Part 1)
Lecture 2 - Introduction to string theory and conformal field theory (Part 2)
Lecture 3 - Perturbation theory for Bosonic strings. Belavin-Knizhnik theorem, Mumford forms, string infinities and the boundary of moduli spaces.
Lecture 4 - The operator formulation of string theory (Part 1)
Lecture 5 - The operator formulation of string theory (Part 2)
Lecture 6 - Virasoro action on moduli space, more general conformal theories, nonpertubative ideas and recent developments.

b) ***Riemann Surfaces and Infinite Grassmannians.*** (6 lectures in English).
Prof. E. ARBARELLO (Università di Roma "La Sapienza", Roma, Italy).

Contents

— Compact Riemann surfaces and their moduli (stable and semi-stable curves). Picard's group of the moduli space Mg. The Riemann-Roch-Grothendieck theorem.
— Kodaira-Spencer theory. Schiffer variations. Calculus of cohomology à la Mayer-Vietoris. The cotangent bundle of moduli space.
— Lie algebras d (Virasoro) and D (Virasoro-Heisenberg) and their relations with moduli space Mg.
— The Lie algebra gl_∞ and the "Boson-Fermion correspondence". Relations with the Lie algebras d and D. Calculation of the relevant cohomology.
— The infinite Grassmannian Gr and its geometry. The central extension of GL_∞. Tautological and determinant bundles; the function τ. Relation with Plücker's coordinates. The K.P. hierarchy.
— Krichever application. Relations between the functions τ and θ. The correlation function. The trisecting formula. Novikov conjecture.
— The sheaf of differential operators of order less or equal to one, acting on sections of the determinant bundle over Gr. Its restriction to moduli space.
— Some known results about moduli spaces of Riemann surfaces.

References

- L. ALVAREZ-GAUME, C. GOMEZ, C. REINA, Loop Groups, Grassmannians and String Theory, Phys. Lett. 190B, 55-62 (1987).
- E. ARBARELLO, M. CORNALBA, The Picard Groups of the Moduli Spaces of Curves, Topology 26(2), 153-171 (1987).
- E. ARBARELLO, C. DE CONCINI, V. KAC, C. PROCESI, Moduli Space of Curves and Representation Theory, preprint, 1987.
- A.A. BEILINSON, YU. I. MANIN, The Mumford Form and the Polyakov Measure in String Theory, Comm. Math. Phys. 107, 359-376 (1986).
- A.A. BEILINSON, YU. I. MANIN, V.V. SCHECHTMAN, Sheaves of the Virasoro and Neveu Schwarz Algebras, Moscow Univ. preprint, 1987.
- E. DATE, M. JIMBO, M. KASHIWARA, T. MIWA, Transformation Groups for Soliton Equations, in "Nonlinear Integrable Systems. Classical Theory and Quantum Theory", World Sci. (Singapore, 1983), pp. 39-119.
- J. HARER, The Second Homology Group of the Mapping Class Group of an Orientable Surface, Inv. Math. 72, 221-239 (1983).
- J. HARER, Stability of the Homology of the Mapping Class Group of Orientable Surfaces, Ann. of Math. 121, 215-249 (1985).
- V.G. KAC, D.H. PETERSON, Spin and Wedge Representations of Infinite Dimensional Lie Algebras and Groups, Proc. Nat. Acad. Sci. U.S.A. 78, 3308-3312 (1981).
- V.G. KAC, Highest Weight Representations of Conformal Current Algebras, in "Geometrical Methods in Field Theory", World Sci. (Singapore, 1986), pp. 3-15.
- N. KAWAMOTO, Y. NAMIKAWA, A. TSUCHIYA, Y. YAMADA, Geometric Realization of Conformal Field Theory on Riemann Surfaces, Nagoya Univ. preprint, 1987.
- YU. I. MANIN, Quantum String Theory and Algebraic Curves, Berkeley I.C.M. talk, 1986.
- E.Y. MILLER, The Homology of the Mapping Class Group, Journ. Diff. Geom. 24, 1-14 (1986).
- D. MUMFORD, Stability of Projective Varieties, L'Enseignement Mathém. 23, 39-110 (1977).
- A. PRESSLEY, G. SEGAL, Loop Groups, Oxford Univ. Press (Oxford, 1986).
- G. SEGAL, G. WILSON, Loop Groups and Equations of KdV Type, Publ. Math. I.H.E.S. 61, 3-64 (1985).
- C. VAFA, Conformal Theories and Punctured Surfaces, preprint, 1987.
- E. WITTEN, Quantum Field Theory, Grassmannians and Algebraic Curves, preprint, 1987.

c) ***The Topology and Geometry of Moduli Spaces.*** (6 lectures in English).
Prof. N.J. HITCHIN (Oxford University, Oxford, UK).

Contents

Topics will include:

(i) Moduli space instantons
(ii) Moduli space of monopoles
(iii) Moduli space of vortices
(iv) Teichmüller space
(v) Moduli spaces related to Riemann surfaces

References

- M.F. ATIYAH, Instantons in 2 and 4 Dimensions, Comm. Math. Phys. 93, 437-451 (1984).
- D. FREED, K. UHLENBECK, Instantons and 4 - Manifolds, Springer Verlag (Berlin, 1984).
- N.J. HITCHIN, A. KALBADE, U. LINDSTROM, M. ROCEK, Hyperkähler Metrics and Supersymmetry, Comm. Math. Phys. 108, 535-589 (1987).
- M.F. ATIYAH, N.J. HITCHIN, Geometry and Dynamics of Magnetic Monopoles, Princeton University Press (Princeton, 1988).
- A. JAPPE, C. TOMBES, Vortices and Monopoles, Birkhäuser (1980).
- N.J. HITCHIN, The Self-Duality Equation on a Riemann Surface, Proc. London Math. Soc. 55, 59-126 (1987).
- M.F. ATIYAH, R. BOTT, The Yang-Mills Equations over Riemann Surfaces, Phil. Trans. Roy. Soc. London, sec. A. 308, 523-615 (1982).

d) ***Differential Algebras in Field Theory.*** (6 lectures in English).
Prof. R. STORA (LAPP, Annecy-le-Vieux, France).

Contents

Lecture 1 - Introduction. The role of locality in perturbative quantum field theory ([1]).

Lecture 2 - The description of continuous symmetries in perturbative quantum field theories: global symmetries and their current algebras ([2]).

Lecture 3 - Current Algebra anomalies: algebraic structure. Local cohomologies of gauge Lie algebras ([3] - [10]).

Lecture 4 - Perturbative quantization of gauge theories, Slavnov-symmetries and the corresponding differential algebras ([6]).

Lecture 5 - More general differential algebras involving diffeomorphisms. Gravitational anomalies ([6], [7], [10]). Application to the first quantized string ([11]).

Lecture 6 - Miscellaneous examples:
(i) "Higher cocycles" in field theory: BRST current algebra versus Schwinger term (gauge Lie algebras extensions) ([12], [13]).
(ii) The "Torino" differential algebras for gravity and supergravity (tentative) ([14]).

References

The brief bibliography contains mainly review articles, from which the overwhelmingly voluminous original literature can be traced back.

[1] H. EPSTEIN, V. GLASER, Ann. Inst. H. Poincaré 19, 211 (1973).

[2] C. BECCHI, A. ROUET, R. STORA, Renormalizable Theories with Symmetry Breaking in "Field Theory Quantization and Statistical Physics", E. Tirapegui ed.; Reidel (Dordrecht), 1981).

[3] B. ZUMINO, in "Relativity, Groups and Topology. II", Les Houches XLIV, 1983; B. De Witt, R. Stora eds.; North-Holland (Amsterdam, 1984).

[4] R. STORA, in "Progress in Gauge Field Theory", Cargése 1983; G. 't Hooft et al. eds.; NATO ASI Series B 115, Plenum (New York, 1984).

[5] J. MANES, R. STORA, B. ZUMINO, Comm. Math, Phys. 102, 157 (1985).

[6] R. STORA, in "New Perspectives in Quantum Field Theories", XVIth GIFT Seminar, 1985; J. Abad et al. eds.; World Sci. (Singapore, 1986).

[7] L. ALVAREZ-GAUME, P. GINSPARG, Ann. Phys. 161, 423 (1985).

[8] M. DUBOIS-VIOLETTE, M. TALON, C.M. VIALLET, Comm. Math. Phys. 102, 105 (1985).

[9] L. BONORA, P. COTTA RAMUSINO, M. RINALDI, J. STASHEFF, CERN-Th 4647/87 and 4750/87 (to appear in Comm. Math. Phys.).

[10] L. ALVAREZ-GAUME, An Introduction to Anomalies, in "Erice 1985", G. Velo, A.S. Wightman eds.; ASI Series B 141, Plenum (New York, 1986).

[11] C. BECCHI, preprint, Genova 1987;
L. BAULIEU, M. BELLON, LPTHE 87-39.

[12] B. ZUMINO, Nucl. Phys. B 253, 477 (1985).

[13] L. BAULIEU, B. GROSSMANN, R. STORA, Phys. Lett. 180B, 95 (1986).

[14] T. REGGE, in "Relativity, Groups and Topology. II", Les Houches XLIV, 1983;
B. DE WITT, R. STORA eds.; North-Holland (Amsterdam, 1984).

Seminars

A number of seminars and special lectures will be offered during the Session.

LIST OF C.I.M.E. SEMINARS

Year	Seminar	Publisher
1954 -	1. Analisi funzionale	C.I.M.E.
	2. Quadratura delle superficie e questioni connesse	"
	3. Equazioni differenziali non lineari	"
1955 -	4. Teorema di Riemann-Roch e questioni connesse	"
	5. Teoria dei numeri	"
	6. Topologia	"
	7. Teorie non linearizzate in elasticità, idrodinamica, aerodinamica	"
	8. Geometria proiettivo-differenziale	"
1956 -	9. Equazioni alle derivate parziali a caratteristiche reali	"
	10. Propagazione delle onde elettromagnetiche	"
	11. Teoria della funzioni di più variabili complesse e delle funzioni automorfe	"
1957 -	12. Geometria aritmetica e algebrica (2 vol.)	"
	13. Integrali singolari e questioni connesse	"
	14. Teoria della turbolenza (2 vol.)	"
1958 -	15. Vedute e problemi attuali in relatività generale	"
	16. Problemi di geometria differenziale in grande	"
	17. Il principio di minimo e le sue applicazioni alle equazioni funzionali	"
1959 -	18. Induzione e statistica	"
	19. Teoria algebrica dei meccanismi automatici (2 vol.)	"
	20. Gruppi, anelli di Lie e teoria della coomologia	"
1960 -	21. Sistemi dinamici e teoremi ergodici	"
	22. Forme differenziali e loro integrali	"
1961 -	23. Geometria del calcolo delle variazioni (2 vol.)	"
	24. Teoria delle distribuzioni	"
	25. Onde superficiali	"
1962 -	26. Topologia differenziale	"
	27. Autovalori e autosoluzioni	"
	28. Magnetofluidodinamica	"
1963 -	29. Equazioni differenziali astratte	"
	30. Funzioni e varietà complesse	"
	31. Proprietà di media e teoremi di confronto in Fisica Matematica	"

1964 -	32. Relatività generale	C.I.M.E.
	33. Dinamica dei gas rarefatti	"
	34. Alcune questioni di analisi numerica	"
	35. Equazioni differenziali non lineari	"
1965 -	36. Non-linear continuum theories	"
	37. Some aspects of ring theory	"
	38. Mathematical optimization in economics	"
1966 -	39. Calculus of variations	Ed. Cremonese, Firenz
	40. Economia matematica	"
	41. Classi caratteristiche e questioni connesse	"
	42. Some aspects of diffusion theory	"
1967 -	43. Modern questions of celestial mechanics	"
	44. Numerical analysis of partial differential equations	"
	45. Geometry of homogeneous bounded domains	"
1968 -	46. Controllability and observability	"
	47. Pseudo-differential operators	"
	48. Aspects of mathematical logic	"
1969 -	49. Potential theory	"
	50. Non-linear continuum theories in mechanics and physics and their applications	"
	51. Questions of algebraic varieties	"
1970 -	52. Relativistic fluid dynamics	"
	53. Theory of group representations and Fourier analysis	"
	54. Functional equations and inequalities	"
	55. Problems in non-linear analysis	"
1971 -	56. Stereodynamics	"
	57. Constructive aspects of functional analysis (2 vol.)	"
	58. Categories and commutative algebra	"
1972 -	59. Non-linear mechanics	"
	60. Finite geometric structures and their applications	"
	61. Geometric measure theory and minimal surfaces	"
1973 -	62. Complex analysis	"
	63. New variational techniques in mathematical physics	"
	64. Spectral analysis	"

1974 - 65. Stability problems		Ed. Cremonese, Firenze
66. Singularities of analytic spaces		"
67. Eigenvalues of non linear problems		"
1975 - 68. Theoretical computer sciences		"
69. Model theory and applications		"
70. Differential operators and manifolds		"
1976 - 71. Statistical Mechanics		Ed. Liguori, Napoli
72. Hyperbolicity		"
73. Differential topology		"
1977 - 74. Materials with memory		"
75. Pseudodifferential operators with applications		"
76. Algebraic surfaces		"
1978 - 77. Stochastic differential equations		"
78. Dynamical systems		Ed. Liguori, Napoli and Birkhäuser Verlag
1979 - 79. Recursion theory and computational complexity		Ed. Liguori, Napoli
80. Mathematics of biology		"
1980 - 81. Wave propagation		"
82. Harmonic analysis and group representations		"
83. Matroid theory and its applications		"
1981 - 84. Kinetic Theories and the Boltzmann Equation	(LNM 1048)	Springer-Verlag
85. Algebraic Threefolds	(LNM 947)	"
86. Nonlinear Filtering and Stochastic Control	(LNM 972)	"
1982 - 87. Invariant Theory	(LNM 996)	"
88. Thermodynamics and Constitutive Equations	(LN Physics 228)	"
89. Fluid Dynamics	(LNM 1047)	"
1983 - 90. Complete Intersections	(LNM 1092)	"
91. Bifurcation Theory and Applications	(LNM 1057)	"
92. Numerical Methods in Fluid Dynamics	(LNM 1127)	"
1984 93. Harmonic Mappings and Minimal Immersions	(LNM 1161)	"
94. Schrödinger Operators	(LNM 1159)	"
95. Buildings and the Geometry of Diagrams	(LNM 1181)	"
1985 - 96. Probability and Analysis	(LNM 1206)	"
97. Some Problems in Nonlinear Diffusion	(LNM 1224)	"
98. Theory of Moduli	(LNM 1337)	"

Note: Volumes 1 to 38 are out of print. A few copies of volumes 23,28,31,32,33,34,36,38 are available on request from C.I.M.E.

1986 -	99. Inverse Problems	(LNM 1225)	Springer-Verlag
	100. Mathematical Economics	(LNM 1330)	"
	101. Combinatorial Optimization	to appear	"
1987 -	102. Relativistic Fluid Dynamics	to appear	"
	103. Topics in Calculus of Variations	(LNM 1365)	"
1988 -	104. Logic and Computer Science	to appear	"
	105. Global Geometry and Mathematical Physics	to appear	"

Vol. 1290: G. Wüstholz (Ed.), Diophantine Approximation and Transcendence Theory. Seminar, 1985. V, 243 pages. 1987.

Vol. 1291: C. Mœglin, M.-F. Vignéras, J.-L. Waldspurger, Correspondances de Howe sur un Corps p-adique. VII, 163 pages. 1987

Vol. 1292: J.T. Baldwin (Ed.), Classification Theory. Proceedings, 1985. VI, 500 pages. 1987.

Vol. 1293: W. Ebeling, The Monodromy Groups of Isolated Singularities of Complete Intersections. XIV, 153 pages. 1987.

Vol. 1294: M. Queffélec, Substitution Dynamical Systems – Spectral Analysis. XIII, 240 pages. 1987.

Vol. 1295: P. Lelong, P. Dolbeault, H. Skoda (Réd.), Séminaire d'Analyse P. Lelong – P. Dolbeault – H. Skoda. Seminar, 1985/1986. VII, 283 pages. 1987.

Vol. 1296: M.-P. Malliavin (Ed.), Séminaire d'Algèbre Paul Dubreil et Marie-Paule Malliavin. Proceedings, 1986. IV, 324 pages. 1987.

Vol. 1297: Zhu Y.-l., Guo B.-y. (Eds.), Numerical Methods for Partial Differential Equations. Proceedings. XI, 244 pages. 1987.

Vol. 1298: J. Aguadé, R. Kane (Eds.), Algebraic Topology, Barcelona 1986. Proceedings. X, 255 pages. 1987.

Vol. 1299: S. Watanabe, Yu.V. Prokhorov (Eds.), Probability Theory and Mathematical Statistics. Proceedings, 1986. VIII, 589 pages. 1988.

Vol. 1300: G.B. Seligman, Constructions of Lie Algebras and their Modules. VI, 190 pages. 1988.

Vol. 1301: N. Schappacher, Periods of Hecke Characters. XV, 160 pages. 1988.

Vol. 1302: M. Cwikel, J. Peetre, Y. Sagher, H. Wallin (Eds.), Function Spaces and Applications. Proceedings, 1986. VI, 445 pages. 1988.

Vol. 1303: L. Accardi, W. von Waldenfels (Eds.), Quantum Probability and Applications III. Proceedings, 1987. VI, 373 pages. 1988.

Vol. 1304: F.Q. Gouvêa, Arithmetic of p-adic Modular Forms. VIII, 121 pages. 1988.

Vol. 1305: D.S. Lubinsky, E.B. Saff, Strong Asymptotics for Extremal Polynomials Associated with Weights on $\mathbb{R}$. VII, 153 pages. 1988.

Vol. 1306: S.S. Chern (Ed.), Partial Differential Equations. Proceedings, 1986. VI, 294 pages. 1988.

Vol. 1307: T. Murai, A Real Variable Method for the Cauchy Transform, and Analytic Capacity. VIII, 133 pages. 1988.

Vol. 1308: P. Imkeller, Two-Parameter Martingales and Their Quadratic Variation. IV, 177 pages. 1988.

Vol. 1309: B. Fiedler, Global Bifurcation of Periodic Solutions with Symmetry. VIII, 144 pages. 1988.

Vol. 1310: O.A. Laudal, G. Pfister, Local Moduli and Singularities. V, 117 pages. 1988.

Vol. 1311: A. Holme, R. Speiser (Eds.), Algebraic Geometry, Sundance 1986. Proceedings. VI, 320 pages. 1988.

Vol. 1312: N.A. Shirokov, Analytic Functions Smooth up to the Boundary. III, 213 pages. 1988.

Vol. 1313: F. Colonius, Optimal Periodic Control. VI, 177 pages. 1988.

Vol. 1314: A. Futaki, Kähler-Einstein Metrics and Integral Invariants. IV, 140 pages. 1988.

Vol. 1315: R.A. McCoy, I. Ntantu, Topological Properties of Spaces of Continuous Functions. IV, 124 pages. 1988.

Vol. 1316: H. Korezlioglu, A.S. Ustunel (Eds.), Stochastic Analysis and Related Topics. Proceedings, 1986. V, 371 pages. 1988.

Vol. 1317: J. Lindenstrauss, V.D. Milman (Eds.), Geometric Aspects of Functional Analysis. Seminar, 1986–87. VII, 289 pages. 1988.

Vol. 1318: Y. Felix (Ed.), Algebraic Topology – Rational Homotopy. Proceedings, 1986. VIII, 245 pages. 1988

Vol. 1319: M. Vuorinen, Conformal Geometry and Quasiregular Mappings. XIX, 209 pages. 1988.

Vol. 1320: H. Jürgensen, G. Lallement, H.J. Weinert (Eds.), Se groups, Theory and Applications. Proceedings, 1986. X, 416 pag 1988.

Vol. 1321: J. Azéma, P.A. Meyer, M. Yor (Eds.), Séminaire Probabilités XXII. Proceedings. IV, 600 pages. 1988.

Vol. 1322: M. Métivier, S. Watanabe (Eds.), Stochastic Analy Proceedings, 1987. VII, 197 pages. 1988.

Vol. 1323: D.R. Anderson, H.J. Munkholm, Boundedly Contro Topology. XII, 309 pages. 1988.

Vol. 1324: F. Cardoso, D.G. de Figueiredo, R. Iório, O. Lopes (Ec Partial Differential Equations. Proceedings, 1986. VIII, 433 pag 1988.

Vol. 1325: A. Truman, I.M. Davies (Eds.), Stochastic Mechanics Stochastic Processes. Proceedings, 1986. V, 220 pages. 1988.

Vol. 1326: P.S. Landweber (Ed.), Elliptic Curves and Modular Form Algebraic Topology. Proceedings, 1986. V, 224 pages. 1988.

Vol. 1327: W. Bruns, U. Vetter, Determinantal Rings. VII,236 pag 1988.

Vol. 1328: J.L. Bueso, P. Jara, B. Torrecillas (Eds.), Ring The Proceedings, 1986. IX, 331 pages. 1988.

Vol. 1329: M. Alfaro, J.S. Dehesa, F.J. Marcellan, J.L. Rubio Francia, J. Vinuesa (Eds.): Orthogonal Polynomials and their Appl tions. Proceedings, 1986. XV, 334 pages. 1988.

Vol. 1330: A. Ambrosetti, F. Gori, R. Lucchetti (Eds.), Mathemat Economics. Montecatini Terme 1986. Seminar. VII, 137 pages. 19

Vol. 1331: R. Bamón, R. Labarca, J. Palis Jr. (Eds.), Dynam Systems, Valparaiso 1986. Proceedings. VI, 250 pages. 1988.

Vol. 1332: E. Odell, H. Rosenthal (Eds.), Functional Analysis. ceedings, 1986–87. V, 202 pages. 1988.

Vol. 1333: A.S. Kechris, D.A. Martin, J.R. Steel (Eds.), Cabal Sem 81–85. Proceedings, 1981–85. V, 224 pages. 1988.

Vol. 1334: Yu.G. Borisovich, Yu. E. Gliklikh (Eds.), Global Anal – Studies and Applications III. V, 331 pages. 1988.

Vol. 1335: F. Guillén, V. Navarro Aznar, P. Pascual-Gainza, F. Pue Hyperrésolutions cubiques et descente cohomologique. XII, pages. 1988.

Vol. 1336: B. Helffer, Semi-Classical Analysis for the Schrödin Operator and Applications. V, 107 pages. 1988.

Vol. 1337: E. Sernesi (Ed.), Theory of Moduli. Seminar, 1985. VIII, pages. 1988.

Vol. 1338: A.B. Mingarelli, S.G. Halvorsen, Non-Oscillation Dom of Differential Equations with Two Parameters. XI, 109 pages. 1988

Vol. 1339: T. Sunada (Ed.), Geometry and Analysis of Manifol Procedings, 1987. IX, 277 pages. 1988.

Vol. 1340: S. Hildebrandt, D.S. Kinderlehrer, M. Miranda (Ed Calculus of Variations and Partial Differential Equations. Proceedir 1986. IX, 301 pages. 1988.

Vol. 1341: M. Dauge, Elliptic Boundary Value Problems on Cor Domains. VIII, 259 pages. 1988.

Vol. 1342: J.C. Alexander (Ed.), Dynamical Systems. Proceedin 1986–87. VIII, 726 pages. 1988.

Vol. 1343: H. Ulrich, Fixed Point Theory of Parametrized Equivar Maps. VII, 147 pages. 1988.

Vol. 1344: J. Král, J. Lukeš, J. Netuka, J. Veselý (Eds.), Poter Theory – Surveys and Problems. Proceedings, 1987. VIII, 271 pag 1988.

Vol. 1345: X. Gomez-Mont, J. Seade, A. Verjovski (Eds.), Holomor Dynamics. Proceedings, 1986. VII, 321 pages. 1988.

Vol. 1346: O. Ya. Viro (Ed.), Topology and Geometry – Ro Seminar. XI, 581 pages. 1988.

Vol. 1347: C. Preston, Iterates of Piecewise Monotone Mappings an Interval. V, 166 pages. 1988.

Vol. 1348: F. Borceux (Ed.), Categorical Algebra and its Applicatic Proceedings, 1987. VIII, 375 pages. 1988.

Vol. 1349: E. Novak, Deterministic and Stochastic Error Bound Numerical Analysis. V, 113 pages. 1988.